Basic
Concepts
of Geometry

A BLAISDELL BOOK IN THE
PURE AND APPLIED MATHEMATICS

Seymour Schuster, Carleton College

CONSULTING EDITOR

WALTER PRENOWITZ
MEYER JORDAN

BROOKLYN COLLEGE

Basic

Concepts

of Geometry

BLAISDELL PUBLISHING COMPANY

A DIVISION OF GINN AND COMPANY

Waltham, Massachusetts · Toronto · London

TO
S

Preface

This book has two main objectives:

(1) To present geometry as a branch of contemporary mathematics involving the interrelated study of many specific geometrical systems, which are characterized and studied by properly chosen postulate systems.

(2) To present, in the framework of (1), a treatment of Euclidean geometry which meets current standards of rigor.

The principal innovation which makes possible the realization of these objectives is the introduction—as a basic unifying idea—of the concept of Incidence Geometry, which was originated by Professor Saul Gorn. An *incidence geometry*, as we use the term, is a system involving three primitive ideas, *point, line, plane*, which essentially satisfies Hilbert's postulates of incidence.

All the familiar kinds of geometry with the exception of spherical (or double elliptic) geometry can be characterized as incidence geometries which satisfy one or more additional postulates concerning parallelism, order, congruence, continuity. The following chart indicates the relation of various familiar types of geometry to incidence geometry.

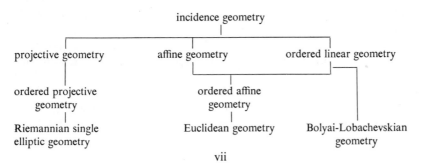

In our experience one of the principal difficulties in teaching the Foundations of Geometry is to convince the student of the need for a rigorous, abstract, deductive treatment of geometry. He may feel that school geometry has a soundness and a logical quality which few other mathematical subjects have, and that to devote much time and subtle intellectual effort to rigorize the subject is to belabor the obvious. A rigorous treatment is apt to be unmotivated for him, and to have little pedagogical impact. Consequently, we have organized the book into two parts: the second is a rigorous deductive treatment of geometry for which the first part is preparatory.

Part I (Chaps. 1–6) is directed towards the creation of dissatisfaction in the student's mind with his present knowledge of geometry, and hopes to broaden his perspective and sharpen his critical sense. In spirit it is discursive, analytic, intuitive, semi-rigorous. Part I begins (Chap. 1) with a critical treatment of deductive reasoning in Euclid and in school geometry. This is introduced with a "proof" that every triangle is isosceles. There follows (Chap. 2) a discussion of Euclid's Parallel Postulate: its role in the theory, and the historical attempts to deduce it from the remaining postulates of Euclidean geometry. This leads naturally into the study, in Chapter 3, of absolute (or *neutral*) geometry where no parallel postulate is assumed, and, in Chapter 4, of non-Euclidean geometries where parallel postulates contrary to Euclid's are introduced. Chapter 4 contains an elementary study of the geometry of Bolyai and Lobachevsky, followed by a discussion, largely descriptive, of Riemann's non-Euclidean geometries. Part I concludes with a treatment of the logical consistency of the non-Euclidean geometries (Chap. 5) and their applicability to physical space (Chap. 6).

Part II, which is essentially independent of Part I, is constructive. It attempts to build a theory of geometry in which the student is not required to pay for rigor of development and the advantages of the postulational method by loss of intuitive insight and conceptual understanding. It tries to achieve this in several ways:

First, by choosing postulates for simplicity, naturalness, and ease of recall, not for logical independence or to yield quick results in artificial ways.

Second, by introducing new concepts gradually. The postulates of incidence are introduced and studied in detail, before the theory of order is mentioned. The postulates of linear order are introduced and the theory of order on the line developed, before planar or spatial order properties are considered. The full complement of Euclidean postulates is introduced only toward the end of the book.

Third, by using similar methods, when possible, in the study of the line, the plane, and space. For example, the basic separation theorems for line, plane, and space are proved in the same way. Indeed, it would not be too hard to extend the methods used to higher dimensional space.

Part II begins with the Theory of Incidence (Chap. 7) based on Hilbert's postulates of incidence, omitting an existence postulate. The elementary consequences of the postulates are deduced abstractly without reliance on diagrams. Chapter 8 is devoted to *Incidence Geometries*, that is, models of the Theory of Incidence. A set of almost thirty models in presented which includes: finite and infinite models; two- and three-dimensional ones; models satisfying the Euclidean, Lobachevskian, Riemannian parallel postulates, and none of these; projective and affine geometries; algebraic models based on the fields of real numbers and of the integers modulo 2 and modulo 3; models on the sphere and hemisphere. The models are discussed and compared and the idea of isomorphism is introduced. The use of models to prove consistency and independence also arises naturally in this context. The models are referred to in later portions of the book, particularly in exercises, to illustrate and illuminate the theory.

In Chapter 9, an Affine Geometry is characterized as a type of incidence geometry, and parallelism properties in space are studied. A measure of unity is achieved by studying the idea of "transverseness" (of two lines, a line and a plane, two planes) and its relation to parallelism.

The concept of *order* is now introduced into the Theory of Incidence (Chaps. 10, 11). Chapter 10 studies the order of points of a line in terms of the concept of betweenness. An innovation is the inclusion, in the set of betweenness postulates, of a "linear" form of Pasch's Postulate, which fosters uniformity of treatment of basic order properties for line, plane, and space. Chapter 11 introduces Pasch's Postulate to complete the postulational characterization of an Ordered Incidence Geometry, and studies the planar and spatial analogues, by strictly similar methods, of much of the material in Chapter 10. An important point is that the student has encountered all the essential difficulties of concept and method in the pedagogically simpler case of the geometry of a line.

Chapter 12 is devoted to the nonmetrical study of angles and the related theory of betweenness of rays. Many of the subtle properties of angles (for example, interiority properties) are intimately related to order properties of rays, which are easily accessible to intuition and bear analogy to order properties of points.

Chapter 13 comes to grips with the relatively deep and difficult problems involved in the study of separation properties of angles and triangles—this is the core of the traditional course in the foundations of geometry. The essential difficulty of the material is mitigated somewhat by exploiting the concept of order of rays (Chap. 12), and the principal theorems are proved by methods which seem more insightful than the conventional ones.

The final two chapters study the theory of congruence. Chapter 14 introduces the theory of congruence into an ordered incidence geometry by using a modification of the School Mathematics Study Group treatment based on the real number system. It culminates in the theorem on the angle sum of a triangle in Euclidean geometry. Chapter 15 gives a purely geometrical treatment of essentially the same material.

The book is intended primarily for an advanced undergraduate or introductory graduate course for mathematics majors. It requires little in the way of specific mathematical preparation, but mathematical maturity and some geometric background are desirable. Portions of the book, in preliminary form, were used successfully in summer institutes for secondary mathematics teachers, and we feel it can be very useful in the training of secondary teachers, since a deeper understanding of the subject equips them better to teach high school geometry.

The material can be treated quite flexibly. Students of high ability and good preparation could be expected to assume the major burden of studying Part I on their own, with class time being used mainly for discussion of ideas. Then Part II could be taught in a more conventional manner. For a class of lower mathematical maturity it would be desirable to cover Part I more slowly to prepare the student for the postulational treatment in Part II, and to study some portions of Part II descriptively instead of working through the details of all proofs. Keeping in mind these considerations and the variety of teachers' objectives we mention the following possible sequences for a one-semester course: Chapters 1–12, 14, 15; Chapters 1–14; Chapters 1–4, 7, 8, 10–12, 14; Chapters 1–8; or even Chapters 1–4, 7, 8. For sufficiently well-prepared students it should be feasible to begin with Part II which is logically complete in itself and refers to Part I merely in exercises and for purposes of illustration.

We have tried to construct a rich and varied collection of exercises. There are graded sequences which often lead to an interesting result, miscellaneous sets, and difficult exercises which may challenge the student's ingenuity. We have included some sequences which introduce attractive material for which there was no room in the text. Difficult exercises are indicated by a dagger.

About twenty years ago a rumor was current that if one found an error in a book of Edmund Landau, he received a quarter. Despite the intervening currency inflation and the care with which we have written this book, we shall not offer a quarter even to the first person who finds an error, but our gratitude and a promise to use the reader's criticism to improve the text.

Our thanks are due to Professor G. W. Booth, Professor I. C. Gentry, Professor B. Greenspan, Professor L. Guggenbuhl, Professor R. A. McCoy, Professor J. M. Osborn, and Professor I. Rose who used the preliminary version in class and criticized it. Also, we are indebted to Professor Booth for a critical reading of a large portion of the final manuscript. We are especially grateful to Jane W. DiPaola who read and criticized both versions of the book and prepared the index. Finally, we wish to thank Edward Prenowitz and Saul Zaveler for their criticism of portions of the final manuscript.

WALTER PRENOWITZ
MEYER JORDAN

Contents

PART II

Basic
Concepts
of Geometry

PART I

Introduction

Our main object in this book is to give a treatment of geometry which meets modern standards of logical precision and has the breadth and sweep which the axiomatic approach has made possible in modern mathematics. Such a treatment must seem remote from geometry as studied in high school or the early years of college. Our conception of geometry is broader and more abstract than the Euclidean geometry of high school for two important reasons. First, Euclidean geometry is not quite the perfect example of deductive reasoning it was supposed to be for twenty centuries. (And we shall have to examine its basic ideas and principles quite deeply in order to construct a treatment which meets current standards of logical precision.) Secondly, Euclidean geometry is only one of many valid geometrical sciences or geometrical systems which interest mathematicians today. We have: the classical non-Euclidean geometries, which differ from Euclidean geometry in the assumption concerning parallel lines; projective geometry, which arose historically in the study of visual perspective, was treated for many years in terms of Euclidean geometry, and is studied as an independent branch of mathematics based on its own postulates; nonmetrical geometries in which positional relations of points and lines are studied but not distance or angular measure or area. There are finite geometrical systems which have only a finite number of points, lines, and planes. There are geometries in which a line is conceived as being closed, with a finite length, so that a point of a line does not separate it into two parts, as is the case in Euclidean geometry. In fact, the term geometry is often used now in the sense of a specific type or example of geometric theory, and we speak of Euclidean *geometries*, projective *geometries*, finite *geometries*, et cetera.

Our formal deductive treatment of geometry is presented in Part II. The object of the present part is, so to speak, to bridge the gap between the classical and modern views of geometry—it is intended to be a preparation for the more abstract treatment of Part II. We try to show why the conventional standards of geometrical proof—which are not too bad and did satisfy mathematicians for twenty centuries—do not meet the requirements of modern mathematics. And we indicate the historical basis for this change of standards in the revolutionary development of non-Euclidean theories of geometry. Concurrently, we try to indicate the richness of geometry. Our treatment in this part is critical, discursive, historical, philosophical. It does not attempt to be mathematically rigorous in a formal sense—but does try to foster a deepened critical sense of proof as well as a broad view of geometry, for the better appreciation of the abstract, formal, deductive treatment of Part II.

CHAPTER 1

Logical Deficiencies in Euclidean Geometry

For over two thousand years the "Elements" of Euclid (300 B.C.) has been considered a model of mathematical reasoning. Since his time, the study of the "Elements" or an equivalent text has been an essential factor in a liberal education. Generation after generation has been instilled with an appreciation of deductive reasoning from the study of Euclidean geometry. At the end of the nineteenth century, mathematicians, probing deeply into the basic ideas of Euclid, found deficiencies in his treatment. It is not considered "rigorous" today. Our object in this chapter is to illustrate and explain certain logical deficiencies in Euclid's treatment and to show why it does not meet current standards of rigor.

Many students of mathematics have the impression that Euclidean geometry is a perfect model of logical reasoning. The development begins with the statement of basic assumptions, then every theorem is proved—that is, logically deduced—from these assumptions. However, there are certain subtle logical gaps in the reasoning, both in Euclid and in the standard high school texts. These gaps arise because certain tacit assumptions, based on visual evidence or spatial intuition, are made. This breaks the orderly sequence of logical inferences. We will make this vivid by showing that the tacit employment of visual evidence in the usual manner of high school geometry may result in the proof of such startling theorems as:

Every triangle is isosceles.
There exists a triangle with two right angles.

We will also discuss several geometric theorems whose proofs in many high school texts are logically deficient, pointing out where tacit assumptions

3

are made, and indicating how these proofs may be revised to make them logically sound.

1. Proof That Every Triangle Is Isosceles

Suppose $\triangle ABC$ is not isosceles. Then the bisector of $\angle B$ is not perpendicular to AC, for in the contrary case $\triangle ABC$ would be isosceles. Thus the bisector of $\angle B$ intersects the perpendicular bisector of AC at a point E which is either inside, on, or outside $\triangle ABC$. (Several diagrams are shown in Figure 1.1. The proof applies equally well to each.) From E drop perpendiculars to AB and BC, meeting AB in F, and BC in G.

In right triangles BFE and BGE we have $\angle FBE = \angle GBE$ (definition of angle bisector), and $BE = BE$ (identity). Hence $\triangle BFE \cong \triangle BGE$ (hypotenuse, acute angle). Therefore, we have

(1) $$BF = BG$$

(corresponding parts of congruent triangles).

In right triangles FAE and GCE we have $AE = CE$ (since E is a point on the perpendicular bisector of AC), and $FE = GE$ (since E is a point on the

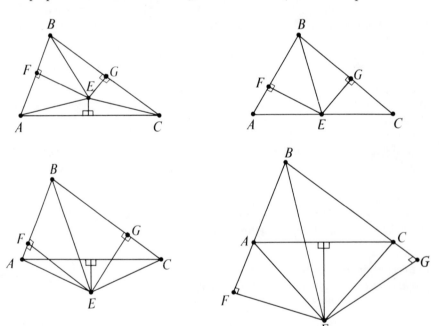

FIGURE 1.1

bisector of $\angle B$). Thus $\triangle FAE \cong \triangle GCE$ (hypotenuse, arm). Therefore,

(2) $$FA = GC$$

(corresponding parts of congruent triangles).

In the case of the first three diagrams, we add (1) and (2) getting

$$BF + FA = BG + GC$$

or $BA = BC$. In the case of the fourth diagram, we subtract (2) from (1), getting

$$BF - FA = BG - GC$$

and again $BA = BC$. In either case $\triangle ABC$ is isosceles.

2. Where Does the Difficulty Lie

This is rather distressing, since the proof seems as sound as many standard proofs in high school geometry and sounder than much of the reasoning encountered in everyday discourse. It may occur to us that an accurately drawn diagram might disclose a clue to the paradox. Suppose we draw one and find in our diagram (Figure 1.2) that point E lies outside $\triangle ABC$, F lies on the extension of AB beyond A, and G lies between B and C. Then, as above, $\triangle BFE \cong \triangle BGE$, $\triangle FAE \cong \triangle GCE$, $BF = BG$, and $FA = GC$. Now we cannot obtain $BA = BC$ from the preceding equations either by addition

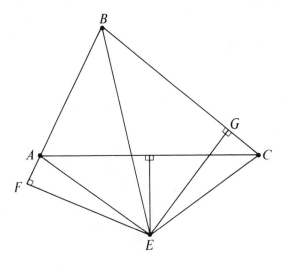

FIGURE 1.2

or by subtraction, for addition yields

$$BF + FA = BG + GC,$$

and although $BG + GC$ does equal BC, $BF + FA$ does not equal BA since BF itself is greater than BA. Similarly, subtraction yields

$$BF - FA = BG - GC,$$

and $BF - FA = BA$, but $BG - GC$ is less than BC.

Does this demolish the paradox? The answer is no. We have shown merely that for the case illustrated by Figure 1.2 the proof fails, and so perhaps not all triangles are isosceles. We do not know, however, that an arbitrary scalene triangle must fall into this case. It may fit into one of the previous cases, and so prove to be isosceles.

We can resolve the paradox by proving that the case of Figure 1.2 is essentially the only case. That is, for any scalene triangle ABC, the bisector of angle B meets the perpendicular bisector of AC in a point E which is outside the triangle, and the perpendiculars from E to AB and BC are such that the foot of one falls on a side of the triangle and so is between two vertices, whereas the foot of the other falls on the extension of a side of the triangle and so is not between two vertices. (This is considered in the exercises at the end of the chapter.)

Our analysis indicates that, for a completely logical treatment of geometry, we must study the concepts of "inside" and "outside" of a triangle, and the concept of a point on a line lying between two other points of the line.

3. A Triangle with Two Right Angles

Consider next a "proof" that there exists a triangle with two right angles:

Let two circles meet in points A, B (see Figure 1.3). Let AC, AD be their respective diameters from A. Let CD meet the respective circles in E, F. Then

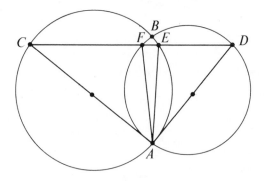

FIGURE 1.3

$\angle AEC$ is a right angle, since it is inscribed in semicircle AEC. Similarly, $\angle AFD$ is a right angle. Thus $\triangle AEF$ has two right angles.

Here, a carefully drawn diagram might *suggest* that CD passes through B. But we cannot *argue* that CD passes through B because it *appears* so in a carefully drawn diagram.

4. Proof of a Theorem

The above results were not taught in high school as theorems; and, of course, they shouldn't be! There are theorems, however, whose proofs *are* logically deficient. Consider, for example, the following proof taken from a textbook in plane geometry.

THEOREM.   The base angles of an isosceles triangle are equal.

FIGURE 1.4

Given: $\triangle ABC$ with $AC = BC$.
To prove: $\angle A = \angle B$.

Proof.

Statements	Reasons
1. Draw the bisector of $\angle C$.	1. Every angle has a bisector.
2. Extend it to meet AB at D.	2. A line may be extended.
3. In $\triangle ACD$ and BCD, $AC = BC$.	3. Hypothesis.
4. $\angle 1 = \angle 2$.	4. Definition of angle bisector.
5. $CD = CD$.	5. Identity.
6. $\triangle ACD \cong \triangle BCD$.	6. SAS.
7. $\angle A = \angle B$.	7. Corresponding parts of congruent triangles are equal.

As we read the proof, and glance at the diagram, we find it most convincing; but is it a valid argument?

To test it, suppose we cover the diagram, and re-examine the argument. There is a difficulty in Step 2. The reason, *A line may be extended*, does not justify extending the line to meet a second line. It may be that the two lines are parallel.

Suppose for a moment that this difficulty has been overcome, and we know that the bisector of angle *C* meets *AB* in a point *D*. Let us continue our analysis of the proof. Steps 3, 4, 5, 6 seem justified. In Step 7, we have $\angle A = \angle B$. This means

(1) $$\angle CAD = \angle CBD,$$

since these are the corresponding angles of the congruent triangles *ACD*, *BCD*. This is not what we set out to prove. The base angles of triangle *ABC* are $\angle CAB$ and $\angle CBA$, and we must prove

(2) $$\angle CAB = \angle CBA.$$

But it seems evident that

$$\angle CAD = \angle CAB \quad \text{and} \quad \angle CBD = \angle CBA,$$

so that (2) follows from (1) by substitution. The phrase "seems evident" means here that the diagram suggests the inference so strongly that the contrary does not even occur to us. This, however, is not the same as a logical proof. In order to assert $\angle CAD = \angle CAB$, we must know that *D* lies on side *AB* of $\angle CAB$; otherwise these angles are supplementary. And to assert $\angle CBD = \angle CBA$ we must know that *D* is on side *BA* of $\angle CBA$. Thus to complete the proof, we need to know that *D* is on *segment AB*, or that *D* is *between A* and *B*. Unfortunately, there is nothing in the *proof* (as distinct

 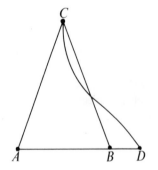

FIGURE 1.5

from the *diagram*) to justify this. As for the proof, D might be on the prolongation of AB beyond B in which case B is between A and D. Consider the adjoining diagrams (Figure 1.5) which illustrate this possibility.

You may be shocked by these and say: "This can't happen. How could the bisector of $\angle ACB$ lie outside the angle; or how could a line through vertex C of $\angle ACB$ get outside the angle once it is inside it? Straight lines don't behave this way." Do you mean straight chalk streaks on a blackboard? Of course they don't behave this way. But we are off the track. We are not making a visual study of chalk streaks, but a *logical* study of whether a certain conclusion can be deduced from our basic assumptions.

On the basis of the postulates which Euclid (or a standard high school geometry text) explicitly assumes, it is not possible to prove that D is between A and B. What Euclid (and the familiar texts) frequently do is to assume implicitly certain properties of geometric figures which are not stated as postulates. Consequently many of the standard "proofs" are not logically satisfactory. To convert them into valid proofs, we must make explicit the properties which have been tacitly (and sometimes unconsciously) assumed from the diagram.

In the present case, the logical deficiency could be eliminated by assuming a new postulate:

The bisector of an angle intersects any segment which joins the sides of the angle.

Then D will exist, and will be between A and B.

The analysis indicates again the need to characterize the concept of betweenness for points on a line, and the concepts of the inside and outside of a geometric figure, particularly an angle. In the standard treatments, the concept of "inside" is not discussed. A point is observed to be inside a given figure just on the basis of unanalyzed visual impressions. This is not logically satisfactory. There are important properties involving "inside" and "outside" which must be included in our treatment of geometry, either as postulates, theorems, or definitions. For example:

A line which contains the vertex of an angle and a point inside the angle intersects any segment joining the sides of the angle. (Compare this with the postulate suggested above concerning the bisector of an angle.)

The theorem that the base angles of an isosceles triangle are equal can be proved validly by using a different and simpler approach. The following proof attributed to Pappus (300 A.D.) is based on the validity of the SAS principle when applied to a triangle and itself.

Given: $\triangle ABC$ with $AC = BC$.

To prove: $\angle BAC = \angle ABC$.

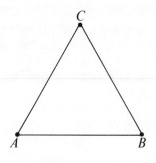

FIGURE 1.6

Proof. Consider *ACB, BCA* as triangles with vertices *A, C, B* corresponding to *B, C, A*, respectively.

1. $AC = BC$.	1. Given.
2. $BC = AC$.	2. Given.
3. $\angle ACB = \angle BCA$.	3. Identity.

Thus the parts *AC, BC,* $\angle ACB$ of $\triangle ACB$ are equal to the corresponding parts *BC, AC,* $\angle BCA$ of $\triangle BCA$.

4. Therefore $\triangle ACB \cong \triangle BCA$.	4. SAS.
5. $\angle BAC = \angle ABC$.	5. Corresponding parts of congruent triangles are equal.

5. Another Flawed Proof

We shall now examine another familiar proof in high school geometry to show how easily logical flaws may creep into an argument.

THEOREM.   The diagonals of a parallelogram bisect each other.

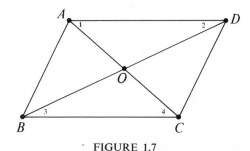

FIGURE 1.7

Given: Parallelogram *ABCD* with diagonals *AC* and *BD*.
To prove: $AO = OC, BO = OD.$

Proof.

Statements	Reasons
1. $AD \parallel BC, AB \parallel DC.$	1. By hypothesis *ABCD* is a parallelogram.
2. $\angle 1 = \angle 4, \angle 2 = \angle 3.$	2. Alternate interior angles of parallel lines are equal.
3. $AD = BC.$	3. Opposite sides of a parallelogram are equal.
4. $\triangle AOD \cong \triangle COB.$	4. ASA.
5. $AO = OC, BO = OD.$	5. Corresponding parts of congruent triangles are equal.

The proof appears to be quite logical, since each step seems carefully justified. The flaw is in the statement of what is to be proved, in which it is tacitly assumed that the diagonals *AC* and *BD* meet in a point *O*. There is nothing in the hypothesis to warrant this assumption. It is, of course, visually evident, and we certainly want it to be a valid statement (postulate or theorem) in our treatment of geometry.

To indicate the difficulties involved in proving that the diagonals of a parallelogram intersect, observe that this property is related to the property that the diagonals of a parallelogram are wholly inside the figure. This is not true for all polygons, nor even for all quadrilaterals. In Figure 1.8, observe that—disregarding endpoints—diagonal *AC* is wholly inside quadrilateral *ABCD*, whereas diagonal *BD* is wholly outside. Clearly these diagonals do not meet.

To avoid a deep and difficult analysis of the concepts of the inside and outside of a parallelogram, we might adopt the following postulate:

The diagonals of a parallelogram intersect.

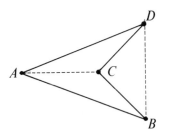

FIGURE 1.8

By using this postulate the proof can easily be validated. But the postulate seems rather specialized. (Should we, for example, have a similar postulate for a trapezoid?) We might wonder if a different approach would enable us to prove the theorem without assuming the postulate. The following analysis suggests a method of doing this.

Analysis: Let M be the midpoint of AC. To show that AC and BD bisect each other is equivalent to showing that M is the midpoint of BD. This is difficult to do directly since we do not even know that M is on BD. This suggests that we extend BM its own length through M to a point D' (see Figure 1.9), and then show that D' is the same point as D. The details are given in the following proof.

Given: Parallelogram $ABCD$.
To prove: AC and BD bisect each other.

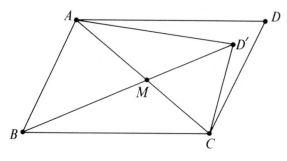

FIGURE 1.9

Proof.

Statements	Reasons
1. Let M be the midpoint of AC, so that $AM = MC$.	1. A line segment has a midpoint.
2. Extend BM its own length through M to D' so that $BM = MD'$.	2. A line segment can be extended its own length.
3. $\angle AMD' = \angle CMB$.	3. Vertical angles are equal.
4. $\triangle AMD' \cong \triangle CMB$.	4. SAS.
5. $\angle MAD' = \angle MCB$.	5. Corresponding parts of congruent triangles are equal.
6. $AD' \parallel BC$.	6. If two lines are cut by a transversal so as to make a pair of alternate interior angles equal, then the lines are parallel.

7. $AD \parallel BC$.

7. By hypothesis $ABCD$ is a parallelogram.

8. Lines AD' and AD coincide, so that D' lies in line AD.

8. Through a point not on a given line there is only one line parallel to the given line.

In a similar way, using triangles CMD' and AMB, we show that lines CD' and CD coincide, so that D' lies in line CD.

9. D' and D coincide.

9. Two lines can intersect in only one point.

10. $BM = MD$.

10. By substituting D for D' in Statement 2.

Thus M is the midpoint of BD, and AC and BD bisect each other, since they have the same midpoint M.

Although the proof validates the theorem without assuming an additional postulate, its methods do not shed light on the general problem of the intersection of the diagonals of a quadrilateral (for example, it does not help us to prove that the diagonals of a trapezoid meet). The problem requires a deep study of positional and ordinal properties of points and lines, which is made later in the book.

6. Summary

The following points should be clear from our discussion: (1) Euclidean geometry, as we conceive it, is not the study of carefully drawn figures—it is not a branch of engineering drawing*; (2) from the point of view of logic, to determine that a point lies between two points by *observation* in a diagram, is just as inadmissible as to determine that two segments are equal by *measurement* in a diagram; (3) the standard treatments of high school geometry are logically deficient in that certain properties are tacitly assumed on the basis of diagrams or spatial intuition. Since we want to be able to make our demonstrations of theorems logically flawless, we must make explicit these tacit assumptions and use them formally in the development of the subject. After devoting several chapters to further critical and historical discussion of Euclidean and non-Euclidean geometry, we shall tackle the problem of creating a basis for Euclidean and other geometries which is free from the objections raised here.

---

* Although diagrams are not to be used to supplant reasons in formal proofs it must be emphasized that they have an important role in clarifying difficulties and discovering geometrical relationships.

7. Implications for Teaching

A final word on the relation of our work to the study of geometry in high school. The severity of our criticism of the conventional treatment might lead one to conclude that we consider it wholly inappropriate for high school students, and that a logically sound, rigorous treatment of the type constructed in this book should be taught to high school students. We hold no such belief. Our criticism is directed toward showing the inadequacy of school geometry as a definitive mathematical treatment of the subject.

Euclidean geometry, like all branches of mathematics, is to be taught at various levels of abstraction; what is appropriate for a tenth grader would not be suitable for a college sophomore or a first year graduate student. At the high school level we believe that geometry should be presented essentially as a deductively organized subject which is suggested by and is applicable to physical space, not as an abstract mathematical science. This will tend to foster the student's feeling that deductive reasoning is a worthwhile human activity—not a mere logical game—in which he reasons about important concrete material, justifying its properties and even discovering new ones. It will better prepare the student for later, more abstract treatments, since he will have an understanding of the concrete material which motivates the abstract theory and gives it point. It would be a grave error to obliterate the history and concrete origin of our subject in teaching at the secondary level (or even at higher levels) under the guise of fostering an appreciation of abstract reasoning or putting the student in contact with modern mathematics. It might well produce antipathy to abstract reasoning.

Similar considerations apply to the question of the level of rigor at which the material is taught. A sense of rigor is more comparable to a personal quality than to subject matter to be mastered in a fixed time. It is a quality of mind that develops and matures, often quite slowly, in response to need and growing insight. An attempt to force it prematurely may yield a mere formalistic or mechanical sense of rigor, not a deep sense of critical alertness.

However, we do feel that an understanding of the rigorous development of geometry as presented in this book is important for teachers. They should make contact with the best treatments of their subject available at higher levels, and in the light of their own understanding, try to make it as interesting and significant as possible to the student at his level.

EXERCISES

These exercises are to be done in the framework of high school geometry, using diagrams. Many are designed to give a deeper insight into certain graphical properties of figures, especially those encountered in Sections 1, 2, 3. For the sake of clarity we use $\overline{AB}$ to denote segment AB.

1. Let two circles meet in points A, B. Let $\overline{AC}$, $\overline{AD}$ be their respective diameters from A. Prove that B is on line CD.

2. From a point P on side BA of $\angle ABC$, a perpendicular is dropped onto side BC (extended if necessary) of the angle. Prove that the foot of the perpendicular is on side BC of $\angle ABC$ or on the extension of side BC according as $\angle ABC$ is acute or obtuse.

3. In $\triangle ABC$ let $\overline{AB} \neq \overline{BC}$. Prove that the bisector of $\angle B$ and the perpendicular bisector of $\overline{AC}$ meet in just one point. (That is, they have a point, but not two points, in common.)

4. In $\triangle ABC$ let $\overline{AB} \neq \overline{BC}$, and let E be the point of intersection of the bisector of $\angle B$ and the perpendicular bisector of $\overline{AC}$. Prove that A, B, C, and E lie on a circle. (Note how strongly this supports the conclusion that E is outside $\triangle ABC$.)

5. In $\triangle ABC$ let the bisector of $\angle B$ meet the circle containing A, B, C in a point E distinct from B. Prove that $\overline{BE}$ is a diameter of the circle if and only if $\overline{AB} = \overline{BC}$.

6. In $\triangle ABC$ let $\overline{AB} \neq \overline{BC}$, and let the bisector of $\angle B$ meet the perpendicular bisector of $\overline{AC}$ in E. Prove that one of the angles BAE, BCE is acute, and the other obtuse.

7. In $\triangle ABC$ let $\overline{AB} \neq \overline{BC}$, and let the bisector of $\angle B$ meet the perpendicular bisector of $\overline{AC}$ in E. From E drop perpendiculars to sides $\overline{AB}$, $\overline{BC}$ (extended if necessary) of the triangle. Prove that one of the feet of the perpendiculars falls on the side to which it was dropped, and the other on an extension of the side.

8. Prove that every triangle has an "interior" altitude (one that meets the side to which it is drawn).

9. How many interior altitudes can a triangle have? Justify your answer.

10. Criticize the following proof that any right angle is obtuse.

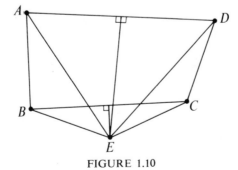

FIGURE 1.10

Let $\angle ABC$ be a right angle. Choose point D such that $\angle BCD$ is obtuse, and $\overline{AB} = \overline{CD}$. We show $\angle ABC = \angle BCD$. Draw $\overline{AD}$. Let E be the intersection of the perpendicular bisectors of $\overline{AD}$ and $\overline{BC}$. Join E to A, B, C, D. Then $\overline{EB} = \overline{EC}$ since E is on the perpendicular bisector of $\overline{BC}$. Similarly $\overline{EA} = \overline{ED}$. Thus $\triangle EAB \cong \triangle EDC$ by SSS, so that

(1) $$\angle EBA = \angle ECD.$$

From $\overline{EB} = \overline{EC}$ it follows that

(2) $$\angle EBC = \angle ECB.$$

Subtracting (2) from (1), we get

$$\angle ABC = \angle BCD.$$

11. Criticize the following proof.

Every point inside a circle, other than the center, lies on the circle.

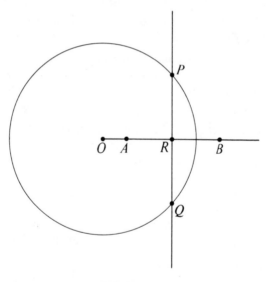

FIGURE 1.11

Given: Circle with center O, radius r. A, any point inside the circle distinct from O.

To Prove: A is on the circle.

Proof. Let B be on the extension of $\overline{OA}$ through A such that

(1) $$OA \cdot OB = r^2.$$

Let the perpendicular bisector of $\overline{AB}$ meet the circle in P and Q; let R be the midpoint of $\overline{AB}$. Then

(2) $$OA = OR - RA, \qquad OB = OR + RA.$$

(1), (2) yield

$$r^2 = (OR - RA)(OR + RA)$$
$$= \overline{OR}^2 - \overline{RA}^2$$
$$= (r^2 - \overline{PR}^2) - (\overline{AP}^2 - \overline{PR}^2)$$
$$= r^2 - \overline{AP}^2.$$

Thus $\overline{AP}^2 = 0$ and $AP = 0$. Hence A coincides with P and is on the circle.

12. Criticize the following proof.

If one pair of opposite sides of a quadrilateral are equal the second pair are parallel.

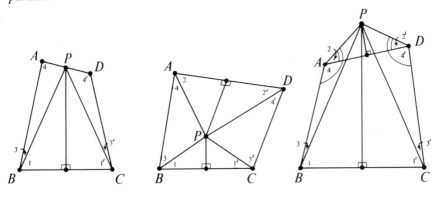

FIGURE 1.12

Given: Quadrilateral $ABCD$ with $\overline{AB} = \overline{CD}$.
To Prove: $AD \parallel BC$.

Proof. If the perpendicular bisectors of $\overline{BC}$ and $\overline{AD}$ are parallel, or coincide, certainly $AD \parallel BC$. Suppose then the perpendicular bisectors of $\overline{BC}$, $\overline{AD}$ meet in a single point P. Either P is on $ABCD$, inside $ABCD$, or outside $ABCD$. In all cases $\overline{PB} = \overline{PC}$ and $\overline{PA} = \overline{PD}$, since P is on the perpendicular bisectors of $\overline{BC}$ and $\overline{AD}$. Suppose P not on $ABCD$. Then

(1) $$\angle 1 = \angle 1',$$

(2) $$\angle 2 = \angle 2'.$$

$\triangle ABP \cong \triangle DCP$ by SSS so that

(3) $\angle 3 = \angle 3',$

(4) $\angle 4 = \angle 4'.$

Suppose P inside $ABCD$. Adding (1) and (3) we obtain

(5) $\angle B = \angle C.$

Similarly, addition of (2) and (4) yields

(6) $\angle A = \angle D.$

Since $\angle A + \angle B + \angle C + \angle D = 360°$, (5) and (6) yield $\angle A + \angle B = 180°$ and we infer $AD \parallel BC$.

If P is outside $ABCD$, (5) and (6) still follow from (1), (2), (3), (4), in one case by addition and in one case by subtraction. If P is on $ABCD$ a similar but simpler argument applies.

13. Criticize the following proof taken from a geometry text. Indicate how it may be improved and find as many "hidden assumptions" as you can.

THEOREM.   Two triangles are congruent if three sides of one are equal, respectively, to three sides of the other.

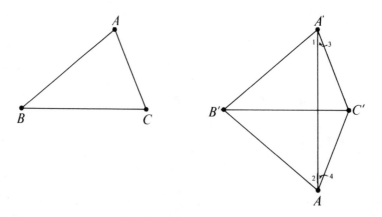

FIGURE 1.13

Given: $\triangle ABC$ and $\triangle A'B'C'$, with $AB = A'B'$, $BC = B'C'$, $AC = A'C'$.
To Prove: $\triangle ABC \cong \triangle A'B'C'$.

Proof.

Statements	Reasons
1. Place $\triangle ABC$ so that BC and $B'C'$ coincide and A and A' lie on opposite sides of $B'C'$.	1. A geometric figure may be moved about without changing size or shape; equal segments can be made to coincide.
2. Draw AA'.	2. Two points determine a line.
3. $\triangle AB'A'$ is isosceles.	3. $AB = A'B'$ by hypothesis.
4. $\angle 1 = \angle 2$.	4. The base angles of an isosceles triangle are equal.
5. $\triangle AC'A'$ is isosceles.	5. $AC = A'C'$ by hypothesis.
6. $\angle 3 = \angle 4$.	6. See 4.
7. $\angle 1 + \angle 3 = \angle 2 + \angle 4$,	7. If equals are added to equals the sums are equal.
8. or $\angle A' = \angle A$.	8. Substitution.
9. $\triangle AB'C' \cong \triangle A'B'C'$, that is $\triangle ABC \cong \triangle A'B'C'$.	9. SAS.

14. Examine some of the proofs in a high school text to see how sound they are.
15. Formulate some of the basic properties of betweenness for points on a line which are implicit in high school geometry.
16. Formulate some of the basic properties of points inside (outside) a triangle. Try to frame a simple definition of the idea that a point is inside (outside) a triangle.
17. The same as Exercise 16, replacing triangle by angle.
18. The same as Exercise 16, replacing triangle by tetrahedron or triangular pyramid.

R E F E R E N C E S

T. L. Heath, *The Thirteen Books of Euclid's Elements*, 2nd ed., 3 vols, Cambridge University Press, New York, 1926. Reprinted by Dover Publications, Inc., 1956.
E. A. Maxwell, *Fallacies in Mathematics*, Cambridge University Press, Cambridge, 1959.
School Mathematics Study Group, *Geometry*, Student's Text, Parts I and II, Yale University Press, New Haven, 1961.

CHAPTER 2

Euclid's Parallel Postulate

In Chapter 1 we have seen that geometric proofs which draw conclusions from diagrams are considered unsatisfactory today. They do not meet current standards of rigor. On the other hand, Euclid, who was a consummate logician, leaned heavily on diagrams in his proofs. How did this change in attitude come about?

It seems likely that the main force which motivated this change was the development of non-Euclidean theories of geometry which contradicted Euclid's Parallel Postulate. So long as mathematicians believed that Euclidean geometry was the only possible theory of space, and that it exactly described the physical world, it never occurred to them that diagrams might mislead them. But when the absolute and unique position of Euclidean geometry was assailed early in the nineteenth century by the discovery of non-Euclidean geometry, mathematicians were badly shaken. A revolution in mathematics occurred, comparable to the Copernican revolution in astronomy or the Darwinian revolution in biology. Ideas about the nature of geometry and the unique position of Euclidean geometry, which had been held by the greatest minds for over two thousand years, were destroyed in the decade of 1820–1830.

In Chapter 4 we shall present an elementary introduction to non-Euclidean geometry. We prepare for it in this chapter by discussing Euclid's famous Parallel Postulate. For two millenia mathematicians were dissatisfied with it and made numerous attempts to deduce it as a theorem from Euclid's other, seemingly simpler, postulates. Right up to the beginning of the nineteenth century, competent mathematicians were convinced that they had

20

solved the problem, only to have flaws discovered in their proofs. The failure of every attempt to prove the parallel postulate led eventually to the realization that the parallel postulate was not indubitable, that Euclid's theory was not sacrosanct, and that other (non-Euclidean) theories of geometry were possible. Later in this chapter we shall present three important and typical attempts to prove Euclid's Parallel Postulate.

1. The Structure of Euclidean Plane Geometry

Euclid's parallel postulate, which is one of the most important single sentences in the history of intellectual controversy, may be stated as follows:

If two lines are cut by a transversal in such a way that the sum of two interior angles on one side of the transversal is less than 180°, the lines will meet on that side of the transversal.

The historical importance of the parallel postulate is based on the important role it plays in Euclid's theory. To make this role evident it is necessary to examine the structure of his theory. Therefore, we begin with a sketch of the Euclidean theory of plane geometry. Our treatment does not follow the details of Euclid's development too closely, but paraphrases his basic ideas in modern terms, and is close to the treatment usually accorded his work in current school texts.

We begin by listing a number of assumptions or postulates for Euclidean plane geometry.

I.   *Things equal to the same thing or to equal things are equal to each other.*
II. *If equals are added to equals, the sums are equal.*
III. *If equals are subtracted from equals, the differences are equal.*
IV. *The whole is greater than any of its parts.*
V.   *A geometric figure can be moved without change of size or shape.*
VI. *Every angle has a bisector.*
VII. *Every segment has a midpoint.*
VIII. *Two points are on one and only one line.*
IX. *Any segment can be extended by a segment equal to a given segment.*
X.   *A circle can be drawn with any given center and radius.*
XI. *All right angles are equal.*

From these postulates we can deduce (with the attendant logical deficiencies pointed out in Chapter 1) a number of elementary theorems. Among these are:

(1)   Vertical angles are equal.
(2)   Congruence properties of triangles (SAS, ASA, SSS).
(3)   The theorem on the equality of the base angles of an isosceles triangle, and its converse.
(4)   The existence of a unique perpendicular to a given line at a given point of the line.
(5)   The existence of a perpendicular to a given line through an external point.
(6)   The construction of an angle equal to a given angle with preassigned vertex and side.
(7)   The construction of a triangle congruent to a given triangle with preassigned side equal to a side of the given triangle.

Now we can prove the Exterior Angle Theorem, a key to later developments. First we introduce a notation which will be used in the proof.

NOTATION: $\overline{AB}$ denotes segment AB.

THEOREM 1.   (EXTERIOR ANGLE THEOREM)   An exterior angle of a triangle is greater than either remote interior angle.

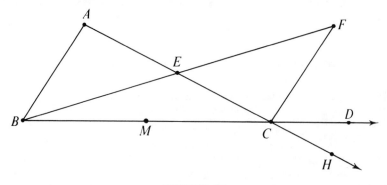

FIGURE 2.1

Proof.   Let ABC be any triangle, and let D be on the extension of $\overline{BC}$ through C. We show first that exterior angle $\angle ACD$ is greater than $\angle A$.

Let E be the midpoint of $\overline{AC}$, and let $\overline{BE}$ be extended its own length through E to F. Then $\overline{AE} = \overline{EC}$, $\overline{BE} = \overline{EF}$, and $\angle AEB = \angle CEF$ (vertical angles are equal). Thus $\triangle AEB \cong \triangle CEF$ (SAS), and $\angle BAE = \angle FCE$ (corresponding parts of congruent triangles are equal). Since $\angle ACD >$ $\angle FCE$ (the whole is greater than any of its parts), we conclude $\angle ACD >$ $\angle BAE = \angle A$.

To show $\angle ACD > \angle B$, extend $\overline{AC}$ through C to H, forming $\angle BCH$. Then show $\angle BCH > \angle B$, using the procedure of the first part of the proof: let M be the midpoint of $\overline{BC}$, extend $\overline{AM}$ its own length through M, et cetera. To complete the proof, observe that $\angle BCH$ and $\angle ACD$ are vertical angles and so are equal.

Note that the criticisms in Chapter 1 apply to this proof. The statement $\angle ACD > \angle FCE$ depends on the diagram.

It is now easy to prove several important results.

THEOREM 2.   If two lines are cut by a transversal so as to form a pair of equal alternate interior angles, then the lines are parallel.

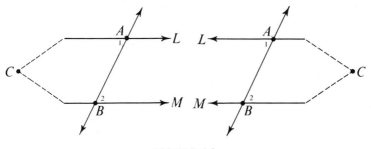

FIGURE 2.2

Proof. Recall that two lines in the same plane are defined to be *parallel* if they do not meet. Let a transversal cut two lines L, M in points A, B to form a pair of alternate interior angles, $\angle 1$ and $\angle 2$, which are equal, and suppose that L and M are not parallel. Then they meet in a point C, forming $\triangle ABC$. C lies on one side of AB or the other. In either case, an exterior angle of $\triangle ABC$ is equal to a remote interior angle. (For example, if C is on the same side of AB as $\angle 2$ then exterior angle $\angle 1$ is equal to the remote interior angle $\angle 2$.) This contradicts the preceding theorem. Therefore L and M are parallel.

COROLLARY 1.   Two lines perpendicular to the same line are parallel.

As an immediate consequence of Corollary 1 we have

COROLLARY 2.   There is only one perpendicular to a given line through an external point.

COROLLARY 3.   (EXISTENCE OF PARALLEL LINES)   If a point P is not on a given line L, then there exists at least one line through P which is parallel to L.

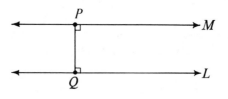

FIGURE 2.3

Proof.   From P drop a perpendicular to L with foot Q, and at P erect line M perpendicular to PQ. Then M is parallel to L by Corollary 1.

THEOREM 3.   The sum of two angles of a triangle is less than 180°.

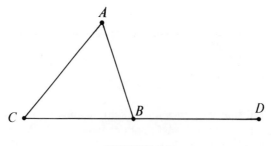

FIGURE 2.4

Proof.   Let $\triangle ABC$ be any triangle. We show $\angle A + \angle B < 180°$.
Extend $\overline{CB}$ through B to D. Then $\angle ABD$ is an exterior angle of $\triangle ABC$. By Theorem 1, $\angle ABD > \angle A$. But $\angle ABD = 180° - \angle B$. By substituting for $\angle ABD$ in the first relation, we have

$$180° - \angle B > \angle A, \quad \text{or} \quad 180° > \angle A + \angle B.$$

Thus $\angle A + \angle B < 180°$, and the theorem is proved.

2. A Substitute for Euclid's Parallel Postulate

In Section 1 we have, in essence, sketched the beginning of Euclid's development of plane geometry. Observe that no use has been made of the parallel postulate. The further development requires it. In current textbooks, however, Euclid's parallel postulate is usually replaced by the following statement:

There exists only one line parallel to a given line through a given point not on the line.

This is called Playfair's Postulate. How is it related to Euclid's parallel postulate? Certainly the two statements are not the same. The former is a statement about parallel lines, the latter about intersecting lines. Yet they play the same role in the logical development of geometry. We say the statements are *logically equivalent* or simply *equivalent*. This means that if the first statement is taken as a postulate (together with all of Euclid's postulates except the parallel postulate), then the second statement can be deduced as a theorem; and conversely, if the second is taken as a postulate (together with all of Euclid's postulates except the parallel postulate), then the first can be deduced as a theorem. It is thus logically immaterial which of the two statements we assume as a postulate, and which we deduce as a theorem.

3. The Equivalence of Euclid's and Playfair's Postulates

We now prove the equivalence of Euclid's parallel postulate and Playfair's postulate.

First, assuming Euclid's parallel postulate, we deduce Playfair's postulate.

Given line L and point P not on L (Figure 2.5). We show there is only one line through P parallel to L.

We know that there exists a line through P parallel to L, and we know how to construct it (see Corollary 3 to Theorem 2). From P we drop a perpendicular to L with foot Q, and at P erect line M perpendicular to PQ. Then M is parallel to L.

Now let N be any line through P distinct from M. We show N meets L. Let $\angle 1$, $\angle 2$ denote the angles line N makes with PQ. Then $\angle 1$ is not a

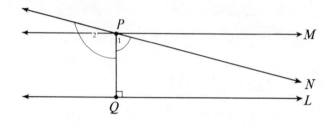

FIGURE 2.5

right angle for otherwise N and M coincide, contrary to assumption. Thus $\angle 1$ or $\angle 2$ is acute, say $\angle 1$.

Summary. Lines L and N are cut by transversal PQ so as to form an acute angle $\angle 1$ and a right angle, which are interior angles on the same side of the transversal. Since the sum of these angles is less than 180°, Euclid's parallel postulate applies and we conclude that N meets L. Thus M is the *only* line through P which is parallel to L, and we have deduced Playfair's postulate from Euclid's parallel postulate.

Now assuming Playfair's postulate, we deduce Euclid's parallel postulate.

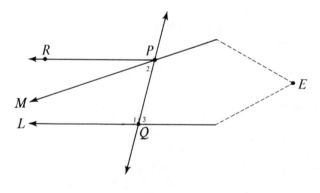

FIGURE 2.6

Let lines L, M be cut by a transversal in points Q, P forming $\angle 1$ and $\angle 2$, a pair of interior angles on one side of the transversal whose sum is less than 180° (Figure 2.6). That is

(1) $$\angle 1 + \angle 2 < 180°.$$

Let $\angle 3$ denote the supplement of $\angle 1$ which lies on the opposite side of PQ from $\angle 1$ and $\angle 2$ (see Figure 2.6). Then

(2) $$\angle 1 + \angle 3 = 180°.$$

Relations (1), (2) imply

(3) $$\angle 2 < \angle 3.$$

At P construct $\angle QPR$ equal to, and alternate interior to, $\angle 3$. Then $\angle 2 < \angle QPR$, so that RP is distinct from M. By Theorem 2, RP is parallel to L. Hence by Playfair's postulate, M is not parallel to L; therefore M and L meet.

Suppose they meet on the opposite side of PQ from $\angle 1$ and $\angle 2$, say in point E. Then $\angle 2$ is an exterior angle of $\triangle PQE$; hence $\angle 2 > \angle 3$, contrary to (3). Consequently, the supposition is false, and so M and L meet on the side of transversal PQ containing $\angle 1$ and $\angle 2$. Thus Euclid's parallel postulate follows from Playfair's postulate and the two postulates are equivalent.

4. The Role of Euclid's Parallel Postulate

Sections 2 and 3 have been a digression devoted to relating Euclid's parallel postulate to its modern counterpart. We shall return now to the topic of Section 1, the structure of the Euclidean theory. The results of Section 1 do not depend on the parallel postulate. Further progress requires it. Most of the important and powerful results considered typical of Euclidean geometry are consequences of the parallel postulate.

By assuming Euclid's parallel postulate (or equivalently Playfair's postulate), the following important results can be justified:

1. *If two parallel lines are cut by a transversal, any pair of alternate interior angles formed are equal.*
2. *The sum of the angles of any triangle is 180°.*
3. *The opposite sides of a parallelogram are equal.*
4. *Parallel lines are everywhere equidistant.*
5. *The existence of rectangles and squares.*
6. *The familiar theory of area in terms of square units.*
7. *The theory of similar triangles, including the existence of a figure of arbitrary size similar to a given figure.*

Now you see why we (and mathematicians from the Greeks onward) have been so concerned over Euclid's parallel postulate. It is the source of so many important results. Without it (or an equivalent) we would not have the familiar theories of area, of similarity, and the powerful Pythagorean relation. Without it school geometry would seem a pallid counterpart of itself. Euclid's parallel postulate may not have seemed so important when you studied geometry in high school, since it is used only once in order to derive the basic result 1 on alternate interior angles, which is then constantly used to derive further results.

The manner in which Euclid arranged his theorems suggests that he was not wholly satisfied with his parallel postulate. He stated it at the beginning of his work but did not use it until he could advance no further without it. Presumably Euclid had the feeling that the parallel postulate did not have the simple, intuitive quality of the other postulates. Such a feeling *was* held by geometers for twenty centuries. They tried to deduce the parallel postulate from the other postulates, or to replace it by a postulate which seemed indubitable. We now discuss three such attempts to "solve the problem" of Euclid's parallel postulate.

5. Proclus' Proof of Euclid's Parallel Postulate

Proclus (410–485) gave a "proof" of Euclid's parallel postulate which we paraphrase as follows:

We assume Euclid's postulates other than the parallel postulate, and deduce Playfair's postulate. Let P be a point not on line L (Figure 2.7). We construct line M through P parallel to L in the familiar way. Let PQ be perpendicular to L at Q, and let M be perpendicular to PQ at P. Now suppose there is another line N through P parallel to L. Then N makes an acute angle with PQ, that lies let us say on the right side of PQ. The part of N to the right of P is then wholly contained in the region bounded by L, M, and PQ. Now let X be any point of M to the right of P. Let XY be perpendicular to L at Y and let it meet N at Z. Then $\overline{XY} > \overline{XZ}$. Let X recede endlessly

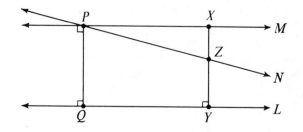

FIGURE 2.7

on M. Then $\overline{XZ}$ increases indefinitely, since $\overline{XZ}$ is at least as large as the segment from X perpendicular to N. Thus $\overline{XY}$ also increases indefinitely. But the distance between two parallel lines must be bounded. Therefore we have a contradiction and our supposition is false. Thus M is the only line through P parallel to L. Hence Playfair's postulate holds, and so does its equivalent, Euclid's parallel postulate.

This argument has a rather modern ring; we do not expect the sophisticated notion of variation to appear in an argument in elementary geometry by a fifth century mathematician. Let us examine it. At face value it involves three assumptions:

(A) *If two lines intersect, the distance to one from a point on the other increases indefinitely, as the point recedes endlessly.*

(B) *The shortest segment joining an external point to a line is the perpendicular segment.*

(C) *The distance between two parallel lines is bounded.*

(A) and (B) can be justified without recourse to Euclid's parallel postulate.* Thus the crux of the proof is the assumption (C). In effect, Proclus tacitly assumed (C) as an additional postulate. Let us call this hidden assumption Proclus' postulate. Then we may assert: *Proclus' postulate is equivalent to Euclid's parallel postulate.* For, as we indicated in result 4 of Section 4, Euclid's parallel postulate implies that the distance between parallel lines is constant, and so bounded. Conversely, it follows from Proclus' argument that Proclus' postulate implies Euclid's parallel postulate.

Thus Proclus merely replaced the parallel postulate by an equivalent postulate, and did not, as he believed, establish the validity of the parallel postulate.

6. Wallis' Solution of the Problem

John Wallis (1616–1703) consciously replaced Euclid's parallel postulate by the following postulate:

There exists a triangle with one side arbitrarily preassigned which is similar to a given triangle.

---

* For (A) see Exercises I,15 at end of Chapter 3. For (B) see Theorem 7-6, p. 206, School Mathematics Study Group, *Geometry*, Student's Text, Part I, Yale University Press, New Haven, 1961.

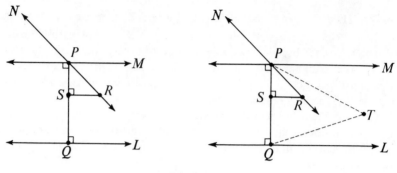

FIGURE 2.8

From this, Playfair's postulate can be deduced as follows:

Let P be a point not on line L. From P drop PQ perpendicular to L, meeting L in Q, and at P erect line M perpendicular to PQ (Figure 2.8). Let N be any line other than M containing P. We show N meets L. Let R be any point on N in the region between L and M. From R drop RS perpendicular to PQ, meeting PQ in S. Now, using Wallis' postulate, we can find a triangle PQT such that $\triangle PQT$ is similar to $\triangle PSR$ and T is on the same side of PQ as R. Then $\angle TPQ = \angle RPS$, and PR and PT coincide. Thus T is on N. Moreover, $\angle PQT = \angle PSR$, so that $\angle PQT$ is a right angle. Since L is perpendicular to PQ at Q, it follows that T is on L. Hence N meets L in T, and there is only one line containing P which is parallel to L.

It is clear therefore that Wallis' postulate implies Euclid's parallel postulate. As we have already observed in result 7 of Section 4, the converse holds. Thus Wallis' postulate is logically equivalent to Euclid's. Wallis apparently felt that his postulate was indubitable, and that he had settled the problem of the parallel postulate for all time.

Is Wallis' postulate really more obvious or simpler than Euclid's? In effect it says that if $\triangle ABC$ and segment $\overline{PQ}$ are given (Figure 2.9), there exists a

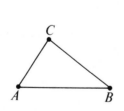
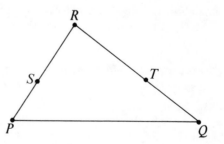

FIGURE 2.9

point R such that $\triangle PQR$ is similar to $\triangle ABC$. How could we obtain such a point R? On a given side of PQ we could construct $\angle QPS = \angle A$ and $\angle PQT = \angle B$. Then R would appear as the intersection of lines PS and QT. Consequently, Wallis' postulate implies that PS and QT must meet. Note that $\angle A + \angle B < 180°$ by Theorem 3, so that $\angle P + \angle Q < 180°$. Thus Wallis' postulate asserts that, in a certain case, if two lines meet a transversal so as to form a pair of angles on one side of the transversal whose sum is less than 180°, then the two lines must meet. This is very similar to Euclid's parallel postulate. But Wallis' postulate asserts much more, since it requires $\angle R = \angle C$ and the proportionality of corresponding sides of the two triangles. It seems then, that Wallis' postulate is no more indubitable than Euclid's, and hardly less complicated.

7. Saccheri's Attempt to Vindicate Euclid

Girolamo Saccheri (1667–1733) made a deep and critical study of geometry in a book, *Euclides Vindicatus*, published in the year of his death. He approached the problem of proving Euclid's parallel postulate in a radically new way. His procedure was equivalent to assuming Euclid's parallel postulate to be false, and arriving at a contradiction by logical reasoning. This would validate the parallel postulate by the principle of the indirect method.

Saccheri's point of departure was the study of quadrilaterals which have two sides which are equal and perpendicular to a third side. Without assuming any parallel postulate he made an exhaustive study of such quadrilaterals, now called *Saccheri quadrilaterals*. Let $ABCD$ be a Saccheri quadrilateral with $\overline{AD} = \overline{BC}$ and right angles at A, B (Figure 2.10). Saccheri proved that $\angle C = \angle D$, and then considered the three possibilities concerning angles C and D:

(1) Hypothesis of the right angle ($\angle C = \angle D = 90°$).
(2) Hypothesis of the obtuse angle ($\angle C = \angle D > 90°$).
(3) Hypothesis of the acute angle ($\angle C = \angle D < 90°$).

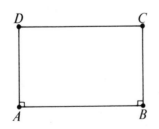

FIGURE 2.10

If Euclid's parallel postulate is assumed, then the hypothesis of the right angle follows (because the parallel postulate implies that the angle sum of any quadrilateral is 360°). Saccheri's basic line of argument is the following:

Show that the hypothesis of the obtuse angle and the hypothesis of the acute angle both lead to contradictions. This establishes the hypothesis of the right angle, which is equivalent to Euclid's parallel postulate.

Saccheri proved in a carefully reasoned sequence of theorems, that the hypothesis of the obtuse angle leads to a contradiction.

He then considered the implications of the hypothesis of the acute angle. Among these are a number of unusual theorems, two of which we state as follows:

The sum of the angles of any triangle is less than 180°.

If L and M are two lines in a plane, then one of the following properties is satisfied:

(a) *L and M intersect, in which case they diverge from the point of intersection;*

(b) *L and M do not intersect but have a common perpendicular, in which case they diverge in both directions from the common perpendicular;*

(c) *L and M do not intersect and do not have a common perpendicular, in which case they converge in one direction, and diverge in the other.*

Saccheri did not arrive at the sought-for contradiction, although he thought he had; and indeed, we know today that Saccheri's theory of the hypothesis of the acute angle is as free from contradiction as Euclidean geometry. Actually, he proved a number of theorems in the non-Euclidean geometry which was developed about a century later by Bolyai and Lobachevsky.

Apparently, Saccheri's faith in Euclid prevented him from making the logical jump to non-Euclidean geometry. His attempt to validate Euclid was a failure, but a great failure, which only a person of unusual ability and rare dedication could have achieved.

EXERCISES

1. Prove Playfair's postulate equivalent to the alternate interior angle theorem: If two parallel lines are cut by a transversal, any pair of alternate interior angles formed are equal.

2. Criticize the following proof of Euclid's parallel postulate suggested by an argument of W. Bolyai (1775–1856):

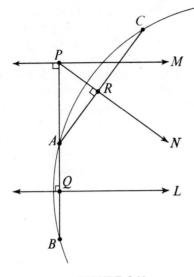

FIGURE 2.11

Given point P not on line L, PQ perpendicular to L at Q, and line M perpendicular to PQ at P. Let N be any line through P distinct from M (Figure 2.11). We show that N meets L so M is the only line through P parallel to L. Let A be any point between P and Q. Extend $\overline{AQ}$ its own length through Q to B. Let AR be perpendicular to N at R and extend $\overline{AR}$ its own length through R to C. Then A, B, C do not lie on a line and determine a triangle. Let Z be the circle which circumscribes this triangle. Then L and N are perpendicular bisectors of chords of Z and so must meet at its center.

3. Criticize the following proof of Euclid's parallel postulate:

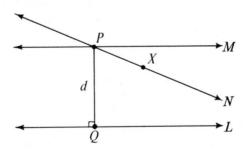

FIGURE 2.12

Given point P not on line L, PQ perpendicular to L at Q. Let d be the distance PQ from P to L. Let M be the line whose points are on the same side of L as P and at a distance d from L. Clearly M contains P and does not meet L. Now let N be any other line through P. We show N meets L. N crosses M and enters the region between M and L, let us say to the right of line PQ. Let X be any point on N to the right of line PQ. Then as X recedes from P, the distance from X to M will increase indefinitely. Eventually it becomes greater than d, the constant distance from a point of M to L. Thus eventually N must cross L, and certainly N meets L. Therefore, M is the only line through P parallel to L.

4. Criticize the proof of the following proposition: Two lines which are not everywhere equidistant must meet.

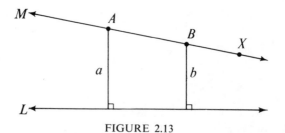

FIGURE 2.13

Suppose line L is not everywhere equidistant from line M. Then there are two points A, B on L with respective distances a, b from M such that $a \neq b$. Let us say $a > b$. Let $c = a - b$ and let d be the distance AB. As a point moves through the distance d from A to B along M, its distance to L decreases by c. Choose X on M in the same direction from A as B, such that the distance AX is $(a/c)d$. Then as a point moves through the distance $(a/c)d$ from A to X along M, its distance to L decreases by $(a/c)c = a$. Hence the distance from X to L is $a - a = 0$, that is M meets L in X.

This implies Playfair's postulate: Let P be a point not on line L; then there can be only one line through P parallel to L, namely the line through P which is everywhere equidistant from L.

5. Criticize the following proof of Euclid's parallel postulate of A. M. Legendre (1752–1833):

Given point P not on line L, PQ perpendicular to L at Q and line M perpendicular to PQ at P. Let N be any line through P distinct from M. We show N meets L so that M is the only line through P parallel to L. Certainly N meets L if N is perpendicular to L, so we may assume N not perpendicular to L. Since N is distinct from M, there must be a point R on N such that $\angle QPR$ is acute. Construct $\angle QPR'$ equal to $\angle QPR$ with

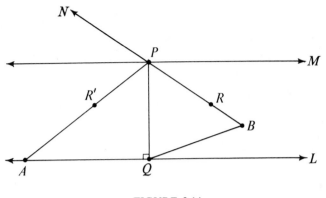

FIGURE 2.14

R' on the opposite side of PQ from R. Then Q is inside $\angle RPR'$. Hence L contains a point inside $\angle RPR'$ and must meet one of its sides. If L meets side PR, certainly N meets L. Suppose then that L meets side PR' in A. Choose point B on side PR such that $\overline{PB} = \overline{PA}$. Then $\triangle PQA \cong \triangle PQB$ (SAS) and $\angle PQB$ is a right angle. Thus B is on L and N meets L.

6. Criticize the following proof of Euclid's parallel postulate:

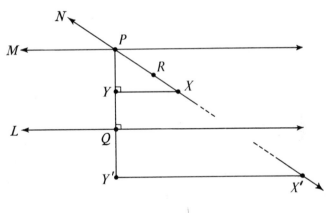

FIGURE 2.15

Given point P not on line L, PQ perpendicular to L at Q and line M perpendicular to PQ at P. Let N be any line through P distinct from M. We show that N meets L so that M is the only line through P parallel to L. Certainly N meets L if N is perpendicular to L, so we may assume N not perpendicular to L. Since N is distinct from M, there must be a point R on N such that $\angle QPR$ is acute. Choose point X arbitrarily on side PR of

$\angle QPR$ and let Y be the foot of the perpendicular from X to side PQ of $\angle QPR$. Let X recede endlessly. Then Y recedes endlessly. Hence there is a position Y' of Y on side PQ such that $\overline{PY'} > \overline{PQ}$. Let X' be the corresponding position of X on side PR. Since $\overline{PY'} > \overline{PQ}$, points P and Y' must be on opposite sides of L. But X' and Y' are on the same side of L, since $Y'X' \parallel L$. Hence P and X' are on opposite sides of L, and line N which joins them must meet L.

7. Euclid's parallel postulate implies that the angle sum of every triangle is *180°*. Do you think the converse is true? Try to prove this, but do not spend an unreasonable amount of time on it, as it is a difficult problem without suitable preparation.

REFERENCES

R. Bonola, *Non-Euclidean Geometry*, Translated by H. S. Carslaw, Dover Publications, Inc., New York, 1955.

G. B. Halsted, *Girolamo Saccheri's Euclides Vindicatus*, The Open Court Publishing Company, LaSalle, Illinois, 1920.

T. L. Heath, *The Thirteen Books of Euclid's Elements*, 2nd Ed., 3 vols. Cambridge University Press, New York, 1926. Reprinted by Dover Publications, Inc., 1956.

Neutral Geometry

The failure of attempts to prove Euclid's parallel postulate foreshadows the development of geometric theories which contradict the parallel postulate. In this chapter we shall study the consequences of Euclid's postulates other than the parallel postulate. This serves several purposes: it helps to clarify the role of the parallel postulate in Euclidean geometry; it suggests and opens the way for our study of non-Euclidean geometry in the next chapter; and it actually yields theorems which are valid in non-Euclidean geometry. The wealth of results obtained (many appear in the exercises at the end of the chapter) is at first sight quite surprising.

1. Introduction

Let us consider what the theory of geometry will be like if the "controversial" parallel postulate is dropped. Can interesting and important theorems be proved? Will light be thrown on the problem of Euclid's parallel postulate?

We call this theory *Neutral Geometry** to indicate that we refrain from assuming any parallel postulate. The theorems of neutral geometry then, are precisely those which can be deduced from Euclid's postulates without employing the parallel postulate. In studying neutral geometry we are treading the path of Saccheri, but not adopting Saccheri's predetermined imperative that Euclid's parallel postulate must be validated. Rather, we explore the

---

* The classical name for the subject is *Absolute Geometry*. This name reflects the historical feeling that Euclid's other postulates are indubitable—only the parallel postulate could be doubted. The current viewpoint is that postulates are suppositions, formulated abstractly, which may or may not be true when interpreted and tested.

possibilities inherent in the other postulates and so deepen our knowledge of geometry.

We begin our study of neutral geometry by observing—like the man in Molière's comedy who discovered he had been speaking prose his whole life—that we already know some of its theorems. For in Chapter 2, Section 1, the consequences (1)–(7) of Euclid's postulates, as well as Theorems 1, 2, 3, and corollaries, are proved *before* the introduction of the parallel postulate, and so are propositions of neutral geometry. The familiar notions involved in measurement of segments and of angles, for example, the idea of right angle and the degree measure of angles, also belong to neutral geometry. We shall appeal to these results in our arguments (often without reference) and in general continue as in Chapter 2, drawing inferences from diagrams when necessary. For convenience our treatment is restricted, as in Chapter 2, to plane geometry.

2. The Sum of the Angles of a Triangle

To prove an important and insufficiently known theorem concerning the angle sum of a triangle we introduce the following lemma.

LEMMA.   Given $\triangle ABC$ and $\angle A$. Then there exists a triangle, $\triangle A_1B_1C_1$, such that $\triangle A_1B_1C_1$ has the same angle sum as $\triangle ABC$, and $\angle A_1 \leq \frac{1}{2}\angle A$.

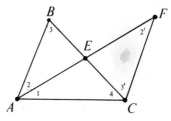

FIGURE 3.1

Proof.   Let E be the midpoint of $\overline{BC}$, and let F be chosen on AE such that $\overline{AE} = \overline{EF}$ and E is between A and F. Then $\triangle BEA \cong \triangle CEF$ and their corresponding angles are equal. We show $\triangle AFC$ is the $\triangle A_1B_1C_1$ we are seeking. By labeling the angles as in the diagram we have $\angle 2 = \angle 2'$, $\angle 3 = \angle 3'$ and

$$\angle A + \angle B + \angle C = \angle 1 + \angle 2 + \angle 3 + \angle 4$$
$$= \angle 1 + \angle 2' + \angle 3' + \angle 4$$
$$= \angle CAF + \angle AFC + \angle FCA.$$

To complete the proof, observe that $\angle A = \angle 1 + \angle 2$ which implies

$$\angle A = \angle 1 + \angle 2'.$$

In this equation, one of the terms on the right-hand side, $\angle 1$ or $\angle 2'$, must be less than or equal to one-half the term on the left, $\angle A$. If $\angle 1 \leq \frac{1}{2}\angle A$, relabel A as A_1; if not, relabel F as A_1. Then relabel the other two vertices of $\triangle AFC$ as B_1 and C_1 and the lemma is proved.

In intuitive terms the lemma says that we can replace a triangle by a "slenderer" one without altering its angle sum. In effect this is proved by cutting off $\triangle ABE$ from $\triangle ABC$ and pasting it back as $\triangle FCE$.

The lemma is less trivial than may appear at first sight, for in neutral geometry we cannot assume that the angle sum is constant for all triangles—this is a Euclidean theorem whose proof depends on the parallel postulate. Hence the lemma is important because it indicates that given a triangle we can construct a noncongruent triangle having the same angle sum. The lemma implies, by an easy argument, the existence of an infinite sequence of noncongruent triangles, all having the same angle sum as a given triangle.

We can now prove a remarkable theorem which is a consequence of Saccheri's result (Ch. 2, Sec. 7), on the falsity of the hypothesis of the obtuse angle. An independent proof was given by A. M. Legendre (1752–1833).

THEOREM 1.   (SACCHERI-LEGENDRE)   The angle sum of any triangle is less than or equal to $180°$.

Proof.   Suppose the contrary. Then there exists a triangle, $\triangle ABC$, with angle sum $180° + p°$, where p is a positive number. Now, applying the preceding lemma, there exists a more "slender" triangle, $\triangle A_1B_1C_1$, with the same angle sum as $\triangle ABC$, $180° + p°$, such that

$$\angle A_1 \leq \tfrac{1}{2}\angle A.$$

By applying the lemma to $\triangle A_1B_1C_1$, we see that there exists a triangle, $\triangle A_2B_2C_2$, with the same angle sum, $180° + p°$, such that

$$\angle A_2 \leq \tfrac{1}{2}\angle A_1 \leq \tfrac{1}{4}\angle A.$$

By continuing in this fashion we construct a sequence of triangles

$$\triangle A_1B_1C_1, \quad \triangle A_2B_2C_2, \quad \triangle A_3B_3C_3, \ldots,$$

each with angle sum $180° + p°$, such that, for every positive integer n,

$$\angle A_n \leq \frac{1}{2^n} \angle A.$$

Clearly, we can select n sufficiently large so that $\angle A_n$ is as small as we please, in particular so that

$$\angle A_n \leqq p°.$$

Since $\angle A_n + \angle B_n + \angle C_n = 180° + p°$, it follows that

$$\angle B_n + \angle C_n \geqq 180°.$$

This contradicts Theorem 3 of Chapter 2. The supposition is therefore false, and the theorem follows.

As a specific example of this theorem, suppose $p = 1$ and $\angle A = 25°$. Thus in our original triangle, $\triangle ABC$, we have $\angle A + \angle B + \angle C = 181°$ and $\angle A = 25°$. By the lemma, there exists a triangle, $\triangle A_1 B_1 C_1$, such that $\angle A_1 + \angle B_1 + \angle C_1 = 181°$ and $\angle A_1 \leqq \dfrac{25°}{2}$. In like fashion there exists a triangle, $\triangle A_2 B_2 C_2$, such that $\angle A_2 + \angle B_2 + \angle C_2 = 181°$ and $\angle A_2 \leqq \dfrac{25°}{4}$. To see the contradiction, apply the lemma three times more to obtain $\triangle A_5 B_5 C_5$ in which $\angle A_5 + \angle B_5 + \angle C_5 = 181°$ and $\angle A_5 \leqq \dfrac{25°}{32} < 1°$. Consequently $\angle B_5 + \angle C_5 > 180°$ which is impossible.

COROLLARY.   The angle sum of any quadrilateral is less than or equal to 360°.

The corollary implies Saccheri's conclusion that the hypothesis of the obtuse angle is false (Ch. 2, Sec. 7). Similarly, the theorem denies that the angle sum of a triangle can exceed 180°. But the possibility that the angle sum of a triangle may be less than 180°, which corresponds to Saccheri's hypothesis of the acute angle, forces itself on our attention.

3. Do Rectangles Exist

In continuing our study of neutral geometry, we become interested in whether a rectangle can exist in such a geometry, and what can be deduced if it does.

The existence of rectangles in a geometry is not a trivial thing. Imagine what Euclidean geometry would be like if we did not have or could not use rectangles. It is hard to see how one could construct a rectangle without assuming Euclid's parallel postulate, or one of its consequences, such as, the angle sum of a triangle is 180°. Consequently, all our theorems in this section have the hypothesis that a rectangle exists. To avoid ambiguity, we formally define the term "rectangle" in the sense in which we shall use it.

DEFINITION.   A quadrilateral is called a *rectangle* if each of its angles is a right angle.

Notice that since we are studying neutral geometry, we cannot *automatically* apply familiar Euclidean propositions such as (a) the opposite sides of a rectangle are parallel, or (b) that they are equal, or (c) that a diagonal divides a rectangle into two congruent triangles. If we want to assert any of these results, we have to prove them from our definition without assuming a parallel postulate. For example, (a) is the immediate result of Corollary 1 of Theorem 2 in Chapter 2.

THEOREM 2.   If one rectangle exists, then there exists a rectangle with an arbitrarily large side.

Restatement.   Suppose a rectangle $ABCD$ exists, and $\overline{XY}$ is any given segment. Then there exists a rectangle with one side greater than $\overline{XY}$.

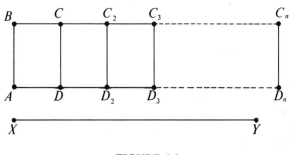

FIGURE 3.2

Proof.   We use $ABCD$ as a "building block" to construct the desired rectangle. Construct a quadrilateral D_2C_2CD congruent to $ABCD$ so that corresponding sides $\overline{C_2D_2}$ and $\overline{BA}$ are on opposite sides of line CD. (A way of doing this is to extend $\overline{BC}$ through C its own length to C_2 and $\overline{AD}$ through D its own length to D_2.) Then D_2C_2CD is a rectangle. Moreover, B,C,C_2 lie on a line since there is a unique perpendicular to CD at C. Similarly, A,D,D_2 lie on a line. Thus $ABCC_2D_2D$ is a quadrilateral ABC_2D_2, and consequently a rectangle. Note that ABC_2D_2 has the property

$$\overline{AD_2} = 2\overline{AD}.$$

In like manner construct $D_3C_3C_2D_2$ congruent to DCC_2D_2 so that $\overline{C_3D_3}$ and $\overline{CD}$ correspond and are on opposite sides of line C_2D_2. It follows easily, as above, that ABC_3D_3 is a rectangle, and that

$$\overline{AD_3} = 3\overline{AD}.$$

By continuing in this fashion, we see that for every positive integer n there exists a rectangle ABC_nD_n such that

$$\overline{AD_n} = n\overline{AD}.$$

Now choose n so large that $n\overline{AD} > \overline{XY}$. Then rectangle ABC_nD_n satisfies the condition of the theorem.

COROLLARY.   If one rectangle exists, then there exists a rectangle with two arbitrarily large adjacent sides.

Restatement.   Suppose a rectangle $ABCD$ exists and $\overline{XY}, \overline{ZW}$ are given segments. Then there exists a rectangle $PQRS$ such that $\overline{PQ} > \overline{XY}$ and $\overline{PS} > \overline{ZW}$.

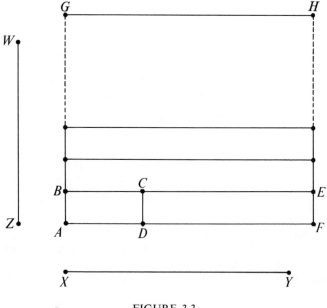

FIGURE 3.3

Proof. By the theorem there exists a rectangle $ABEF$ with $\overline{AF} > \overline{XY}$. By placing successive replicas of $ABEF$ on top of each other, we eventually construct (by the method of the theorem) a rectangle $AFHG$ with $\overline{AG} > \overline{ZW}$. Since $\overline{AF} > \overline{XY}$, $AFHG$ is a rectangle $PQRS$ that satisfies the corollary.

THEOREM 3.   If one rectangle exists, then there exists a rectangle with two adjacent sides equal to preassigned segments $\overline{XY}, \overline{ZW}$.

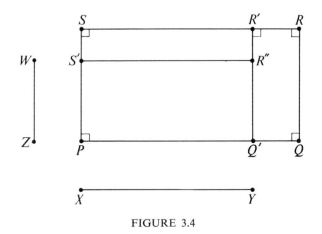

FIGURE 3.4

Proof. Our method is that of a tailor: by the preceding corollary we get rectangle $PQRS$ such that $\overline{PQ} > \overline{XY}$ and $\overline{PS} > \overline{ZW}$; then we cut it down to fit.

There is a point Q' on $\overline{PQ}$ such that $\overline{PQ'} = \overline{XY}$. From Q' drop a perpendicular to line RS with foot R'. We show that $PQ'R'S$ is a rectangle. It certainly has right angles at P, R', and S. We show that $\angle PQ'R'$ is also a right angle. Suppose $\angle PQ'R' > 90°$. Then the sum of the angles of quadrilateral $PQ'R'S$ is greater than $360°$, contrary to Corollary 1 of Theorem 1. Suppose $\angle PQ'R' < 90°$. Then $\angle QQ'R' > 90°$ and the angle sum of quadrilateral $QQ'R'R$ is greater than $360°$. Thus the only possibility is $\angle PQ'R' = 90°$, and $PQ'R'S$ is a rectangle.

In the same way, there is a point S' on $\overline{PS}$ such that $\overline{PS'} = \overline{ZW}$. Drop a perpendicular from S' to line $Q'R'$ with foot R''. Then, as above, $PQ'R''S'$ is a rectangle. Its adjacent sides $\overline{PQ'}$ and $\overline{PS'}$ equal $\overline{XY}$ and $\overline{ZW}$, respectively, which completes the proof.

THEOREM 4.   If one rectangle exists, then every right triangle has an angle sum of $180°$.

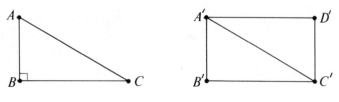

FIGURE 3.5

Proof. Our procedure is to show first that every right triangle is a replica of a triangle formed by the splitting of a rectangle by a diagonal; and second, that the latter type of triangle has an angle sum of 180°.

Let $\triangle ABC$ be a right triangle with right angle at B. By Theorem 3, there exists a rectangle $A'B'C'D'$ with $\overline{A'B'} = \overline{AB}$ and $\overline{B'C'} = \overline{BC}$. Draw $\overline{A'C'}$. Then $\triangle ABC \cong \triangle A'B'C'$, hence $\triangle ABC$ and $\triangle A'B'C'$ have the same angle sum. Let p be the angle sum of $\triangle A'B'C'$, and q the angle sum of $\triangle A'C'D'$. Then

(1) $p + q = 4 \cdot 90° = 360°.$

We show $p = 180°$. By Theorem 1, $p < 180°$ or $p = 180°$. Suppose $p < 180°$. Then by (1), $q > 180°$, contrary to Theorem 1. Thus $p = 180°$, completing the proof.

THEOREM 5.   If one rectangle exists, then every triangle has an angle sum of 180°.

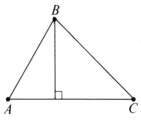

FIGURE 3.6

Proof.   Any triangle ABC can be split into two right triangles by a suitably chosen altitude (Ch. 1, Exercise 8). Each of these triangles has an angle sum of 180° by Theorem 4. It easily follows that the same property is true for $\triangle ABC$.

This is a rather striking result. The existence of one puny rectangle with microscopic sides inhabiting a remote portion of space guarantees that every conceivable triangle has an angle sum of 180°. Since this is a typically Euclidean property, we are tempted to say that if a neutral geometry contains a rectangle then it must be Euclidean. The statement is correct but not yet fully justified, since to characterize a geometry as Euclidean we must show that it satisfies Euclid's parallel postulate. This will be done in the next chapter.

4. Again, the Sum of the Angles of a Triangle

Our results on rectangles can now be used to sharpen Theorem 1, the Saccheri-Legendre Theorem on the angle sum of a triangle. This is easily done since, as suggested by Theorem 5, the existence of a triangle with an angle sum of 180° is equivalent to the existence of a rectangle.

THEOREM 6.   If there exists one triangle with an angle sum of 180°, then there exists a rectangle.

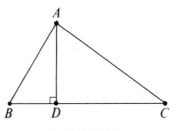

FIGURE 3.7

Proof.   Suppose $\triangle ABC$ has an angle sum of 180°. We show first that there is a right triangle with an angle sum of 180°. Split $\triangle ABC$ by a suitably chosen altitude, say $\overline{AD}$, into two right triangles whose angle sums are say, p and q. Then

$$p + q = 2 \cdot 90° + 180° = 360°.$$

We show $p = 180°$. By Theorem 1, $p \leq 180°$. If $p < 180°$, then $q > 180°$, contrary to Theorem 1. Thus there is a right triangle, say $\triangle ABD$, with right angle at D, whose angle sum is 180°.

Now we put two such right triangles together to form a rectangle.

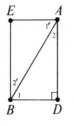

FIGURE 3.8

Construct $\triangle BAE \cong \triangle ABD$ with E on the opposite side of line AB from D, and $\overline{BE}$ corresponding to $\overline{AD}$ (Figure 3.8). Since the angle sum of $\triangle ABD$ is 180°, it follows that

$$\angle 1 + \angle 2 = 90°.$$

Since

$$\angle 1 = \angle 1', \qquad \angle 2 = \angle 2'$$

we have

$$\angle 1 + \angle 2' = 90°, \qquad \angle 1' + \angle 2 = 90°.$$

But $\angle 1 + \angle 2' = \angle EBD$, and $\angle 1' + \angle 2 = \angle EAD$. Thus $\angle EAD = \angle EBD = 90°$, proving $ADBE$ is a rectangle.

COROLLARY 1.   If one triangle has an angle sum of 180°, then *every* triangle has angle sum 180°.

Proof.   By Theorems 6 and 5.

COROLLARY 2.   If one triangle has an angle sum which is less than 180°, then *every* triangle has an angle sum less than 180°.

Proof.   Suppose $\triangle ABC$ has angle sum less than 180°. Consider any triangle, $\triangle PQR$. By Theorem 1, its angle sum p must satisfy $p = 180°$ or $p < 180°$. Suppose $p = 180°$. Then by Corollary 1 above, $\triangle ABC$ has angle sum 180°, contrary to hypothesis. Thus $p < 180°$.

By comparing Corollaries 1 and 2 we observe an important fact not contained in the Saccheri-Legendre Theorem. A neutral geometry is "homogeneous" in the sense that either all of its triangles have angle sums of 180°, or else they all have angle sums less than 180°. The first type of neutral geometry is, as you may guess, merely Euclidean geometry. The second type arose historically in the study of non-Eucidlean geometry. Both are considered in the next chapter.

We include a reference list of propositions of Plane Neutral Geometry which may be assumed in solving the exercises below. We have already proved some of them, the proofs of many can be found in the SMSG *Geometry*, Yale University Press, 1961.

Propositions of Plane Neutral Geometry

Two distinct lines meet in at most one point.
Every segment has exactly one midpoint.

Every angle has exactly one bisector.

Supplements (complements) of equal angles are equal.

Vertical angles are equal.

Congruence of triangles: SAS, ASA, SSS.

If two sides of a triangle are equal, the angles opposite them are equal.

If two angles of a triangle are equal, the sides opposite them are equal.

There exists one and only one line perpendicular to a given line at a given point of the line.

There exists one and only one line perpendicular to a given line through a given external point.

A point is in the perpendicular bisector of a segment if and only if it is equidistant from the endpoints of the segment.

If two sides of a triangle are not equal, then the angles opposite them are not equal, and the greater angle is opposite the greater side.

If two angles of a triangle are not equal, then the sides opposite them are not equal, and the greater side is opposite the greater angle.

The shortest segment joining a point to a line is the perpendicular segment.

The sum of the lengths of two sides of a triangle is greater than the length of the third side.

If two sides of one triangle are equal, respectively, to two sides of a second triangle, and the included angle of the first triangle is greater than the included angle of the second, then the third side in the first triangle is greater than the third side in the second.

If two sides of one triangle are equal, respectively, to two sides of a second triangle, and the third side of the first triangle is greater than the third side of the second, then the angle opposite the third side in the first triangle is greater than the angle opposite the third side in the second.

An exterior angle of a triangle is greater than either remote interior angle.

The sum of two angles of a triangle is less than $180°$.

If two lines are cut by a transversal so as to form a pair of equal alternate interior angles, then the lines are parallel.

Two lines perpendicular to the same line are parallel.

There is at least one line parallel to a given line through a external point.

Suppose a line contains a point whose distance to the center of a circle is less than the radius. Then the line meets the circle in two points.

A line is tangent to a circle if and only if it is perpendicular to a radius at its outer endpoint.

Given $\triangle ABC$ and $\overline{PQ}$ such that $\overline{PQ} = \overline{AB}$. Then there exists a point R on a given side of line PQ such that $\triangle PQR \cong \triangle ABC$.

A circle can be inscribed in any triangle.

E X E R C I S E S   I

1. Prove: Two triangles are congruent if two angles and the side opposite one of them in the first triangle are equal to the three corresponding parts of the second triangle.
2. Prove: Two right triangles are congruent if the hypotenuse and a leg of one are equal to the hypotenuse and a leg of the other.
3. Prove: If a transversal makes equal alternate interior angles with two lines then the two lines have a common perpendicular.

DEFINITION.   Quadrilateral $ABCD$ is called an *isosceles birectangular quadrilateral* or a *Saccheri quadrilateral* (see Ch. 2, Sec. 7) if $\angle B = \angle C = 90°$, and $\overline{AB} = \overline{DC}$. $\overline{BC}$ is the *base* of the Saccheri quadrilateral, $\overline{AB}$ and $\overline{DC}$ its *sides* or *legs*, $\overline{AD}$ its *summit*, and $\angle A$, $\angle D$ its *summit angles*.

4. Prove: The summit angles of a Saccheri quadrilateral are equal and not obtuse.
5. Prove: The line joining the midpoints of summit and base of a Saccheri quadrilateral is perpendicular to its summit and base. Infer that summit and base of a Saccheri quadrilateral are parallel.
6. Prove: The perpendicular bisector of the base of a Saccheri quadrilateral is also the perpendicular bisector of its summit. (See Exercise 5.)
7. Prove: Two lines have a common perpendicular if and only if there are two points of one line which are equidistant from the other line and on the same side of it. (See Exercise 5.)
8. In quadrilateral $ABCD$, let $\angle B = \angle C = 90°$. Prove that $\overline{AB} > \overline{DC}$ implies $\angle D > \angle A$ and conversely.
9. In quadrilateral $ABCD$, let $\angle B = \angle C = 90°$. Prove $\angle A = \angle D$ implies $\overline{AB} = \overline{DC}$. (See Exercise 8.)
10. Prove: The summit of a Saccheri quadrilateral is greater than or equal to its base.
11. Prove: The segment joining the midpoints of summit and base of a Saccheri quadrilateral is smaller than or equal to a leg of the quadrilateral.
12. Prove: If a quadrilateral has three right angles, a side adjacent to the fourth angle is greater than or equal to its opposite side.
13. Prove: If two lines have a common perpendicular the shortest segment joining them is the common perpendicular segment.
14. Given a right triangle, form a new right triangle which has an acute angle in common with the given triangle and hypotenuse double its hypotenuse. Prove that the side opposite the acute angle is at least

doubled. What do you think happens to the adjacent side? Try to justify your answer.

15. Given two lines L, M intersecting at O; P is between O and Q on L; $PP' \perp M$ at P'; $QQ' \perp M$ at Q'. Prove $\overline{QQ'} > \overline{PP'}$. That is, as a point of L recedes from O, its distance to M increases.

16. In Exercise 15 show that as $\overline{OP}$ increases indefinitely so does $\overline{PP'}$. This justifies property (A) (Ch. 2, Sec. 5) that as a point of L recedes endlessly from O its distance to M increases indefinitely. [*Hint.* Apply Exercise 14.]

EXERCISES II

1. Prove: If perpendiculars are drawn to the line joining the midpoints of two sides of a triangle from the endpoints of the third side, a Saccheri quadrilateral is formed. More precisely, if M, N are the midpoints of sides $\overline{AB}$, $\overline{AC}$ of $\triangle ABC$ and $BP \perp MN$ at P, $CQ \perp MN$ at Q then $BPQC$ is a Saccheri quadrilateral. (Note that there are several cases.)

DEFINITION.   Suppose a triangle and a Saccheri quadrilateral are related as in Exercise 1—that is, the base of the quadrilateral lies on the line joining the midpoints of two sides of the triangle and its summit is the third side of the triangle. Then we say that the triangle and the Saccheri quadrilateral are *associated*, or the triangle (quadrilateral) is an *associated triangle* (*quadrilateral*) of the quadrilateral (triangle). Note that a triangle has three associated Saccheri quadrilaterals.

2. Prove: The segment joining the midpoints of two sides of a triangle is less than or equal to half the third side; moreover, it is parallel to the third side. [*Hint.* Study the associated Saccheri quadrilateral.]

DEFINITION.   Two polygons p, q are *equivalent by decomposition* (or *by addition*) if p can be decomposed into triangles $p_1, p_2, \ldots, p_n$ and q into triangles $q_1, q_2, \ldots, q_n$ such that p_i is congruent to q_i for $i = 1, 2, \ldots, n$.

DEFINITION.   Two polygons p, q are *equivalent* (a) if they are equivalent by decomposition; or (b) if there exist polygons p', q' which are equivalent by decomposition such that p' is decomposable into p and a set of triangles $p_1, p_2, \ldots, p_n$ and q' into q and a set of triangles $q_1, q_2, \ldots, q_n$ where p_i is congruent to q_i for $i = 1, 2, \ldots, n$. Assume that the relation equivalence of polygons is *transitive*; that is, if *p* is equivalent to *q* and *q* is equivalent to *r* then *p* is equivalent to *r*.*

3. Prove: A triangle is equivalent to each associated Saccheri quadrilateral. Moreover, its angle sum is the sum of the summit angles of each associated Saccheri quadrilateral. (Note that there are several cases.)
4. Prove: If two triangles have a common associated Saccheri quadrilateral they are equivalent and have the same angle sum.
5. Prove: If the summit of a Saccheri quadrilateral is one side of a triangle and the line of its base bisects a second side of the triangle, then it bisects the third side and the quadrilateral is an associated Saccheri quadrilateral of the triangle.
6. Let a Saccheri quadrilateral be given. Prove that there exists an associated triangle with one side of preassigned length *x*, provided *x* is at least twice the length of a leg of the quadrilateral.
7. Let $\triangle ABC$ be given. Prove that there exists an equivalent triangle which has the same angle sum as $\triangle ABC$ and has one side of preassigned length *x*, provided *x* is greater than the length of one side of $\triangle ABC$.
8. Prove: Every Saccheri quadrilateral has an associated isosceles triangle. Infer that any $\triangle ABC$ is equivalent to and has the same angle sum as an isosceles triangle with base $\overline{AB}$.

E X E R C I S E S   I I I**

Characterizations of Euclidean Geometry

1. Prove: If in a neutral geometry there exists a Saccheri quadrilateral whose summit equals its base, the geometry is Euclidean.
2. Prove: If in a neutral geometry the segment joining the midpoints of two sides of a triangle always equals half the third side, the geometry is Euclidean. (Is it necessary to assume this hypothesis for all triangles?)
3. Prove: If in a neutral geometry every triangle can be circumscribed by a circle, the geometry is Euclidean. (See Ch. 2, Exercise 2.)
4. Prove: If in a neutral geometry any line through a point inside an angle meets the angle, then the geometry is Euclidean. (See Ch. 2, Exercise 5.)
5. Prove: If in a neutral geometry the sum of the angles of a triangle is constant, the geometry is Euclidean.

* See Theorem 2.1, p. 258, Henry G. Forder, *The Foundations of Euclidean Geometry*, Cambridge University Press, New York, 1927; reprinted Dover Publications, New York.
** Assume where necessary that a neutral geometry which contains a rectangle is Euclidean (see p. 71, Cor. 4).

6. Prove: A neutral geometry is Euclidean if it contains two noncongruent similar triangles.

†7. Prove: If in a neutral geometry there exists a triangle such that the segment joining the midpoints of a certain pair of its sides is equal to half the third side, then the geometry is Euclidean.

8. Prove: If the Pythagorean theorem holds in a neutral geometry, the geometry is Euclidean.

9. In a neutral geometry let quadrilateral $ABCD$ have right angles at A and B with $\overline{AD} \neq \overline{BC}$ and let the perpendicular bisector of $\overline{AB}$ bisect $\overline{CD}$. Prove that the geometry is Euclidean.

EXERCISES   IV

Miscellaneous

1. Prove: If the opposite sides of a quadrilateral are equal, then its opposite angles are equal.

2. In quadrilateral $ABCD$ let the angles at A and B be right angles. Prove that $ABCD$ is a Saccheri quadrilateral if one of the following holds:

   (i) the perpendicular bisector of $\overline{AB}$ is perpendicular to CD.

   (ii) the perpendicular bisector of $\overline{CD}$ is perpendicular to AB.

   (iii) the perpendicular bisector of $\overline{CD}$ bisects $\overline{AB}$.

3. Prove: The opposite sides of a rectangle are equal.

4. Prove: The diagonals of a rectangle bisect each other.

5. Prove: If the diagonals of a Saccheri quadrilateral bisect each other, the figure is a rectangle.

6. Prove: In a Saccheri quadrilateral, the line joining the midpoint of the base to the midpoint of the summit passes through the intersection of the diagonals.

7. In a Saccheri quadrilateral, prove that the line joining the midpoints of the legs is bisected perpendicularly by the line joining the midpoints of the base and summit.

8. Prove: The line joining the midpoints of two sides of a triangle is perpendicular to the perpendicular bisector of the third side.

9. Prove: The perpendicular bisectors of the sides of a triangle meet in a point, provided two of them meet. Infer (i) the three perpendicular bisectors concur or are parallel; (ii) either a circle can be circumscribed about a triangle or the perpendicular bisectors of its sides are parallel.

10. Prove: The perpendicular bisectors of the sides of a triangle are the altitudes of the triangle formed by joining the midpoints of its sides.

11. Prove: Any right triangle is equivalent to a trirectangular quadrilateral, and conversely.

DEFINITION.   A *parallelogram* is a quadrilateral which has two equal opposite sides and two adjacent angles which are supplementary and adjacent to the equal sides. Its *base* is the side joining the vertices of the two supplementary angles. Note that any Saccheri quadrilateral is a parallelogram.

12. Prove: The opposite sides of a parallelogram are parallel.
13. Prove: Any triangle is equivalent to a parallelogram; its angle sum is the angle sum of the parallelogram minus 180°.
14. Given a right triangle, form a new right triangle which has an acute angle in common with the given triangle and adjacent leg double its adjacent leg. Prove that the area is more than doubled. [*Hint.*   Try to reproduce the smaller triangle twice within the larger.]
†15. In a Saccheri quadrilateral, prove that the line joining the midpoints of the legs bisects each diagonal.

Introduction to Non-Euclidean Geometry

Saccheri died in 1733. His work seems to have had little influence on the course which geometry took, for his successors up to the beginning of the nineteenth century continued to try to prove Euclid's parallel postulate. At the turn of the century attempts were made by mathematicians of the caliber of Gauss (1777–1855) and Legendre (1752–1833). However, the failures of twenty centuries seemed finally to strike a spark of doubt in the minds of mathematicians so that by 1830, J. Bolyai (1802–1860), a Hungarian army officer, N. I. Lobachevsky (1793–1856), a Russian professor of mathematics at the University of Kazan, and the great Gauss himself had independently developed theories of geometry based on a contradiction of Euclid's parallel postulate. Specifically, they assumed that there is more than one line parallel to a given line through an external point. Gauss, fearful of controversy, was reluctant to publish his ideas, so that Bolyai and Lobachevsky are usually credited as the discoverers or creators of the new theory. Later, in 1854, the eminent German mathematician B. Riemann (1826–1866) introduced a different non-Euclidean theory of geometry based on the assumption that there are no parallel lines. In this chapter our object is to give an elementary introduction to the theories—classic by now—of Bolyai and Lobachevsky, and of Riemann.

1. Lobachevskian Geometry

We now introduce the non-Euclidean geometry of Bolyai and Lobachevsky as a formal theory based on postulates. We call the theory *Lobachevskian geometry* for the sake of simplicity and also to signalize the lifetime of work which Lobachevsky devoted to it. Lobachevskian geometry could be characterized as the type of neutral geometry (Ch. 3, Sec. 4) in which every triangle has an angle sum of less than 180°. We prefer, however, to follow the historical development and study the subject directly in terms of its relation to the Euclidean parallel postulate. Thus, to characterize Lobachevskian geometry we merely assume the postulates of Euclidean geometry but replace the Euclidean parallel postulate by the following postulate.

LOBACHEVSKIAN PARALLEL POSTULATE.   *There are at least two lines parallel to a given line through a given point not on the line.*

Clearly, Lobachevskian geometry is a type of neutral geometry. In effect, we are continuing our study of neutral geometry with an added restriction. Thus the theorems of neutral geometry hold for Lobachevskian geometry, and may be used in our proofs. We shall proceed to derive theorems of Lobachevskian plane geometry, continuing in the same manner and at the same level of rigor as in the study of neutral geometry in Chapter 3.

2. A Nonmetrical Theorem

Our first theorem of Lobachevskian geometry is an elementary one which does not involve *metrical* ideas like distance, perpendicularity, or area. It states a positional or graphical property of lines.

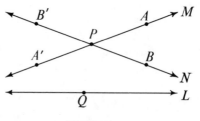

FIGURE 4.1

THEOREM 1.   Any line is contained wholly in the interior of some angle.

Proof.   Let L be a line. Choose point P not on L. By the Lobachevskian parallel postulate there exist distinct lines M, N which contain P and are parallel to L. Lines M and N separate the plane into four regions, each of which is the interior of an angle. These regions may be specifically labeled as the interiors of $\angle APB$, $\angle APB'$, $\angle A'PB$, $\angle A'PB'$, where P is between A and A' on M, and P is between B and B' on N. Let Q be any point on L. Since L does not meet M or N, Q is not on M or N. Thus Q is in the interior of one of the four angles listed above, say $\angle A'PB$ (Figure 4.1). Now where can L lie? Since one of its points Q is in the interior of $\angle A'PB$, and since L does not meet the sides PA', PB of the angle, clearly L is trapped inside $\angle A'PB$—that is, L is contained wholly in the interior of $\angle A'PB$.*

COROLLARY.   There are infinitely many lines parallel to a given line through a given point not on the line.

Proof. Let L be the given line and P the given point. Use the notation of the theorem and let R be any point in the interior of $\angle APB$ (Figure 4.1). Then line PR (excluding point P) is wholly contained in the interiors of $\angle APB$ and $\angle A'PB'$ and cannot meet L which is contained in the interior of $\angle A'PB$. Thus $PR \parallel L$. The corollary follows since there are infinitely many such lines PR.

It is interesting to compare Theorem 1 with the Euclidean situation where only part of a line can be contained in the interior of an angle. For in the Euclidean plane a line through an interior point of an angle meets the angle in two points or in one point (Figure 4.2). In the first case, only a segment of the line is contained in the interior of the angle, in the second, only a ray or half-line.

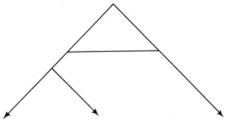

FIGURE 4.2

The theorem indicates a marked difference between Euclidean and Lobachevskian geometries with regard to nonmetrical properties. This should not

* It is interesting that Legendre proved Euclid's parallel postulate on the assumption that a line which contains a point inside an angle must meet the angle (see Exercise 5 at the end of Chapter 2).

be unduly surprising since the Euclidean parallel postulate (in Playfair's form) and the Lobachevskian parallel postulate are themselves markedly different graphical properties. Note the inevitability of the result, despite its apparent strangeness, once the Lobachevskian postulate is assumed.

3. An Objection

You may object that Theorem 1 is valid abstractly, but does not correspond to physical reality—that the conclusion follows logically from the assumption of the Lobachevskian parallel postulate, but that the assumption is palpably false. As you make such a statement you begin to follow the path of the non-Euclidean geometers. For surely when they started to develop their theories they must have had doubts about the empirical validity of the new parallel postulate. All that one needs to think mathematically is a set of assumptions (postulates) from which conclusions (theorems) can be derived by logical reasoning. The validity of a mathematical argument does not depend on the truth or falsity of the basic assumptions.

Nevertheless, does it make sense to choose assumptions that will be false when applied to the physical world? The answer may seem obvious but this is really a difficult and complex question that does not admit of a simple "yes" or "no" answer. Several points must be made. Firstly, the mathematician should be free to choose postulates and study their consequences independent of considerations of practical utility and empirical validity. For example, the study of non-Euclidean geometry is justified by the light it may throw on Euclidean geometry or the insight it may yield into the possible nature of physical space, regardless of its empirical validity.

Secondly, a mathematical proposition is abstract; to test it empirically we must interpret its basic terms. Though it seems false in one interpretation, it may be true in another. For example, a postulate that is false when "line" is interpreted as "taut string" might be true if it is reinterpreted as "ray of light."

Finally, let us not forget that the determination of the empirical truth of a geometrical statement is not our professional concern as mathematicians— it is not a mental experiment to be conducted in an arm chair. It lies in the domain of experimental or observational science and is carried out by physicists, astronomers, surveyors. It is often a difficult problem to determine the empirical truth of a statement in other than an approximative or statistical sense. As a classic example consider Euclid's parallel postulate: it has been used by generations of scientists and engineers; it has stood the test of time. We feel certain that it is an empirical fact. By the same token we are certain that the Lobachevskian parallel postulate is empirically false. Let us think

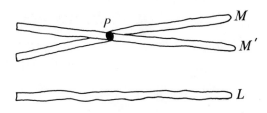

FIGURE 4.3

about this for a minute—just what is involved in these statements? Are we asserting that, given a physical line L and a physical point P not on L, there exists a physical line M through P that is coplanar with L and never meets it? How would this be tested? Would we use ropes, rulings on a blackboard, rays of light? Consider how much more difficult it would be to verify empirically that *only one* such line exists. Suppose that one such line M is known to exist (Figure 4.3). Do we really know enough of the properties of physical space to assert the uniqueness of such a line? Suppose M' is a physical line that contains P and makes a very minute angle with M; can we assert as a *physical* fact that M' must meet L?

The question of the empirical truth of our postulates is a difficult one and will be reexamined in Chapter 6. At this point we are content if we have managed to cast doubt on the belief that, in an empirical sense, Euclid's parallel postulate is certainly true and Lobachevsky's is certainly false. We hope this is sufficient to dispel the feeling that Lobachevskian geometry is merely an abstract mental exercise divorced from physical reality.

4. The Angle Sum of a Triangle in Lobachevskian Geometry

Theorem 1 indicates how positional or nonmetrical properties in a non-Euclidean geometry may differ from our Euclidean presuppositions. We shall show in Theorem 2 how a metrical property, the angle sum of a triangle, is altered when we change the parallel postulate.

We begin with two lemmas which are actually valid in neutral geometry, and could have been included in the last chapter. We postponed their introduction since they are used only to establish Theorem 2. The first lemma is hardly more than a restatement of the Saccheri-Legendre Theorem (Ch. 3, Th. 1).

LEMMA 1. The sum of two angles of a triangle is less than or equal to their remote exterior angle.

Proof. Consider $\triangle ABC$. By the Saccheri-Legendre Theorem

$$\angle A + \angle B + \angle C \leq 180°.$$

By subtracting $\angle C$ from both sides of this inequality we get

$$\angle A + \angle B \leq 180° - \angle C.$$

The lemma follows since the exterior angle at C equals $180° - \angle C$.

The second lemma (which is not trivial in itself) is a special case of the intuitive idea that if one side of a triangle is fixed and the other two are stretched out endlessly, their included angle approaches zero.

LEMMA 2.   Let L be a line, P a point not on L, and Q a point on L; let a side of line PQ be given. Then there exists a point R of L, on the given side of PQ, such that $\angle PRQ$ is as small as we please.

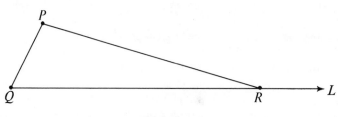

FIGURE 4.4

Proof. Let a be any given angle (not matter how small). We must show (Figure 4.4) that there is a point R of L, on the given side of PQ, such that $\angle PRQ < a$.

First we set up a procedure for obtaining a sequence of angles,

$$\angle PR_1Q, \angle PR_2Q, \ldots,$$

each term of which is no larger than half its predecessor.

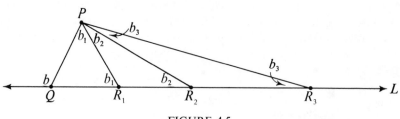

FIGURE 4.5

Let R_1 be a point of L on the given side of PQ such that $\overline{QR_1} = \overline{PQ}$ (Figure 4.5). Draw $\overline{PR_1}$. Then $\triangle PQR_1$ is isosceles, and

$$\angle QPR_1 = \angle PR_1Q = b_1.$$

Let b be the exterior angle of $\triangle PQR_1$ at Q. By Lemma 1

$$b_1 + b_1 = 2b_1 \leq b,$$

so that

(1) $b_1 \leq \tfrac{1}{2}b.$

Now we construct a new triangle and repeat the argument. Extend $\overline{QR_1}$ through R_1 to R_2, making $\overline{R_1R_2} = \overline{PR_1}$. Draw $\overline{PR_2}$. Then $\triangle PR_1R_2$ is isosceles and

$$\angle R_1PR_2 = \angle PR_2R_1 = \angle PR_2Q = b_2.$$

Thus, by Lemma 1,

$$b_2 + b_2 = 2b_2 \leq b_1,$$

so that

$$b_2 \leq \tfrac{1}{2}b_1.$$

This with Equation (1) implies

$$b_2 \leq \frac{1}{2^2} b.$$

Repeating this "bisection" process n times, we obtain a point R_n of L, on the given side of PQ, such that

$$b_n = \angle PR_nQ \leq \frac{1}{2^n} b.$$

The result follows quickly. Choose n so large that $\frac{1}{2^n}b < a$. Then $\angle PR_nQ < a$. Thus the theorem holds with $R = R_n$.

THEOREM 2.   There exists a triangle whose angle sum is less than 180°.

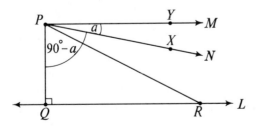

FIGURE 4.6

Proof. Let L be a line and P a point not on L. We obtain a line M through P parallel to L, in the usual way: Let PQ be perpendicular to L at Q and let M be perpendicular to PQ at P. By the Lobachevskian parallel postulate, there is another line N through P parallel to L. One of the angles that N makes with PQ must be acute. Let X be a point of N such that $\angle QPX$ is acute. Let Y be a point of M on the same side of line PQ as X. Let $a = \angle XPY$. Then $\angle QPX = 90° - a$.

Now apply Lemma 2. Let R be a point of L, on the side of PQ containing X, such that $\angle PRQ < a$. Consider $\triangle PQR$. We have

$$\angle PQR = 90°,$$
$$\angle QRP < a,$$
$$\angle RPQ < \angle XPQ = 90° - a.$$

By adding, we obtain

$$\angle PQR + \angle QRP + \angle RPQ < 90° + a + 90° - a = 180°,$$

so that $\triangle PQR$ has an angle sum of less than 180°, and the proof is complete.

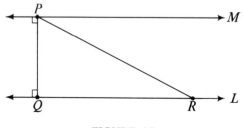

FIGURE 4.7

The basic thread of the proof is quite simple. To appreciate it better, consider first an analogous situation in Euclidean geometry. Let L and M be perpendicular to PQ at Q and P, respectively (Figure 4.7), and let R be an arbitrary point of L on a given side of PQ. Then as R recedes endlessly $\angle QRP$ approaches 0° and $\angle QPR$ approaches 90°. The Lobachevskian situation is a bit different. We still have L and M perpendicular to PQ at Q and P, so that M is parallel to L. But now (as we saw in proving Theorem 2) there exists another line, PX, parallel to L, such that $\angle QPX < 90°$ (Figure 4.8). Let R be an arbitrary point of L on the same side of PQ as X. Then as R recedes endlessly $\angle QRP$ approaches 0° as in the Euclidean case. But $\angle QPR$ does *not* approach 90°, for $\angle QPR$ is always less than $\angle QPX$. Thus, if R is far enough out, $\triangle PQR$ will have an angle sum of less than 180°. For example if $\angle QPX = 89°$ we need merely locate R so that $\angle QRP < 1°$.

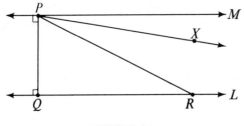

FIGURE 4.8

Finally, you may object that we have not really justified that $\angle QPR <$ $\angle QPX$; that is, that the ray $\overrightarrow{PR}$* lies inside $\angle QPX$. To take care of this, observe that rays $\overrightarrow{PR}$ and $\overrightarrow{PX}$ are distinct and both lie on the same side of line PQ. Consequently, one of them has to fall inside the angle formed by $\overrightarrow{PQ}$ and the other. Suppose $\overrightarrow{PX}$ fell inside $\angle QPR$. Then $\overrightarrow{PX}$ would meet $\overline{QR}$ and so meet L. Since this is impossible, $\overrightarrow{PR}$ must be inside $\angle QPX$.

The following theorem is an important and direct consequence of Theorem 2.

THEOREM 3.   The angle sum of every triangle is less than 180°.

Proof.   By Theorem 2, there exists a triangle whose angle sum is less than 180°. Hence the same is true of every triangle (Ch. 3, Th. 6, Cor. 2).

COROLLARY 1.   The angle sum of every quadrilateral is less than 360°.

COROLLARY 2.   No rectangles exist.

Although Theorem 3 is in striking contrast with the corresponding Euclidean result, you may be tempted to assume that the angle sum of a

---

* We are formally using the symbol $\overrightarrow{PR}$ to denote the *ray* or *half-line PR*, that is the portion of line *PR* consisting of all points that are on the same side of *P* as *R*.

triangle is constant, as in Euclidean geometry. This is not so. There exists a triangle whose angle sum is any preassigned value between 0° and 180°.*

Exercise.   Prove that there are two triangles with different angle sums. Can you find more than two? [*Hint.*   Use the indirect method.]

5. Do Similar Triangles Exist in Lobachevskian Geometry

We show next that similar triangles cannot exist in Lobachevskian geometry, except of course for the trivial case of congruent triangles. This follows readily from the following theorem.

THEOREM 4.   Two triangles are congruent if their corresponding angles are equal.

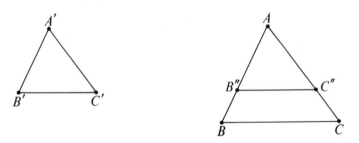

FIGURE 4.9

Proof.   Suppose the theorem false. Then there exist two triangles, $\triangle ABC$ and $\triangle A'B'C'$, such that $\angle A = \angle A'$, $\angle B = \angle B'$, $\angle C = \angle C'$, but the triangles are not congruent. Then $\overline{AB} \neq \overline{A'B'}$ (otherwise the triangles are congruent by ASA). Similarly $\overline{AC} \neq \overline{A'C'}$ and $\overline{BC} \neq \overline{B'C'}$. Consider the triples of segments $\overline{AB}, \overline{AC}, \overline{BC}$ and $\overline{A'B'}, \overline{A'C'}, \overline{B'C'}$. One of the triples must contain two segments which are larger than the two corresponding segments of the other triple. Consequently, it is not restrictive to suppose $\overline{AB} > \overline{A'B'}$ and $\overline{AC} > \overline{A'C'}$.

---

* This is not easy to prove without preparation. It can be shown using Theorem 4 p. 324 by the method of Theorem 1 p. 340 of Edwin Moise, *Elementary Geometry From An Advanced Standpoint*, Addison-Wesley Publishing Company, Reading, 1963.

We can thus find B'' on $\overline{AB}$ and C'' on $\overline{AC}$ such that $\overline{A'B'} = \overline{AB''}$ and $\overline{A'C'} = \overline{AC''}$. Consequently, $\triangle AB''C'' \cong \triangle A'B'C'$, so that

$$\angle AB''C'' = \angle B' = \angle B.$$

Hence $\angle BB''C''$ is supplementary to $\angle B$. Similarly $\angle B''C''C$ is supplementary to $\angle C$. Therefore quadrilateral $BB''C''C$ has an angle sum of 360°, which contradicts Corollary 1 of Theorem 3.

We have here a striking contrast with Euclidean geometry. In view of Theorem 4, there cannot be in Lobachevskian geometry a nontrivial theory of similar figures based on the usual definition; for if two triangles are similar their corresponding angles are equal, and hence the triangles must be congruent. In general, two similar figures would be congruent, and so have the same size. In a Lobachevskian world, pictures and statues would have to be life size to avoid distortion, so a proud father of a newborn child could always say truthfully "Of course the snapshot doesn't do him justice."

Query. What light does our discussion of Wallis' postulate (Ch. 2, Sec. 6) throw on the existence of a theory of similarity in Lobachevskian geometry?

6. Lobachevskian Theory of Area

Now let us consider the question of measurement of area. For the sake of simplicity, we shall restrict ourselves to triangles. The Euclidean procedure of measuring area in terms of square units clearly will not apply, since squares do not exist in Lobachevskian geometry. It is hard to see how to set up a simple procedure for measuring area. We could, of course, use the ideas of integral calculus and employ a method of successive approximations. Instead of trying to construct a measurement process, let us clarify the problem by examining the essential characteristics of an area measure for triangles. Regardless of how area is defined, surely it should possess the following properties:

(*a*) Positivity. *To each triangle there corresponds a uniquely determined positive real number, called its area.*

(*b*) Invariance Under Congruence. *Congruent triangles have equal areas.*

(*c*) Additivity. *If a triangle T is split into two triangles T_1 and T_2 by a segment joining a vertex to a point of the opposite side, then the area of T is the sum of the areas of T_1 and T_2.*

In effect, any process for measuring area determines a real-valued function defined for all triangles which satisfies properties (a), (b), and (c). This suggests that we define the concept of an area measure or area function for triangles by means of these properties, independently of any particular process of measurement. Thus we adopt the definition.

DEFINITION.   Consider a function that assigns to each triangle a specific real number in such a way that properties (a), (b), (c) above are satisfied. Then the function is called an *area function* or *area measure* (for triangles). If μ is such a function and ABC a triangle, $\mu(ABC)$ denotes the value which μ assigns to triangle ABC, and is called the *area* or *measure* of $\triangle ABC$ *determined by* μ.

This definition, of course, is not restricted to Lobachevskian geometry; it applies to any neutral geometry. In fact, in Euclidean geometry the familiar formula (area $= \frac{1}{2}bh$) for the area of a triangle readily yields an area function: We merely assign to each triangle as its measure half the product of a base by the corresponding altitude.

We continue by observing that the additivity property (c) of an area function, is extendible to a finite number of terms:

THEOREM 5.   (FINITE ADDITIVITY)   Let a triangle $\triangle$ be decomposed into a finite set of nonoverlapping triangles $\triangle_1, \triangle_2, \ldots, \triangle_n$. Then for any area function μ,

$$\mu(\triangle) = \mu(\triangle_1) + \cdots + \mu(\triangle_n).$$

The result is equally important in Euclidean and Lobachevskian geometry —actually it is a theorem of neutral geometry. We dispense with the proof which is subtle and tedious and has no special relevance to our treatment of Lobachevskian geometry.*

We introduced the idea of area function in Lobachevskian geometry abstractly, without presenting a specific example. There is one example that is just as important as the familiar area formula of Euclidean geometry, but it is most naturally expressed in terms of the angles of a triangle. We state it formally in the following definition.

DEFINITION.   The *defect* of $\triangle ABC$ is $180 - (\angle A + \angle B + \angle C)$. Here $\angle A, \angle B, \angle C$ are taken as the *degree measures* of the indicated angles, so that the defect of a triangle is simply a real number, not a number of degrees. Note that the defect of a triangle indicates the amount by which its angle sum falls short of $180°$.

The defect of a triangle behaves like a measure of area:

* K. Borsuk and W. Szmielew, *Foundations of Geometry*, North–Holland Publishing Company, Amsterdam, 1960, p. 281. Theorem 10 gives a proof of this for a particular area function $\triangle$, which is valid for any area function.

THEOREM 6.   The defect is an area function for triangles.

Proof.   Property (a) follows from Theorem 3. Property (b) holds since congruent triangles have equal corresponding angles, hence equal angle sums and equal defects.

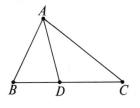

FIGURE 4.10

To establish property (c), let $\triangle ABC$ be given, and let D be a point on $\overline{BC}$, so that $\overline{AD}$ splits $\triangle ABC$ into $\triangle ABD$ and $\triangle ADC$. The sum of the defects of the last two triangles is:

$$180 - (\angle BAD + \angle B + \angle BDA) + 180 - (\angle CAD + \angle C + \angle CDA).$$

By rearranging, and observing that $\angle BDA + \angle CDA = 180$, we have that the sum of the defects of $\triangle ABD$ and $\triangle ADC$ is

$$180 - (\angle BAD + \angle CAD + \angle B + \angle C)$$
$$= 180 - (\angle BAC + \angle B + \angle C),$$

which is the defect of $\triangle ABC$.

The theorem tells us that one area function exists. We naturally wonder if there are other area functions, and how great their variety. A trivial method of constructing new area functions is given by the following theorem, which is an immediate consequence of the definition of area function.

THEOREM 7.   Any multiple of an area function by a positive constant is also an area function.

Multiplication of an area function by a positive constant in effect changes the unit of measure (that is, any triangle whose measure is 1) but not the ratio of the measures of triangles. In the case of the defect this has a simple geometrical significance, for the specific form of our definition of defect depends on the basic convention to measure angles in terms of degrees. If we adopt a different unit for the measure of angles and define "defect" in the natural manner, we obtain a constant multiple of the defect as we originally defined it. For example, suppose we change our unit of angle measurement from degrees to minutes. This would entail two simple changes in the above discussion: (1) each angle measure would have to be multiplied by 60; (2) the key number 180 would have to be replaced by 60 times 180 or 10,800. Thus the appropriate definition of "defect" would be 60 times the defect as we originally defined it.

The last theorem, unfortunately, does not resolve our question on the possible variety of area functions. We are concerned over the possibility of an area function which is *not* a constant multiple of the defect. We may feel that the defect was pulled out of a hat and may not be a "typical" area function—that another area function may be conjured up which is not proportional to the defect. If this happens there may be two triangles which have equal areas as determined by one area function and unequal areas as determined by another. In practical affairs this would be quite distressing: the cost of a house might depend on the system used to measure it. Fortunately, no such thing can happen in Lobachevskian geometry. We have

THEOREM 8.   Any two area functions are proportional.

We shall omit the proof, which is rather difficult and lies partly in the domain of real analysis.*

In view of Theorems 6 and 8, it is perfectly reasonable to define the area of a triangle to be its defect; disregarding a proportionality factor, it is the only possible definition.

It is interesting to note that in Euclidean three–dimensional geometry, the sum of the angles of a spherical triangle is greater than 180°, and the area of a spherical triangle is defined to be its "excess," that is, the sum of the degree measures of its angles minus 180.

We conclude with the observation that Theorem 8 is also true in Euclidean geometry and is needed to validate the familiar Euclidean theory of area.

7. Parallelism and Equidistance of Lines

In Euclidean geometry, an important characteristic of parallel lines is that they are everywhere equidistant. That this is not the case in Lobachevskian geometry is seen in the following theorem.

THEOREM 9.   No two parallel lines are everywhere equidistant.

* See Edwin Moise, *Elementary Geometry From An Advanced Standpoint*, Addison–Wesley Publishing Company, Reading, 1963, p. 345, Theorem 1.

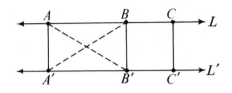

FIGURE 4.11

Proof. We shall show that for any two parallel lines L, L' there do not exist as many as three points on L which are equidistant from L'. Let A, B, C be distinct points on L, with B between A and C. From A, B, C drop perpendiculars to L', meeting L' in A', B', C', respectively. Suppose $\overline{AA'} = \overline{BB'} = \overline{CC'}$. From $\overline{AA'} = \overline{BB'}$, $\angle AA'B' = \angle BB'A'$, $\overline{A'B'} = \overline{B'A'}$, it follows that $\triangle AA'B' \cong \triangle BB'A'$. Hence $\overline{AB'} = \overline{BA'}$. Since $\overline{BB'} = \overline{AA'}$ and $\overline{BA} = \overline{AB}$ we have $\triangle AB'B \cong \triangle BA'A$. Consequently,

$$\angle A'AB = \angle B'BA;$$

that is, the "summit angles" of quadrilateral $AA'B'B$ are equal.

The same reasoning applied to quadrilateral $CC'B'B$ yields

$$\angle C'CB = \angle B'BC.$$

Addition of the last two equations gives

$$\angle A'AB + \angle C'CB = \angle B'BA + \angle B'BC = 180°.$$

Thus the angle sum of quadrilateral $AA'C'C$ is 360°, contradicting Corollary 1 of Theorem 3. The supposition is therefore false and the theorem follows.*

We conclude this section with a discussion of types of parallel pairs of lines. In view of the proof of the theorem, if two lines are parallel only two cases can arise: (a) there exist two points on one line that are equidistant from the other line; (b) no two points of one line are equidistant from the other line. The first case arises if and only if the lines have a common perpendicular (see Ch. 3, Exercises I, 7). In this case the lines diverge indefinitely on both sides of their common perpendicular (see below, Exercises I, 10). In the second case it can be shown that the lines are *asymptotic*; that is, the distance from a point of one to the other approaches zero as the point recedes in one direction on the line.

Asymptotic lines have the important and subtle property of *boundary parallelism*, which we shall proceed to define. Suppose point P is not on line L,

---

* This result sheds light on Exercise 3 at the end of Chapter 2.

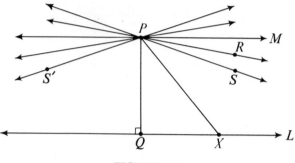

FIGURE 4.12

$PQ \perp L$ at Q, and $M \perp PQ$ at P (Figure 4.12). Suppose line $PR \parallel L$ and $\angle QPR$ is acute. The lines through P intersect L or are parallel to L. Let X be on L on the same side of PQ as R; then we may think of line PX as a typical "intersector" of L through P. Now let X recede endlessly on L. Then $\angle QPX$ is always increasing and is bounded, since $\angle QPX < \angle QPR$. Hence $\angle QPX$ approaches a limit, say $\angle QPS$. Then PX will approach PS as X recedes endlessly. It can be shown that PS does not intersect L, but is a boundary of the lines PX which do intersect L. It is said to be *boundary parallel* to L. Symmetrically there is a line PS', which also is boundary parallel to L, such that $\angle QPS' = \angle QPS$ and S' is on the opposite side of PQ from S. To summarize: PS and PS' are parallel to L and separate the lines through P which intersect L from those lines through P, distinct from PS and PS', which do not intersect L.

This concludes our exposition of the elements of Lobachevskian geometry.

8. Are There Other Neutral Geometries

We have completed our introduction to Lobachevskian geometry, and we naturally inquire about the possibility of other geometrical theories that differ from Euclid's. Specifically, we ask whether there are neutral geometries different from both Euclidean and Lobachevskian geometry. The answer seems to be no. This answer is correct, but its justification requires more than a word. First let us see that there is a problem here, that the question is not trivial. Observe that although Euclidean geometry and Lobachevskian geometry are mutually contradictory, there may possibly exist a neutral geometry which contradicts both, one which is non-Euclidean *and* non-Lobachevskian. There may be a neutral geometry which is "partly Euclidean" and "partly Lobachevskian."

To make this precise let us say, in a neutral geometry, that line L and point P satisfy the *Euclidean parallel property* if there is exactly one line through P parallel to L; and that L and P satisfy the *Lobachevskian parallel property* if there exist at least two lines through P parallel to L. In a neutral geometry we know that there is at least one line through an external point parallel to a given line (Ch. 2, Th. 2, Cor. 3). Hence each pair formed of a line and an external point satisfies either the Euclidean or the Lobachevskian parallel property. But we do not know that a neutral geometry is "homogeneous" in this respect: possibly some of its "point-line" pairs satisfy the Euclidean and others the Lobachevskian parallel property. This calls to mind the homogeneity of a neutral geometry as regards the angle sum of a triangle, mentioned at the end of Section 4 of Chapter 3. It suggests that "parallel property homogeneity" is reducible to "angle sum homogeneity." This is so. To justify it we shall begin by considering implications of the Euclidean parallel property.

THEOREM 10.   In a neutral geometry, let there exist a line and a point which satisfy the Euclidean parallel property. Then there exists a rectangle.

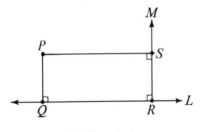

FIGURE 4.13

Proof.   Let L and P be the given line and point. Let PQ be perpendicular to L at Q. Let R be a point on L, distinct from Q. Construct line M perpendicular to L at R. Finally, let PS be perpendicular to M at S. Then $PQRS$ has right angles at Q, R, S. We show $PQRS$ is a rectangle. Since PS and L are both perpendicular to M, they are parallel (Ch. 2, Th. 2, Cor. 1). Since L and P satisfy the Euclidean parallel property, PS is the only line through P parallel to L. But the line perpendicular to PQ at P must also be parallel to L. Hence PS coincides with this line and $PQRS$ is a rectangle by definition (Ch. 3, Sec. 3).

COROLLARY.   In a neutral geometry, let there exist a line and a point which satisfy the Euclidean parallel property. Then every triangle has angle sum 180°.

Proof. The theorem implies the existence of a rectangle and the result follows by Theorem 5 of Chapter 3.

Now let us consider the implications of the Lobachevskian parallel property.

THEOREM 11.   In a neutral geometry, let there exist a line and a point which satisfy the Lobachevskian parallel property. Then there exists a triangle whose angle sum is less than 180°.

Proof. In essence, this has already been proved. Recall the proof of Theorem 2, that in *Lobachevskian geometry* there exists a triangle whose angle sum is less than 180°. By examining the proof closely, you will find that the full force of the Lobachevskian parallel postulate is not employed: it is applied only to a single line *L* and a single point *P*; that is, the argument merely requires *L* and *P* to satisfy the Lobachevskian parallel property.

COROLLARY.   In a neutral geometry, let there exist a line and a point which satisfy the Lobachevskian parallel property. Then the angle sum of every triangle is less than 180°.

Proof. By this theorem there exists a triangle whose angle sum is less than 180°, and so every triangle has the same property (Ch. 3, Th. 6, Cor. 2).

Now we can quickly obtain our main result on the homogeneity of a neutral geometry as regards the parallel property.

THEOREM 12.   In a neutral geometry let there exist a line and a point which satisfy the Euclidean parallel property. Then every line and each external point satisfy the Euclidean parallel property; that is, the geometry is Euclidean.

Proof. Suppose the theorem false. Then there exist a line and a point which satisfy the Lobachevskian parallel property. By the last corollary every triangle has an angle sum of less than 180°. But the hypothesis implies (Th. 10, Cor.) that every triangle has an angle sum of 180°. This contradiction yields the theorem.

COROLLARY 1.   In a neutral geometry, let there exist a line and a point which satisfy the Lobachevskian parallel property. Then every line and each external point satisfy the Lobachevskian parallel property; that is, the geometry is Lobachevskian.

Proof. Suppose line L and point P satisfy the Lobachevskian parallel property. Let L' be any line and P' an external point. Then L' and P' cannot satisfy the Euclidean parallel property; otherwise there would be a contradiction of the theorem.

We immediately infer the following important *classification* property of neutral geometries:

COROLLARY 2. Every neutral geometry is either Euclidean or Lobachevskian.

COROLLARY 3. A neutral geometry is Euclidean or Lobachevskian according as it contains a triangle whose angle sum is, or is less than, 180°.

Proof. In a neutral geometry, let there exist a triangle whose angle sum is 180°. Then the geometry cannot be Lobachevskian and so must be Euclidean by Corollary 2. The other case is similar.

We now can tie into our chain of deductions a loose thread remaining from Chapter 3. We proved in Chapter 3 (Th. 5) that if a rectangle exists in a neutral geometry, every triangle has an angle sum of 180°, and conjectured that the geometry had to be Euclidean. By Corollary 3 the conjecture takes the status of an inference and we may make the assertion:

COROLLARY 4. A neutral geometry which contains a rectangle is Euclidean.

We conclude this section with some remarks on two important ideas that appeared in it. First, consider our basic result in Corollary 2 above, that every neutral geometry is Euclidean or Lobachevskian. This is a *classification* principle. It asserts that all neutral geometries fall into two well-characterized classes. It illustrates an important kind of contemporary mathematical problem: the classification of all mathematical systems which satisfy a given set of postulates. (For example, the classification in modern algebra of all commutative groups, or in geometry of all finite projective planes.) This is somewhat different from the classification problems of traditional mathematics. There we classify in a given branch of mathematics all entities which have certain properties; for example, all curves that satisfy a quadratic equation in two variables, or all functions that have positive derivatives.

Secondly, in proving Theorem 11 we introduced a technique called "proof analysis." This is the process of scrutinizing the proof of a theorem to determine whether it uses the full force of the hypothesis. If it does not, a stronger or "improved" theorem can be asserted, namely, that the conclusion follows from a weaker hypothesis. Proof analysis often yields uninteresting results, such as theorems with bizarrely weakened hypotheses—but sometimes it yields an essentially new theorem or a new insight into a situation. In the case above, the analysis of the proof of Theorem 2 produced a huge reduction in the hypothesis from the Lobachevskian parallel postulate to the Lobachevskian parallel property for a single line and a single point. This yielded several important results, including the classification of neutral geometries into Euclidean and Lobachevskian geometries.

9. Riemann's Non-Euclidean Theory of Geometry

Once the ice had been broken by Bolyai's and Lobachevsky's successful challenge to Euclid's parallel postulate, mathematicians were stimulated to set up other non-Euclidean theories of geometry. The first and best known of these was proposed by Riemann in 1854. Riemann's theory contradicts Euclid's parallel postulate by assuming the following principle:

RIEMANN'S PARALLEL POSTULATE. *There are no parallel lines.*

Riemann's theory requires abandoning not only Euclid's parallel postulate but other postulates as well. For we have shown, without assuming any parallel postulate, that parallel lines exist (Ch. 2, Th. 2, Cor. 3); the existence of parallel lines is, then, a theorem of neutral geometry. In other words, Riemann's postulate, that there are no parallel lines, is *inconsistent* with the postulates of neutral geometry. Consequently, we will have to discover which postulate or postulates of neutral geometry imply the existence of parallel lines, and also drop these from our list.

A natural procedure for doing this is to analyze the proof of the Existence of Parallel Lines (Ch. 2, Th. 2, Cor. 3) to see upon which properties it depends. Glancing at the proof, we see that it follows directly from the following property:

(A) *Two lines perpendicular to the same line are parallel* (Ch. 2, Th. 2, Cor. 1).

Property (A) is a direct consequence of the Exterior Angle Theorem, so we have to determine which postulates the Exterior Angle Theorem depends upon. But the proof of the Exterior Angle Theorem is complex and involves the tacit acceptance of graphic properties from a diagram. Consequently, it is quite difficult to determine which crucial properties are to be dropped. However, there is an alternate proof of Property (A) which is simple and does not require the Exterior Angle Theorem. We present it and analyze it to derive the crucial properties.

THEOREM.   Two lines perpendicular to the same line are parallel.

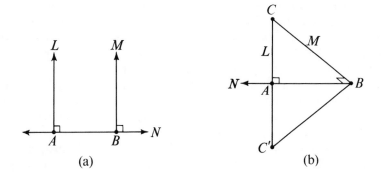

(a)   (b)

FIGURE 4.14

Given.   Two (distinct) lines *L*, *M* which are perpendicular to line *N* (Figure 4.14(a)).
To prove: *L* is parallel to *M*.

Proof.   Suppose *L* parallel to *M* is false. Then *L* and *M* will meet in a point *C* (Figure 4.14(b)). Let *L*, *M* meet *N* in *A*, *B*, respectively.

1. Extend $\overline{CA}$ its own length through   1. A segment may be doubled.
   A to *C'*.
2. Draw *C'B*.   2. Two points determine a line.
3. $\triangle ABC \cong \triangle ABC'$.   3. SAS.
4. $\angle ABC = \angle ABC'$.   4. Corresponding parts.

Thus $\angle ABC'$ is a right angle since $\angle ABC$ is a right angle; and BC and BC' are perpendicular to AB.

5. BC and BC' coincide.

5. There is only one line perpendicular to a given line at a given point of the line.

Thus AC and BC or L and M have points C and C' in common.

6. Therefore L and M coincide.

6. Two points determine a line.

This contradicts our hypothesis that L and M are distinct lines. Thus our supposition is false and the theorem holds.

If Riemann's parallel postulate is to hold, this theorem must go by the board. Thus we must discard (in addition to the Euclidean parallel postulate) one of the principles used in the proof. Certainly we want to retain the basic properties of congruent triangles and perpendicular lines—we won't tinker with these! With this in mind let us analyze the proof. The crucial point seems to be Step 6, that L and M coincide since they have the distinct points C and C' in common. This step (and so the proof) will fail if C and C' are not distinct, that is, if they coincide. How could they coincide? Rather, we should ask how we know that they are distinct. This crucial point in the proof is not formally justified, but seems indubitable from the diagram. Can we find a geometrical *principle* to justify it?

To answer this, observe that Euclid tacitly assumes that every line "separates" the plane into two opposite sides. Stated more precisely: if L is a given line, the points of the plane not on L form two figures or sets of points, called the sides of L. These sides have no point in common, and have the property that every segment which joins a point of one side to a point of the other or opposite side meets L. In view of this *separation property*, the construction in Step 1 of the proof (to extend $\overline{CA}$ its own length to C') guarantees that C and C' are on opposite sides of N, and so are distinct points. Without the separation property the distinctness of C from C' has no formal justification, and the proof fails. This suggests that we can set up a "Riemannian" theory of geometry by discarding the postulate that every line separates the plane.

If you feel that to give up the separation principle is too much, we can manage to retain it provided we pay for it by sacrificing something else. If the separation principle is accepted, C and C' must be distinct points; but we can still avoid the contradiction in Step 6, if we abandon the principle that two points determine a line, and permit two lines to intersect in two points.

At first sight this may seem an exorbitant payment, but it leads to an interesting and rather simple geometrical theory.

Summary. There are two geometrical theories which assume Riemann's parallel postulate. In the first, any two lines intersect in exactly one point, but no line separates the plane. In the second, two lines intersect in exactly two points, and every line separates the plane. These theories are called, respectively, *single elliptic geometry* and *double elliptic geometry.* (The terms "single" and "double" indicate the intersection properties of two lines in the geometries; and the term "elliptic" is used rather subtly in the sense of a classification based on projective geometry in which Euclidean and Lobachevskian geometries are called *parabolic* and *hyperbolic*.)

It is worth noting that Riemann introduced a radically different kind of geometric theory which builds up the properties of space in the large by studying the behavior of distance between points which are close together. This theory, called *Riemannian Geometry*, is useful in applied mathematics and physics, and is the mathematical basis of Einstein's general theory of relativity.

10. Lines as Closed Figures

In these two elliptic geometries another familiar and important property is discarded, namely, that a line is an unbounded open figure which is separated into two parts (rays or half-lines) by each of its points.

First consider single elliptic geometry. By examining the situation indicated by Figure 4.14(b) in the proof of the theorem that two lines perpendicular to the same line are parallel, we see that if a theory of single elliptic geometry is to be possible at all, point C' must coincide with point C. Thus in extending $\overline{CA}$ its own length to C' we have *returned* to point C. In other words, we have traced out the entire line CA which consists of segment $\overline{CA}$ and its extension. Consequently, a line is to be conceived of as a closed figure. It follows that a point does not separate a line into two parts; but two points of a line separate it into two segments, and so determine on the line not one but two segments of which they are the common endpoints.

This conception of line may be motivated in double elliptic geometry in the following way. Let line L be given and let A be a point of L (Figure 4.15). Let M be perpendicular to L at A. Then L and M meet in a second point B. Whatever our concept of line, we require A and B to be the endpoints of one

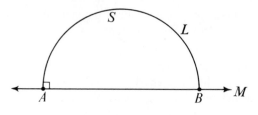

FIGURE 4.15

segment, at least, contained in line L. Let S be a segment that joins A and B, contained in L. Since M separates the plane and M meets L in exactly two points, S must be (apart from endpoints) entirely on one side of M.

Next we want to justify that every point of L on the given side of M is on S. Our concept of line requires that any point of L which is not on segment S must be on an extension of S beyond one of its endpoints A or B. But in the process of extending S beyond A or B, line L "crosses" M, and enters the side of M opposite to S. Thus any point of L on the same side of M as S must be on S, and we infer that S constitutes the portion of L on the given side of M.

We would like to argue that there is a corresponding segment S', contained in line L, that joins A and B on the other side of M and constitutes the portion of L on that side of M. For this purpose, recall a cardinal idea of Euclidean plane geometry (in fact of neutral geometry) that any figure F can be "reflected" (perpendicularly) in a given line to produce a symmetrical figure F'. We want this theory of symmetry to be preserved in double elliptic geometry. Thus there will exist a figure S' symmetrical to segment S, that joins A and B on the other side of M from S. Since S is a segment, S' also is a segment. Since S is perpendicular to M at A, the symmetrical segment S' is also perpendicular to M at A. Since S and S' are segments perpendicular to the same line at the same point, they must lie in a line; that is, S' is contained in L. By the argument of the last paragraph any point of L on the same side of M as S' must be on S'. We conclude that L is constituted by segments S and S'. Thus we are led to conceive of a line as a closed figure, as in single elliptic geometry.

11. Representation on a Euclidean Sphere

At first encounter the elliptic geometries may seem rather strange geometrical theories, but they can be represented faithfully in terms of Euclidean concepts. The representation involves Euclidean spherical geometry and is especially simple for double elliptic geometry. The following table lists some

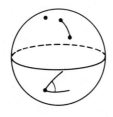

FIGURE 4.16

of the basic concepts of double elliptic geometry and the corresponding representation on a Euclidean sphere (Figure 4.16).

Double Elliptic Geometry	Euclidean Representation
point	point on a sphere S
line	great circle of S
plane	sphere S
segment	arc of great circle of S
distance between two points	length of shortest arc of great circle of S joining two points
angle (formed by two lines)	spherical angle (formed by two great circles)
measure of angle	measure of spherical angle

Observe that Riemann's parallel postulate is satisfied in this representation: each two "lines" (great circles) meet, and in fact in exactly two points. Further, the separation postulate is satisfied, since each great circle separates the sphere into two hemispheres. For example, the equator separates a globe into the northern and southern hemispheres so that any arc of a great circle joining a point on one hemisphere to a point on the other necessarily meets the equator. Note finally that each "line" appears as a closed figure.

Beware the pitfall of thinking that Riemann's double elliptic geometry is merely Euclidean spherical geometry with new names, so that we merely *call* a great circle a line, an arc of a great circle a segment, et cetera. Quite the contrary. Riemann has set up a new abstract theory of how lines behave. We might say, a new theory of straightness which contradicts Euclid's theory at several points. Consequently, Riemannian lines cannot be represented faithfully by Euclidean lines on a Euclidean plane, and it is quite remarkable that they can be represented faithfully by great circles on a Euclidean sphere.

The representation of single elliptic geometry is derived from that of double geometry by a rather clever device. A great circle on a sphere does

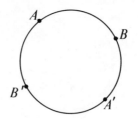

FIGURE 4.17

not represent properly a line of single elliptic geometry, since two great circles always intersect in two diametrically opposite points. To try to circumvent this difficulty, suppose that we consider any two opposite points of a sphere to be the same; or as we say, we agree to "identify" any point and its opposite point. Then we can represent single elliptic geometry essentially as we represented double elliptic geometry. Thus a line of single elliptic geometry is represented by a great circle (with the agreement that opposite points are to be identified). A segment is represented by a *minor* arc of a great circle (since a major arc or a semicircle now represents an entire line). To determine the distance between two points represented by A and B, remember that A and its opposite A' represent the same point, similarly for B and its opposite B' (Figure 4.17). Thus the distance is taken to be the length of the smaller of the minor arcs $\overset{\frown}{AB}$, $\overset{\frown}{AB'}$ (or equivalently the minor arcs $\overset{\frown}{A'B}$, $\overset{\frown}{A'B'}$). Angles and their measures are represented as in double elliptic geometry.

Observe that in this representation, as in the preceding one, a "line" is a closed figure and Riemann's parallel postulate is satisfied. But now, since opposite points are identified, two "lines" meet in just one point. It also follows that the postulate, *two points determine a line*, is satisfied. Moreover, it is not hard to see that the separation postulate fails.

12. Critique

You may feel that the basis of the representation of single elliptic geometry—the identification of a point and its opposite point—is unsound. You may argue: If things are identical there is no need to "identify" them; if they are distinct it is impossible to identify them. This seems impossible to answer. Without trying to answer it, let us examine the problem of representing single elliptic geometry and see what led us to introduce the idea of "identification." A "single elliptic" point is not represented by a unique point of a sphere, as is a "double elliptic" point. We must think of it as being represented equally well by either one of a pair of diametrically opposite points. But we would like each "single elliptic" point to have a unique representation. This led us to agree that a point of the sphere and its opposite were to be identified, and so involved us in the difficulty above. Can we find a unique representation for a point of single elliptic geometry? Very easily, provided we give up the idea that a point of single elliptic geometry must be represented by a *point* on the sphere. We merely represent a point of single elliptic geometry by a *point pair*, consisting of two opposite points of the sphere. Of course the representation of line, plane, segment must be modified to correspond. For example, a line is represented by the set of pairs of opposite points which are contained in a great circle. In effect, this justifies the original, intuitively simpler representation by putting it on a basis that is logically unassailable. Observe that in taking two opposite points A and A' and forming the pair (A, A') we have constructed a single entity of the two, which might be described, in a certain sense, as a process of identification.

13. Representation of Mathematical Systems

Before continuing our discussion of elliptic geometry we shall devote a few lines to the general idea of representation of mathematical systems. The representation of the elliptic geometries in Euclidean geometry may seem a trivial or isolated thing, like a photograph made with a distorting lens; but this view is not justified. We have introduced these representations to make elliptic geometry more accessible and to give a graphic picture in familiar terms of an unfamiliar (and incompletely described) theory. But the notion of representation of one mathematical system in another is an intrinsically important idea of modern mathematics. Considering that double elliptic geometry and Euclidean geometry are contradictory theories, it is astonishing that every property of double elliptic geometry can be faithfully pictured by a corresponding property of Euclidean spherical geometry. This indicates a deep interrelation between double elliptic geometry and Euclidean geometry, which might not have seemed possible. We may note here that Lobachevskian geometry also can be represented faithfully in Euclidean geometry, as will be indicated in Chapter 5.

Once we have established a representation of one mathematical system in another, we can interpret propositions in the first by means of corresponding propositions in the second and are thus afforded deeper insight into each system than is obtainable by studying them independently. This process of studying a system by its representation is not really unfamiliar. The study of (Euclidean) analytic geometry is based on a representation of Euclidean geometry in the system of real numbers which pictures geometrical properties of points and lines by corresponding algebraic properties of ordered pairs (x, y) of real numbers and linear equations $ax + by + c = 0$.

In Chapter 5 we shall deal with the problem of consistency of the non-Euclidean geometries, using representations of them in Euclidean geometry.

14. Difficulties Encountered in a Formal Treatment of Riemann's Theory

It would be fortunate if we could present a formal treatment of Riemann's theory comparable to the one we gave for Bolyai's and Lobachevsky's. But this is not feasible. There are difficulties that make an elementary introductory treatment impossible. Recall that in Lobachevskian geometry we had as a base all the familiar results of neutral geometry. The familiar Euclidean congruence, graphical and separation properties were retained, and only the parallel postulate was changed. In Riemann's theory, points and lines behave quite differently than in neutral geometry. As we have seen (Sec. 10), a line is closed and two of its points separate it into two segments. It is harder to define *angle*, since we do not have available the notion of ray or half-line as in neutral geometry. Even the question of framing a suitable definition of *triangle* is a problem. For instance let A, B, C be three non-collinear points and let $\overline{AXC}$, $\overline{AYC}$ be the two segments of line AC determined by A and C (Figure 4.18). Then if segments $\overline{AB}$, $\overline{BC}$, and $\overline{AXC}$ form a triangle, can we also consider that $\overline{AB}$, $\overline{BC}$, and $\overline{AYC}$ form a triangle? If so, will the SAS principle be valid? In Euclidean or Lobachevskian geometry difficulties of this type do not arise, since different triangles cannot have the same vertices. In single elliptic geometry another disturbing possibility suggests itself: since a line does not separate the plane, will a triangle necessarily separate the plane? Will it necessarily have an interior?

These difficulties can be resolved. In fact, the existence of the spherical representations of the elliptic geometries gives an important clue for resolving them. However, our discussion indicates that a formal treatment of elliptic geometry would require a careful preliminary study of the graphical prop-

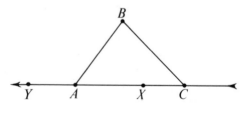

FIGURE 4.18

erties of points and lines, and of the nature of angles and triangles. Such a study is quite out of place here, and we shall conclude our introduction to Riemann's theory with an informal discussion of several important properties.

15. Polar Properties in Elliptic Plane Geometry

In elliptic plane geometry (as in Euclidean and Lobachevskian plane geometry) there is only one line perpendicular to a given line through a given point, if the point is on the line. However, if the point is not on the line this property may fail, since any two lines perpendicular to the same line must meet. The property fails in a rather interesting way, typical of elliptic geometry: namely, for each line L there exists a "polar" point P such that *all* lines through P are perpendicular to L, just as all the great circles on a globe through its north pole are perpendicular to its equator.

To see why this should be so, let us consider an elliptic plane geometry (of either type). Let L be an arbitrary line and let lines M, N be perpendicular to L at the distinct points A and B (Figure 4.19). By Riemann's parallel postulate M and N meet at a point P, so that a triangle PAB is formed with sides $\overline{PA}$, $\overline{PB}$, and $\overline{AB}$. Since $\triangle PAB$ has two equal angles, it is isosceles. Specifically, $\overline{PA} = \overline{PB}$. Let C be the midpoint of segment $\overline{AB}$. Then, as in neutral geometry, congruent triangles $\triangle PAC$, $\triangle PBC$ are formed, with $\overline{PA}$, $\overline{PB}$ as corresponding sides and a segment $\overline{PC}$ as common side. It follows that line PC is perpendicular to AB. By the argument above $\triangle PAC$ and $\triangle PBC$ are isosceles. Thus

$$\overline{PC} = \overline{PA} = \overline{PB}.$$

Clearly, the argument can be repeated by bisecting the third side of $\triangle PAC$ (or of $\triangle PBC$) and we can find as many points as we wish on L that are joined to P by segments that are equal, and perpendicular to L. This suggests the following:

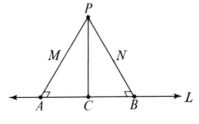

FIGURE 4.19

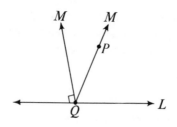

FIGURE 4.20

POLAR PROPERTY.   *Let L be any line. Then there exists a point P, called a pole of L, such that*
   (a) *every segment that joins P to a point of L is perpendicular to L, and*
   (b) *P is equidistant from all points of L.*

We consider some consequences of the Polar Property. First, it should be noted that since two points are joined by more than one segment, the distance between two points is taken to be the length of the shortest joining segment.

Next we observe that if *P* is a pole of *L*, each line perpendicular to *L* passes through *P*. For suppose *M* perpendicular to *L* at point *Q* on *L* (Figure 4.20). There certainly is a line *M'* that passes through *P* and *Q*. By the Polar Property, *M'* is perpendicular to *L* at *Q*. Since *L* has a unique perpendicular at *Q*, *M* and *M'* coincide and *M* must pass through *P*.

Now we introduce the term *polar distance* to denote the constant distance from *P* to the points of *L*. Let line *M* join *P* to a point *Q* of *L*. We show that there is a segment of *M* which joins *P* and *Q* and has length equal to the polar distance of *P* from *L*. By the Polar Property, *M* is perpendicular to *L* at *Q*, and is the only line that joins *P* and *Q* since there is only one perpendicular to *L* at *Q*. Thus there are just two segments that join *P* and *Q*, the segments into which *P* and *Q* separate *M*. The length of the shorter of these is the distance between *P* and *Q* and must be the polar distance from *P* to *L*.

We consider now a construction which suggests that a line may have two poles. Let *P* be a pole of line *L*, and *Q* a point of *L* (Figure 4.21). Let $\overline{PQ}$ be a *polar segment*, that is, a segment joining *P* and *Q* whose length is the polar distance from *P* to *L*. Extend $\overline{PQ}$ its own length through *Q* to *P'*. Considerations of symmetry suggest that *P'* is also a pole of *L*, and that the polar distances of *L* from *P* and *P'* are the same. Let us take this for granted. May we then infer that every line has at least two poles? No, since we have no right to assume from the diagram that *P'* and *P* are distinct points.

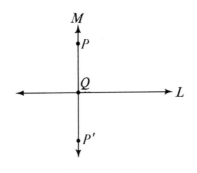

FIGURE 4.21

Examining the situation more closely, we consider first the case of single elliptic geometry. Suppose P' and P do not coincide. It then follows from the Polar Property (as we saw above) that two lines perpendicular to L would meet in the distinct points P and P'. Since this is impossible, P and P' must coincide. Thus in extending $\overline{PQ}$ its own length to P', we have traced out the entire line PQ, and we see that the length of line PQ is twice the polar distance from P to L.

Now consider the double elliptic case. By recalling that L separates the plane, we see that P and P' are on opposite sides of L and cannot coincide. Thus every line has at least two poles. A line cannot have more than two poles since, as we have seen, all perpendiculars to it pass through its poles.

Continuing, we examine the structure of line PQ. In extending $\overline{PQ}$ its own length to P' we have formed a segment $\overline{P'Q}$ which is symmetrical to $\overline{PQ}$ with respect to L. $\overline{PQ}$ and $\overline{P'Q}$ have only Q in common and constitute a segment $\overline{PQP'}$ whose length is twice the polar distance from P to L.

This ties down the relation of the poles P and P' to L and Q. But L meets line PQ in a second point Q' (Figure 4.22). How is Q' related to P, Q, and P'? First we note that Q' is not on $\overline{PQP'}$; for if it were, the distance from P or P' to Q' would be less than the polar distance. Thus P and P' separate line

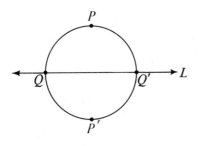

FIGURE 4.22

PQ into $\overline{PQP'}$ and a segment $\overline{PQ'P'}$ that contains Q'. Let $\overline{PQ'}$, $\overline{P'Q'}$ be the segments into which Q' separates $\overline{PQ'P'}$. We assert that $\overline{PQ'}$ is a polar segment, for the polar segment on line PQ that joins P and Q' cannot be the segment complementary to $\overline{PQ'}$, since the latter contains $\overline{PQ}$ and so has length greater than the polar distance. Similarly, $\overline{P'Q'}$ is a polar segment. Thus line PQ is separated by P, Q, P', Q' into four polar segments, and its length is four times the polar distance from P to L.

Finally, we remark that in an elliptic geometry of either type the polar distance is constant and so is the length of a line.

16. Further Remarks on Elliptic Geometry

In elliptic geometry the angle sum of any triangle is greater than 180°. This is evidenced by the existence of triangles with two right angles in our discussion above. It follows that the angle sum of a quadrilateral is greater than 360°. Furthermore, the theory of similarity is "degenerate" as in Lobachevskian geometry. That is, we can prove AAA, that two triangles are congruent if their corresponding angles are equal. Essentially, the proof in Lobachevskian geometry applies. Finally, we observe that area of triangles can be defined essentially as in Lobachevskian geometry: we take the area of a triangle to be its "excess," the difference between the degree measure of its angle sum and 180. This is, of course, the familiar method of measuring the area of a spherical triangle in Euclidean spherical geometry.

17. Conclusion

In its further development non–Euclidean geometry is at least as complex as Euclidean geometry. There is a Lobachevskian and a Riemannian solid geometry, a trigonometry and an analytic geometry. Problems in mensuration of curves, surfaces, solids and problems involving local properties such as tangency and curvature, require the use of integral and differential calculus.

Looking back over the geometrical theories we have examined, we are confronted by the question of which theory is correct. This chapter will not be prolonged to discuss this difficult problem but we shall devote the next two chapters to two aspects of it: the question of the logical consistency of the non–Euclidean geometries and that of their empirical validity.

In conclusion we remark on a common misconception that Euclidean geometry is the correct theory of *straightness* and that the non–Euclidean geometries really study curved lines. Thus two Lobachevskian parallels that have a common perpendicular and are divergent (see below, Exercises I, 9) clearly are curved lines, since parallel lines must be everywhere equidistant. And a Riemannian line is obviously curved since, as everyone knows, straight lines don't close up.

All three theories are theories of straightness, but they don't agree completely on the properties of straightness. It is manifestly unfair to declare a theory incorrect because it is not consistent with the one we were brought up with. From the Lobachevskian viewpoint two Euclidean parallels that are everywhere equidistant cannot both be straight, since as we saw in proving Theorem 9 two parallel lines cannot be equidistant at more than two points.

To facilitate comparison of these three interesting and complex views of the behavior of points and lines we introduce the following table.

Comparison Table for Euclidean and Non-Euclidean Plane Geometry

	Euclidean	Lobachevskian	Riemannian	
Two distinct lines intersect in	at most one	at most one	one (single elliptic) two (double elliptic)	point points
Given line L and point P not on L, there exist	one and only one line	at least two lines	no lines	through P parallel to L
A line	is	is	is not	separated into two parts by a point
Parallel lines	are equidistant	are never equidistant	do not exist	
If a line intersects one of two parallel lines, it	must	may or may not	—	intersect the other
The valid Saccheri hypothesis is the	right angle	acute angle	obtuse angle	hypothesis
Two distinct lines perpendicular to the same line	are parallel	are parallel	intersect	
The angle sum of a triangle is	equal to	less than	greater than	180°
The area of a triangle is	independent	proportional to the defect	proportional to the excess	of its angle sum
Two triangles with equal corresponding angles are	similar	congruent	congruent	

In solving the problems of the following sets use freely the results of the exercises on neutral geometry in Chapter 3. The underlying geometry is Lobachevskian unless the contrary is indicated.

E X E R C I S E S   I

1. Prove: The summit of a Saccheri quadrilateral is greater than the base; the summit angles are acute; the segment joining the midpoints of summit and base is less than a leg.
2. Prove: If two Saccheri quadrilaterals have equal bases and equal legs, they are congruent.
3. Prove: If two Saccheri quadrilaterals have equal bases and equal summit angles, they are congruent.
4. Prove: If two Saccheri quadrilaterals have equal summits and equal summit angles, they are congruent.
5. Prove: If two triangles have the same defect, and a side of one is equal to a side of the other, they are equivalent. [*Hint.* Construct the associated Saccheri quadrilaterals and use the preceding exercise.]
6. Prove: If two triangles have the same defect, they are equivalent.* [*Hint.* Use Exercise 5 and Ch. 3, Exercises II, 7.]
7. (Symmetry of Parallelism) Suppose $PQ \perp L$ at Q, $PR \parallel L$ and $\angle QPR$ acute. Let R' be on the opposite side of line PQ from R and satisfy $\angle QPR' = \angle QPR$. Prove $PR' \parallel L$.
8. Prove: The segment joining the midpoints of two sides of a triangle is less than half the third side.
9. Let A, B, C be points on line L, with B between A and C. Let A', B', C' be points on L' such that AA', BB', CC' are perpendicular to L' and AA' is perpendicular also to L. Prove $\overline{AA'} < \overline{BB'} < \overline{CC'}$. Infer that if two lines have a common perpendicular, they diverge on both sides of the perpendicular.
10. In the preceding exercise, prove that $\overline{BB'}$ increases indefinitely with $\overline{AB}$. Infer that if two lines have a common perpendicular they diverge indefinitely on both sides of the common perpendicular. [*Hint.* Proclus' proof Ch. 2, Sec. 5.]

---

* This asserts that if two triangles have the same defect they occupy the "same amount of space" in a natural geometric sense. The converse also is valid but its proof involves the idea of the defect of a polygon which we have not introduced.

EXERCISES II

1. Prove that there is an upper bound for the areas of all triangles.
2. Prove that there exists a triangle with defect less than a preassigned positive number.
3. If point P is inside $\triangle ABC$, prove

defect $(\triangle ABC)$ = defect $(\triangle PAB)$ + defect $(\triangle PBC)$ + defect $(\triangle PAC)$.

4. If points P, Q, R are on sides $\overline{AB}$, $\overline{BC}$, $\overline{CA}$ of $\triangle ABC$, prove

defect $(\triangle ABC)$ = defect $(\triangle APR)$ + defect $(\triangle BQP)$
$$+ \text{ defect } (\triangle CRQ) + \text{ defect } (\triangle PQR).$$

†5. If P, Q, R are noncollinear points inside $\triangle ABC$, prove that the defect of $\triangle PQR$ is less than the defect of $\triangle ABC$.
†6. If the sides of a triangle are sufficiently small, prove that its defect is arbitrarily small. (Given $e > 0$, there exists $d > 0$ such that, whenever each side of a triangle has length less than d, the defect of the triangle will be less than e.) [Hint. Use Exercises 2 and 5.]
7. Given $\triangle ABC$. Prove that there exists a triangle, $\triangle A'B'C'$, which is equivalent to $\triangle ABC$ and has the same angle sum, such that $\angle A' \leqq \frac{1}{2}\angle A$ and $\angle A' + \angle B' = \angle A$.
   [Hint. Analyze the proof of the Lemma of Ch. 3, Sec. 2.]
8. Given $\triangle ABC$, prove that there exists a triangle, $\triangle A'B'C'$, which is equivalent to $\triangle ABC$ and has the same angle sum, such that $\angle A' + \angle B'$ is smaller than a preassigned angle.

EXERCISES III

1. Prove: Any right triangle can be "doubled;" that is, there exists a triangle whose area (defect) is twice that of the given triangle.
2. Prove: Any triangle can be "halved;" that is, there exists a triangle whose area (defect) is half that of the given triangle.
3. Given two triangles, prove that there exists a triangle whose defect is the average of their defects. [Hint. Convert them into equivalent isosceles triangles with equal bases.]
4. Given two intersecting lines, prove that the foot of the perpendicular from a point on the first to the second is bounded as the point recedes indefinitely. Compare Chapter 2, Exercise 6. [Hint. Assume the contrary and use Ch. 3, Exercises IV, 15 to show that there exists a triangle with arbitrarily large area.]

5. Given any angle, prove that there exists a line which is wholly contained in the interior of the angle. Compare Chapter 2, Exercise 5. [*Hint.* Use Exercise 4.]

†6. Prove: (i) If the base of an isosceles triangle is a fixed segment $\overline{AB}$, and its altitude increases indefinitely, its area approaches a limit L.

   (ii) L is an upper bound for the area of all triangles with side $\overline{AB}$. (See Ch. 3, Exercises II, 8.)

   (iii) If $\overline{A'B'} > \overline{AB}$ and L' is the corresponding limit for isosceles triangles with base $\overline{A'B'}$, then $L' > L$.

   (iv) If $\overline{A'B'} > \overline{AB}$, there exists a triangle with side $\overline{A'B'}$ which has a larger area than any triangle with side $\overline{AB}$. (In other words, there are triangles "too big" to have base $\overline{AB}$.)

†7. Prove: There exists a triangle whose defect is greater than 90, which incidentally cannot be "doubled." See Exercise 1 for the meaning of "doubled." [*Hint.* By Exercises II, 2 there exists a triangle with defect less than 90. Apply Exercises II, 8 to find a triangle with the same defect and one obtuse angle. Then construct a triangle whose defect is at least double the given defect. Repeat the process if necessary.]

8. Prove: A neutral geometry is Lobachevskian if it contains two parallel nonequidistant lines.

†9. Prove: A neutral geometry is Euclidean if through each point in the interior of some particular angle a line can be passed which meets both sides of the angle.

10. Make up an exercise which characterizes Euclidean geometry (Lobachevskian geometry) as a type of neutral geometry.

E X E R C I S E S   I V

†1. Suppose $AB \perp L$ at A. Let P be an arbitrary point of AB on the same side of L as B. Consider the circle K with center P and radius $\overline{PA}$. Prove that as P recedes endlessly K does not approach line L as a limit, contrary to the Euclidean situation. [*Hint.* Choose C on L, $C \neq A$. Let P' be the intersection of K with BC. Show that as P recedes the area of $\triangle PAP'$ increases. Infer that $\angle PAP'$ can not approach $90°$, and so that P' can not approach C. See Figure 4.23.]

Note. The limit of K is a curve called a *horocycle* or *limit curve* and is important in the deeper study of Lobachevskian geometry.

2. Given line L, let C be the set of points on a given side of L which have a fixed distance from L. Prove that C is a convex curve in the sense that any line meets C in at most two points.

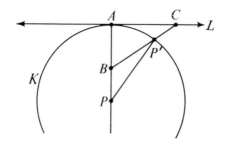

FIGURE 4.23

DEFINITION.   The curve C in Exercise 2 is called an *equidistant curve**; L is its *base*, each line perpendicular to L through a point of C is an *axis* of the equidistant curve.

3. Prove that the line perpendicular to an axis of an equidistant curve at its point of intersection with the curve meets the curve only in this point. The line is said to be *tangent* to the equidistant curve.

4. Prove that an axis M of an equidistant curve C is actually an *axis of symmetry* for C. That is, if P is a point of C and M is the perpendicular bisector of $\overline{PP'}$, then P' is also a point of C.

†5. Prove: If one leg of a right triangle is fixed and the second leg approaches zero, then the defect approaches zero.

6. Prove: If one leg of a right triangle is fixed and the second leg approaches zero, then the angle opposite the first leg approaches a right angle. (See preceding exercise.)

†7. Let a neutral geometry contain two distinct lines such that the foot of the perpendicular from a point of the first to the second is unbounded as the point recedes indefinitely in either direction. Prove the geometry is Euclidean. (See Exercises III, 4.)

REFERENCES

K. Borsuk, and W. Szmielew, *Foundations of Geometry*, North–Holland Publishing Company, Amsterdam, 1960.

Edwin Moise, *Elementary Geometry from an Advanced Standpoint*, Addison–Wesley Publishing Company, Reading, 1963.

* More commonly an equidistant curve is defined as the locus of points in a plane at a given distance from a given line L, and consists of two equidistant curves in our sense.

CHAPTER 5

The Logical Consistency of the
Non-Euclidean Geometries

In this chapter we shall discuss the logical consistency of the non-Euclidean geometries.

1. Is Lobachevskian Geometry Consistent

At this point in our discussion a confirmed partisan of Euclidean geometry might raise the following objection: "I admit that Euclidean geometry in its usual presentation is not the perfect example of a deductive system I thought it was, but presumably this can be remedied. Nevertheless I am convinced that Euclidean geometry is the only possible logical system of geometry and that the Lobachevskian theory, if carried far enough, will yield a contradiction. After all, Euclidean geometry has a history of freedom from contradiction of more than 2000 years. I predict that if the non-Euclidean geometries are studied for a few centuries, contradictions will be discovered."

This argument states the unshaken belief that motivated fruitless attempts to prove Euclid's parallel postulate for twenty centuries. How can the argument be answered? Certainly not on the basis of the evidence adduced in our earlier chapters. We can say only that both the Euclidean and the non-Euclidean geometries have an "empirical" or "probabilistic" presumption of freedom from contradiction, since no inconsistency has been discovered in so long a period of time.

We shall answer the objection by showing that if Lobachevskian geometry is inconsistent, Euclidean geometry is necessarily inconsistent. We use an interesting method employed by Felix Klein (1849–1925). First we construct a representation of Lobachevskian geometry in Euclidean geometry comparable to the spherical representations of the elliptic geometries (Ch. 4, Sec. 11). By means of the representation we establish a process for interpreting every Lobachevskian proposition (postulate or theorem) as a corresponding Euclidean proposition. Then we argue that any inconsistency in the propositions of Lobachevskian geometry must induce a corresponding inconsistency in the propositions of Euclidean geometry. Hence if Euclidean geometry is logically consistent, Lobachevskian geometry must also be consistent.

2. The Representation

We consider the notions point, line, plane, et cetera, of Lobachevskian plane geometry as abstract ideas which we specifically interpret in Euclidean geometry. Let a circle O be given in a Euclidean plane (Figure 5.1). We represent (or interpret) the Lobachevskian plane by the interior of circle O.

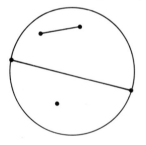

FIGURE 5.1

Thus a Lobachevskian point is represented by a point inside circle O, or as we shall call it an O-*point*. We represent a Lobachevskian line by a chord of circle O that is open in the sense of excluding its endpoints. We call such a chord an O-*chord*. A Lobachevskian segment is represented by a Euclidean segment whose endpoints are interior to circle O, called an O-*segment*. Although the description of our representation is not yet complete, we summarize it in a table and begin to discuss its properties.*

* A different Euclidean representation of Lobachevskian geometry has been given by Poincaré and used to prove the consistency of Lobachevskian geometry, see E. Moise, *Elementary Geometry From An Advanced Standpoint*, Addison-Wesley, 1963, Chs. 9, 25.

Lobachevskian Geometry	Euclidean Representation
point	O-point: point inside O
line	O-chord: open chord of O
plane	the interior of O
segment	O-segment: segment joining two O-points.

3. The Conversion Rule

The function of our representation is to enable us to interpret Lobachevskian propositions in Euclidean terms. The process for doing this is simple but very important, and sometimes quite complicated in its application. So, to lessen the chance of confusion, we present the process of interpreting Lobachevskian propositions in a very formal way. We do this by introducing the following rule for interpreting Lobachevskian statements as, or as we shall say "converting" them into, Euclidean statements.

Conversion Rule: *Replace each Lobachevskian term, "point," "line," "plane," "segment" in a given Lobachevskian statement by the corresponding Euclidean term in the table above.*

For example, consider the Lobachevskian postulate:

Two points are on one and only one line.

This is converted into the Euclidean statement:

Two O-points are on one and only one O-chord.

Observe that the latter is a *valid* Euclidean statement, that is, a proposition of Euclidean geometry. Thus the Lobachevskian postulate that two points determine a line is satisfied in the representation.

Consider next the Lobachevskian statement: *Any segment can be extended to form another segment.* (For example, given segment $\overline{AB}$ there exists a segment $\overline{AC}$ which contains $\overline{AB}$, $C \neq B$.) This is converted into: *Any O-segment can be extended to form another O-segment,* which is a proposition of Euclidean geometry.

We naturally wonder whether the Lobachevskian parallel postulate is satisfied in the representation. Thus we must convert: *If point P is not on*

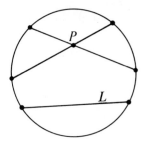

FIGURE 5.2

line L, there are at least two lines through P that do not meet L. This becomes:
If O-point P is not on O-chord L, there are at least two O-chords through P that do not meet L. The latter is clearly a proposition of Euclidean geometry (see Figure 5.2).

Our discussion may suggest the following assertion: *Every Lobachevskian proposition is transformed into a Euclidean proposition by the Conversion Rule.* This statement is true. How can we justify it? We cannot justify it by listing all Lobachevskian propositions and proceeding as above, since it is very doubtful that all Lobachevskian propositions can be listed, and surely there may be some undiscovered theorems of Lobachevskian geometry. This suggests that we consider first the *postulates* of Lobachevskian geometry, and show that they are converted into Euclidean propositions. Then we might expect the same to be true for the theorems.

4. An Objection

Now you may object that our procedure is bound to fail, for the Lobachevskian plane (like the Euclidean plane) is infinite, but it is represented by a circle interior which is finite. This merits discussion. Note first that there are two senses (at least) in which the term infinite can be used in geometry. A geometrical figure may be infinite in the sense of being unbounded; Euclidean lines or planes are unbounded, in contrast to bounded figures such as segments, circular arcs, spheres and triangles. Secondly, a figure may be infinite in a metrical sense, in that it cannot be measured in the usual way by a number. More precisely, we can find portions of the figure whose measures are as large as we please. In this metrical sense a Euclidean (or Lobachevskian) line, ray or plane is infinite.

On analysis, the objection seems to be based on the idea that an O-chord is a bounded Euclidean figure and so cannot be given a metric with respect to which it is infinite. The objection is not valid, but it points up an important incompleteness in our representation. In order to give a faithful picture of Lobachevskian geometry, our representation must tell how the length of a Lobachevskian segment is to be represented, or more simply how equal segments are to be represented. As it stands, it is like a geographical map which seems to be badly distorted, but in which the scale of distance is not given. The map may be a faithful representation of the earth's surface provided a suitable scale is specified. For example, in a Mercator projection of the earth's surface the scale varies in a definite way with the distance from the equator. Our representation of Lobachevskian geometry is like a map for which the scale is not known.

5. The Scale of the Representation

We shall now remedy this deficiency by describing how equal Lobachevskian segments are to be pictured in the representation. Our procedure is to associate with every O-segment a positive number, called its "test number." Then we represent equal Lobachevskian segments by O-segments whose test numbers are equal.

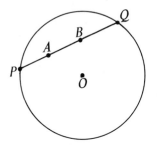

FIGURE 5.3

Let $\overline{AB}$ be any O-segment. Let the Euclidean line containing A and B meet the circle O in points P, Q labeled so that A is between P and B (Figure 5.3). Then the test number for segment $\overline{AB}$, denoted $n(\overline{AB})$, is defined to be $\dfrac{AP}{AQ} \Big/ \dfrac{BP}{BQ}$, where the symbols AP, AQ, et cetera, denote Euclidean distances. In other words $n(\overline{AB})$ is the ratio in which A divides the directed segment $\overline{PQ}$, divided by the ratio in which B divides $\overline{PQ}$, which is the so called "cross-ratio" of the four points P, A, B, Q. Finally, if $\overline{AB}$ and $\overline{CD}$ are O-segments for which $n(\overline{AB}) = n(\overline{CD})$ we say $\overline{AB} \overset{\text{O}}{=} \overline{CD}$, read $\overline{AB}$ is O-*equal* to $\overline{CD}$, or

$\overline{AB}$ is *pseudo-equal* to $\overline{CD}$. Then equal Lobachevskian segments are represented by pseudo-equal O-segments. Thus in the Representation Table of Section 2, equality of Lobachevskian segments is to correspond to pseudo-equality of O-segments. The Conversion Rule of Section 3 is extended similarly: in converting a Lobachevskian statement the term "equal segments" shall be replaced by "pseudo-equal O-segments."

There is an apparent difficulty here that we must discuss before continuing. The definition of the test number $n(\overline{AB})$ for segment $\overline{AB}$ involves the order in which the endpoints A, B are stated, since P is chosen so that A is between P and B. Consequently, $n(\overline{BA})$ might not equal $n(\overline{AB})$ and a given segment might have either of two test numbers, depending on the order in which its vertices are written. Let us calculate $n(\overline{BA})$. Applying the definition above, we note that the roles of P and Q are interchanged since B is between Q and A. Hence

$$n(\overline{BA}) = \frac{BQ}{BP} \bigg/ \frac{AQ}{AP} = n(\overline{AB}),$$

and we conclude that an O-segment has exactly one test number. Note that $n(\overline{AB})$ indicates, so to speak, how the scale of the Lobachevskian map varies; it should not be assumed that it represents Lobachevskian length.

Now we are prepared to show that the Lobachevskian postulates involving equality of segments are satisfied in the representation. For example, it can be seen that the familiar equality properties (*a*) $\overline{AB} = \overline{AB}$; (*b*) if $\overline{AB} = \overline{CD}$ then $\overline{CD} = \overline{AB}$; (c) if $\overline{AB} = \overline{CD}$ and $\overline{CD} = \overline{EF}$ then $\overline{AB} = \overline{EF}$, are satisfied since these properties hold for psuedo-equality.

6. The Property that a Lobachevskian Line is Metrically Infinite

We consider next the Lobachevskian property that a line is metrically infinite (Sec. 4). To prove a line metrically infinite it is sufficient to show that any integral multiple of a given segment of the line can be found on the line. This will follow if we can verify the following familiar postulate.

EXTENSION POSTULATE. *Any segment can be extended by a segment equal to a given segment* [Ch. 2, Sec. 1, IX].

For, using the postulate, a given segment can be doubled; applying the postulate again, we can obtain a triple of the given segment. Continuing in this way, we can obtain any integral multiple of the given segment. Thus we shall show that the Extension Postulate is verified in the representation.

Applying the Conversion Rule to it we obtain: *Any O-segment can be extended by an O-segment pseudo-equal to a given O-segment.*

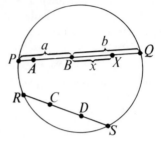

FIGURE 5.4

We prove that this is a Euclidean theorem. We are given O-segments $\overline{AB}$ and $\overline{CD}$ (Figure 5.4) and we are to prove that $\overline{AB}$ can be extended by an O-segment which is pseudo-equal to $\overline{CD}$. That is, we are to prove the existence of an O-point X, such that $\overline{BX} \overset{O}{=} \overline{CD}$, and B is between A and X. In order to enable us to apply the definition of pseudo-equality, the intersections of AB with circle O are labeled P and Q, so that A is between P and B, and those of CD with circle O, R and S, so that C is between R and D (Figure 5.4).

We begin with an analysis of the problem: We assume the existence of an O-point X such that

(1) $$\overline{BX} \overset{O}{=} \overline{CD}, \quad (B \text{ between } A \text{ and } X)$$

and determine the location of X on $\overline{BQ}$. The first relation in (1) yields

(2) $$n(\overline{BX}) = n(\overline{CD}),$$

so that

(3) $$\frac{BP}{BQ} \bigg/ \frac{XP}{XQ} = k,$$

where $k = n(\overline{CD})$. Letting $BP = a$, $BQ = b$, $BX = x$ (Figure 5.4) we have

(4) $$XP = a + x, \quad XQ = b - x.$$

Substitution in (3) yields

(5) $$\frac{a}{b} \bigg/ \frac{a + x}{b - x} = k.$$

This is to be solved for x. We have

$$a(b - x) = kb(a + x),$$
$$ab - ax = kba + kbx,$$
$$x(a + bk) = ab(1 - k),$$

(6)
$$x = \frac{ab(1 - k)}{a + bk}.$$

Thus assuming the existence of O-point X satisfying (1), we have determined x, its distance from B.

To complete the discussion we must prove the *existence* of an O-point X that satisfies (1). Naturally we try to retrace the steps of our analysis. Thus we locate a point X on line PQ (Figure 5.4) so that B is between A and X and $BX = x$ as given by (6). Then our aim is to retrace steps to (1).

Wouldn't it be convenient if valid mathematical arguments always could be retraced automatically? At the very beginning we have a minor but essential point to consider. Since BX is to be a Euclidean (undirected) distance, we must show that x as determined by (6) is positive. Further, we must show $x < b$ (Figure 5.4) to ensure that X will lie between B and Q and be an O-point. Thus we must establish

(7)
$$0 < \frac{ab(1 - k)}{a + bk} < b.$$

Since $a, b, k > 0$, this is equivalent to

(8)
$$0 < a(1 - k) < a + bk,$$

and so to the simultaneous inequalities

(8′)
$$0 < a(1 - k)$$

and

(8″)
$$a(1 - k) < a + bk.$$

Relation (8″) is trivial since $a - ak < a + bk$. To establish (8′) we show $k < 1$. Recalling that $k = n(\overline{CD})$ we have (Figure 5.4)

$$k = \frac{CR}{CS} \Big/ \frac{DR}{DS} = \frac{CR}{DR} \cdot \frac{DS}{CS} < 1.$$

Thus (8) is verified and (7) follows.

Now it is not hard to justify the process of retracing steps from (6) to (1). Thus we may consider the Lobachevskian property, *a line is metrically infinite*, to be verified in the representation.

7. Incompleteness of the Discussion

Our discussion illustrates a method of showing that the postulates of Lobachevskian geometry are verified in the Euclidean representation. The discussion is not complete.† In the first place, our representation is still incomplete, since we have not told how equality of Lobachevskian angles is to be represented. Secondly, we do not have a complete list of postulates for Lobachevskian geometry, just as we do not yet have one for Euclidean geometry. It is not our intention to formulate such postulate sets at this point, for we are not studying Lobachevskian geometry just for its own sake. Indeed, we are arguing that the existence of Lobachevskian geometry as an apparently consistent mathematical system is one of the strongest motivations for putting Euclidean geometry (and other geometrical theories) on a sound postulational basis.

8. Verification of Lobachevskian Theorems

To continue our discussion let us assume that the postulates of Lobachevskian geometry are verified in the representation. Then we assert that all Lobachevskian theorems are verified in the representation. To justify this in somewhat specific terms let us say that $P_1, \ldots, P_{10}$ are the postulates of Lobachevskian geometry. Then $P_1, \ldots, P_{10}$ are abstract statements, independent of any representation. Let $P_1{}^*, \ldots, P_{10}{}^*$ be the statements into which $P_1, \ldots, P_{10}$ are converted by the Conversion Rule. Then $P_1{}^*, \ldots, P_{10}{}^*$ are statements of Euclidean geometry, and by our assumption they are *valid* Euclidean statements—that is postulates or theorems of Euclidean geometry. Let T be any theorem of Lobachevskian geometry and T^* the Euclidean statement into which it is converted. Then T is deducible from $P_1, \ldots, P_{10}$ *abstractly;* that is, independently of any specific representation for the basic terms "point," "line," "segment," et cetera. Hence we argue that T^* is deducible from $P_1{}^*, \ldots, P_{10}{}^*$. But $P_1{}^*, \ldots, P_{10}{}^*$ are Euclidean propositions. Consequently, T^* is a Euclidean proposition. Thus any Lobachevskian theorem T is verified in the representation.

We can make the discussion above very formal and very specific. Let T be a Lobachevskian theorem. Write out a proof of T in "statement-reason" form based solely on the Lobachevskian postulates $P_1, \ldots, P_{10}$ without using any theorems. Replace each "statement" in the proof by the corresponding Euclidean statement, and replace each of the "reasons" $P_1, \ldots, P_{10}$ when it

---

† A detailed treatment is given in K. Borsuk and W. Szmielew, *Foundations of Geometry*, North-Holland, 1960, Ch. IV, Sec. 7.

occurs by the corresponding Euclidean proposition $P_1^*, \ldots, P_{10}^*$. This automatically yields a proof of T^* as a theorem of Euclidean geometry. To summarize, we apply the Conversion Rule to each "statement" and to each "reason" in the proof of Lobachevskian theorem T and automatically obtain a proof of T^* as a Euclidean theorem.

A final point: in order not to interfere with the main line of the argument we have omitted reference to defined terms in Lobachevskian geometry, for example, triangle, quadrilateral, midpoint. How are these to be represented? The answer is simple: merely write out the definition of the defined term in terms of the basic terms, "point," "line," "segment," et cetera. Then replace each basic term by its Euclidean representation. The result gives the representation of the defined term. For example, the Lobachevskian notion of triangle is represented by the figure consisting of three O-points that are not on an O-chord together with the O-segments they determine in pairs; that is, a Euclidean triangle whose vertices are O-points.

9. Consistency Proof for Lobachevskian Geometry

Now we are prepared to present the consistency proof for Lobachevskian geometry; in fact, so well prepared that the proof is quite short. Let us suppose Lobachevskian geometry is inconsistent, that is, that two Lobachevskian propositions contradict each other. To be specific, let us suppose that there are two theorems T_1, T_2 such that T_1 contradicts T_2. Applying the Conversion Rule to T_1, T_2 we get T_1^*, T_2^*. Note that T_1^* has the same form as T_1, and T_2^* the same as T_2, since the conversion process merely replaces certain geometrical terms by others. Thus since T_1 is the contradictory of T_2, it follows that T_1^* must be the contradictory of T_2^*. But as we saw in the last section, T_1^* and T_2^* are Euclidean propositions. Thus two *Euclidean* propositions contradict each other, that is, Euclidean geometry is *inconsistent*. In summary, since Lobachevskian geometry has a representation in Euclidean geometry, the malady of inconsistency is automatically infectious. This must be sad satisfaction to a mathematical Samson who attempts, by a charge of inconsistency, to tear down the edifice of Lobachevskian geometry.

Our discussion shows that if Euclidean geometry is logically consistent then Lobachevskian geometry must be logically consistent. Since it is hard to doubt the consistency of Euclidean geometry it becomes equally hard to doubt the consistency of Lobachevskian geometry. However, a disturbing possibility suggests itself: could Lobachevskian geometry be consistent and Euclidean geometry inconsistent? Our argument does not exclude this possibility. (An interesting and ironic note is that imputation of inconsistency is a many-edged weapon—it is now turned against *Euclidean* geometry.) Actually the possibility cannot occur, for a rather complicated representation of Euclidean geometry can be constructed in Lobachevskian geometry.* Then the argument above shows that Euclidean geometry is consistent if Lobachevskian geometry is consistent. *Thus the Euclidean and Lobachevskian theories stand or fall together in the matter of consistency; either both are consistent or both are inconsistent.*

This conclusion is one of the great achievements of nineteenth century mathematics. First, it definitely settles the "problem of Euclid's parallel postulate." For suppose, as was believed for two thousand years, that Euclid's parallel postulate is deducible from the remaining postulates. Then the Lobachevskian parallel postulate necessarily contradicts the remaining postulates, so Lobachevskian geometry is inconsistent. By the above conclusion, Euclidean geometry is also inconsistent. (We are fortunate indeed that the historical attempts to deduce Euclid's parallel postulate failed, for a successful attempt would have destroyed Euclidean geometry as a valid mathematical theory!)

The conclusion is important also because it fortified the right, assumed early in the nineteenth century by the non-Euclidean geometers, to challenge beliefs that had been considered authoritative for ages. Hereafter mathematicians and scientists would not have to apologize for studying theories at variance with the conventional ones, for there existed the possibility of providing logical respectability to a new or controversial theory by proving it "just as consistent" as a familiar theory. Of course it does not follow that theories which are equivalent as regards consistency are equally interesting, important or useful.

10. Further Remarks on Consistency

Similar considerations apply to the Riemann non-Euclidean geometries (Ch. 4, Sec. 9), since they have representations on a Euclidean sphere (Ch. 4, Sec. 11). In fact it can be shown, as for Lobachevskian geometry, that each of these geometries is consistent, if and only if, Euclidean geometry is consistent.

The type of argument we have used to show that the consistency of one geometric system implies the consistency of another is not restricted in its applicability to geometric theories. It is applicable whenever we can construct a representation of one mathematical theory in terms of the concepts of

---

* See pp. 55–56, p. 84, D. M. Y. Sommerville, *The Elements of Non-Euclidean Geometry*, Bell and Sons, 1914, reprinted Dover Publications; also p. 123, S. Kulczycki, *Non-Euclidean Geometry*, Pergamon, 1961.

another. An interesting and important example is suggested by elementary analytic geometry. Analytic methods are effective in Euclidean geometry precisely because they enable us to represent each notion of Euclidean geometry by a corresponding notion of algebra. Specifically, for Euclidean plane geometry we represent point by ordered pair (x, y) of real numbers, and line by the set of ordered pairs (x, y) of real numbers that satisfy an equation $ax + by + c = 0$ where a, b, c are real numbers and a, b are not both zero. To complete the representation we have to specify how to represent the other basic Euclidean ideas, for example, segment and equality of segments. In any case this can be done in detail, and it can be shown that every postulate of Euclidean geometry is faithfully rendered in the real number system. In other words, there exists a representation of Euclidean geometry in the real number system. Thus we may conclude that Euclidean geometry is consistent provided the real number system is consistent.*

Finally, we observe that the consistency proofs we have described are *relative* proofs of consistency: A mathematical theory is proved consistent on the assumption that another theory is consistent. It may be considered desirable to construct *absolute* proofs of consistency, but this is very difficult if not impossible. A point to be noted is that any consistency proof of a mathematical theory would have to assume the consistency of the underlying theory of logical inference used to formalize the process of deduction in the mathematical theory. In any case, relative consistency proofs should be appreciated for what they are. It is always an advance in knowledge to prove relative consistency, since we learn that two systems are consistent on the hypothesis that one is. Historically such a proof settled definitively the ancient problem of Euclid's parallel postulate. In modern mathematics many apparently unrelated theories can be bound together by long chains of relative consistency proofs. Note finally that a relative consistency proof can show that an unfamiliar and seemingly bizarre theory is as logically satisfactory as a familiar classical theory, and so can help foster the introduction and investigation of new ideas.

EXERCISES

The following exercises refer to the representation of Lobachevskian plane geometry in the interior of a Euclidean circle O (Sec. 2). If an exercise involves analytic geometry the circle is taken to be $x^2 + y^2 = 1$.

1. Prove: If pseudo-equal segments are added to pseudo-equal segments, the results are pseudo-equal.

* See E. Moise, *Elementary Geometry From An Advanced Standpoint*, Addison-Wesley, 1963, Ch. 26.

2. Prove that pseudo-equality of segments is transitive: If $\overline{AB} \overset{O}{=} \overline{CD}$, $\overline{CD} \overset{O}{=} \overline{EF}$ then $\overline{AB} \overset{O}{=} \overline{EF}$.

3. Let A, B, C be on an O-chord with B and C on the same side of A. Prove that $\overline{AB} \overset{O}{=} \overline{AC}$ implies $B = C$.

4. Find two O-segments $\overline{AB}$, $\overline{CD}$ such that the Euclidean distance between A and B is equal to the Euclidean distance between C and D, but $n(\overline{AB}) \neq n(\overline{CD})$.

5. Prove: $0 < n(\overline{AB}) < 1$, for any O-segment $\overline{AB}$.

6. Prove: If A, B, C are O-points on an O-chord and B is between A and C then

$$n(\overline{AB}) \cdot n(\overline{BC}) = n(\overline{AC}).$$

7. Prove: If A, B, C are O-points on an O-chord and B is between A and C, then

$$n(\overline{AB}) + n(\overline{BC}) \neq n(\overline{AC}).$$

8. Let A be a fixed and B a variable point on O-chord $\overline{PQ}$. Prove

   (i) if B approaches A, $n(\overline{AB})$ approaches 1.

   (ii) if B approaches P or Q, $n(\overline{AB})$ approaches 0.

9. Find the "pseudo-midpoint" of the O-segment that has endpoints $(0, 0)$, $(0, \frac{1}{2})$: that is the point of the segment which makes pseudo-equal segments with the endpoints.

10. Prove that every O-segment has a unique pseudo-midpoint. (See Exercise 9.)

11. Let A_n be the point $[(3^n - 1)/(3^n + 1), 0]$, $n = 0, 1, 2, \ldots$. Prove that $A_0, A_1, A_2, \ldots$ form a pseudo-equally-spaced set of points, that is A_{n+1} is always between A_n and A_{n+2} and

$$\overline{A_0 A_1} \overset{O}{=} \overline{A_1 A_2} \overset{O}{=} \cdots \overset{O}{=} \overline{A_{n-1} A_n} \overset{O}{=} \cdots .$$

Try to find another such set of points.

In the Euclidean representation we introduce the following

DEFINITIONS.   If A, B, C are O-points, then O-$\angle ABC$ (read O angle ABC and called an O-*angle*) is the set of points common to $\angle ABC$ and the interior of circle O; a *side* BA (or BC) of O-$\angle ABC$ is the set of points common to the corresponding side BA (or BC) of $\angle ABC$ and the interior of the circle. O-$\angle ABC$ is *pseudo-equal* to O-$\angle A'B'C'$ (written O-$\angle ABC \overset{O}{=}$ O-$\angle A'B'C'$) if there exist points P, Q on sides BA, BC of O-$\angle ABC$, and points P', Q' on sides $B'A'$, $B'C'$ of O-$\angle A'B'C'$ such that $\overline{BP} \overset{O}{=} \overline{B'P'}$, $\overline{BQ} \overset{O}{=} \overline{B'Q'}$, and

$\overline{PQ} \overset{O}{=} \overline{P'Q'}$. If O-$\angle ABC$ is pseudo-equal to a supplement (defined in a natural way) we say O-$\angle ABC$ is a *pseudo-right angle*, or an O-*right angle*. Two O-chords are *pseudo-perpendicular* or O-*perpendicular* if they form a pseudo-right angle.

We extend the Euclidean representation of Lobachevskian geometry by interpreting the Lobachevskian notions angle, right angle, equality of angles, perpendicularity as O-angle, O-right angle, pseudo-equality of O-angles, pseudo-perpendicularity.

12. Prove that the following statements are verified in the representation:
  (i) The base angles of an isosceles triangle are equal.
  (ii) The line joining the vertex of an isosceles triangle and the midpoint of its base is perpendicular to the base.
  (iii) Every angle has a bisector.
13. Prove that a pair of vertical angles formed by the O-chords that lie on the lines $y = 0$ and $y = x - \frac{1}{2}$ are pseudo-equal.
14. Prove: An O-chord is O-perpendicular to a diameter of the circle, if it is perpendicular to it in the Euclidean sense.
15. Find the equation of the locus of points whose O-perpendicular segments to $y = 0$ are pseudo-equal to the segment from $(0, \frac{1}{2})$ to $(0, 0)$. Assume that there is only one O-perpendicular through any O-point to any O-chord.

DEFINITION. The *pseudo-distance* or O-*distance* from A to B, denoted O(A, B) is defined as follows:

(a) O$(A, A) = 0$; (b) if $A \neq B$, O$(A, B) = \log \dfrac{1}{n(\overline{AB})}$, where the base of the logarithm is a fixed number greater than 1.

16. Prove that pseudo-distance has the following properties, where A, B, C, D are O-points:
  (i) O$(A, B) \geqq 0$; O$(A, B) = 0$ if and only if $A = B$.
  (ii) O$(A, B) = $ O(B, A).
  (iii) If A, B, C are on an O-chord and B is between A and C then O$(A, B) + $ O$(B, C) = $ O(A, C).
  (iv) O$(A, B) = $ O(C, D) if and only if $\overline{AB} \overset{O}{=} \overline{CD}$.
  (v) Let A, B, C be on an O-chord with B and C on the same side of A. Then O$(A, B) = $ O(A, C) implies $B = C$.

(vi) Let A be a fixed and B a variable point on O-chord $\overline{PQ}$. Then if B approaches A, O(A, B) approaches 0; if B approaches P or Q, O(A, B) increases indefinitely.

(vii) Let point A on O-chord $\overline{PQ}$ and a positive real number x be given. Then there exists one and only one point B of $\overline{PQ}$, on a given side of A, such that O(A, B) = x.

17. Let L, M be the O-chords on the lines $x + y = 1$, $y = 0$. Let P be an arbitrary point on L, P' the foot of the O-perpendicular from P to M. Then as P recedes to the right O(P, P') approaches 0; as P recedes to the left O(P, P') increases indefinitely. (Assume there is only one O-perpendicular through any O-point to any O-chord. See Exercise 14.) Note that L is boundary parallel to M in the sense of Chapter 4, Section 7.

REFERENCES

K. Borsuk and W. Szmielew, *Foundations of Geometry*, North-Holland Publishing Company, Amsterdam, 1960.

S. Kulczycki, *Non-Euclidean Geometry*, Pergamon Press, New York, 1961.

Edwin Moise, *Elementary Geometry From An Advanced Standpoint*, Addison-Wesley Publishing Company, Reading, 1963.

D. M. Y. Sommerville, *The Elements of Non-Euclidea, Geometry*, G. Bell and Sons, New York, 1914. Reprinted by Dover Publications, Inc., 1958.

CHAPTER 6

The Empirical Validity of the Non-Euclidean Geometries

In this chapter we shall consider the question of the physical truth of the non-Euclidean theories of geometry. We shall not, however, try to consider this question in its broadest aspect; rather we are concerned mainly with the historical (and still widely current) view that Euclidean geometry is applicable to the physical world in a unique manner, whereas the non-Euclidean geometries are merely interesting abstract theories of space. Thus our basic question is not whether some way can be found of applying the non-Euclidean geometries to the physical world, as Riemann's double elliptic geometry, for example, can be applied to a Euclidean sphere. It is rather this: *Are the non-Euclidean geometries applicable to the physical world in the same sense and in the same way as Euclidean geometry; could engineers, architects, physicists, or astronomers use them instead of Euclidean geometry as the geometric underpinning of their craft; could the non-Euclidean geometries meet the usual practical tests of Euclidean geometry?*

The problem lies in the domain of empirical knowledge and philosophy of science, so you must not expect answers of the precise type appropriate to a purely mathematical question. We replace the familiar sharp, dogmatic answers of "white," "black" by somewhat hazy grays: "off-white," "off-black." From the standpoint of empirical validity, Euclidean geometry is not as good as some have believed—and the non-Euclidean geometries certainly not as bad as many think.

Stated a bit imprecisely, our conclusions are: (1) Euclidean geometry cannot be verified empirically with the kind of certainty we seem to attain for the arithmetical proposition $2 + 3 = 5$. Rather, it "works" in an approximative or statistical sense. (2) Lobachevskian and Riemannian geometries as abstract mathematical theories approximate, in a certain sense, Euclidean geometry. Suitably formulated they also should "work" in an appropriate approximative sense.

We shall concentrate on Lobachevskian geometry as an example of a non-Euclidean geometry and indicate briefly why our conclusions also apply to the Riemannian geometries.

1. How is Geometry Applied Practically

One might be tempted to settle the matter in this way: (1) Euclidean geometry is applicable to the physical world; (2) Lobachevskian geometry contradicts Euclidean geometry; (3) therefore Lobachevskian geometry is not applicable to the physical world. This argument read quickly seems very convincing. To help evaluate it, let us consider how mathematical theories are applied in practice.

As a first example, recall that when engineers, navigators and surveyors deal with a small portion of the earth's surface, they treat it as a plane. They use Euclidean plane geometry, not spherical geometry, in making their calculations. Do bridges fall, do ships collide, because of the "error" in underlying theory? Clearly not, for what is required in the practical application of a theory is that the predictions of the theory be verifiable to a degree of precision appropriate to the problem. The theory must "work" in a practical sense, but it hardly seems neccessary to require it to provide a perfect description of physical reality. In view of this it is not remarkable that contradictory theories like Euclidean plane and Euclidean spherical geometry may both be applicable to the same physical situation.

As a second example consider the problem of building a tunnel through the base of a mountain. The engineer makes plans based on Euclidean geometry. He sets shifts of workers to begin construction at points that are miles apart on opposite sides of the mountain. When the tunnel is holed through the two cylindrical segments which compose it and may be ten feet thick, are found to coincide with an error of one inch. Is the engineer chagrined? Does he consider himself a failure and turn in his slide rule? Quite the contrary. He has a champagne party to celebrate the achievement of such a good, though not quite perfect, result, based on geometric theory, geography and the practice of engineering. But, you may say, the discrepancy is not to be attributed to Euclidean geometry, it is due to errors in measurement. This is a difficult issue to resolve. The total process of planning and constructing a tunnel yields a certain error. What basis do we have for attributing the error to one factor in the process rather than another?

2. Attempt to Verify a Postulate

Now let us consider specifically how to test a given geometrical discipline, say Euclidean or Lobachevskian geometry, as a theory of physical space. Naturally we would first attempt to test the postulates, for instance, *two points determine a line*. Recall that a geometrical theory as a branch of mathematics asserts merely that the postulates imply the theorems abstractly, and that diagrams are to be aids in reasoning, but do not validate the processes of reasoning. Thus we must *interpret* the statement, *two points determine a line*, before we can test it. You say this is very easy: "Interpret it on a blackboard using chalk dots to represent points and a ruler to draw streaks representing lines. Then we obviously conclude that two chalk dots determine such a ruled chalk streak." The accompanying diagram indicates that the

FIGURE 6.1

verification of this property is not as simple as one may think. It suggests that we may not always be able to distinguish precisely between streaks which pass through dots and streaks which are in close contact with dots. You may answer by saying: "Make the dots circular and the width of the streaks exactly equal to the diameter of the dots." You have been led into the fallacy of attributing to physical experience a precision which it does not have, a precision which is more typical of *concepts* that we abstract from physical experience. A chalk streak does not have a fixed width, and it is most doubtful that a truly circular chalk dot can be constructed. If you examine closely what happens when you draw a chalk streak you will find that the width tends to increase as you draw it, since the piece of chalk gradually becomes more blunt. Observe also that the most careful machinists and lens grinders do not assert that their products are perfectly circular, but pride themselves on the achievement of approximate circularity with an error, let us say, not greater than one part in a million.

Thus we conclude that if chalk dots and chalk streaks are constructed according to appropriate specifications, it is not at all certain that we can always construct a chalk streak passing through two given chalk dots, or that if one such chalk streak exists that we could not construct two. Of course you may say: "For all practical purposes, two chalk dots do determine a chalk streak, because we can always get a chalk streak that fits the dots to a high degree of approximation, and if we do happen to find *two* chalk streaks that fit the dots, they are practically identical." Note that this is a far cry from the precision which, in our abstract theory, we ascribe to the statement "two points determine a line"; there is no approximative notion present there, no concession to mundane practicality.

Further, you may argue that the difficulty lies in our *interpretation* of the terms point and line. We have made the dots too big and the streaks too wide. It is hard to believe that the situation will be radically different if, for example, we use draftsman's instruments on a drawing board instead of chalk and rulers on a blackboard.

Finally, you may say that our postulate can be confirmed exactly in a limiting sense: as we make the dots smaller and smaller and the lines thinner and thinner, we approach as a limit the situation in which two points determine a line. It is hard to see how this can be verified in view of the molecular constitution of matter. It may be that this attempt to confirm the postulate is hardly more than an abstractly conceived experiment which is incapable of verification. In any case can we say that we have *achieved* a verification of the postulate?

3. Can the Parallel Postulate be Verified

The same kind of difficulty will appear in an attempt to verify the Euclidean parallel postulate or the Lobachevskian counterpart as we have already indicated (Ch. 4, Sec. 3). Indeed, how could one verify empirically the very existence of parallel lines? How could we find two physical models of endless lines and be sure that they had no common point? It seems inadvisable, then, to try to verify the parallel postulate directly. This situation is not at all unfamiliar to a scientist who tests a theory. It is often impracticable, if not impossible, to make a direct verification of the hypotheses of a theory. For example, a physicist who wants to test a theory of the internal constitution of the sun does not poke a probe 93,000,000 miles long into the sun. Rather, he tests the theory indirectly by finding a consequence of its hypotheses which *is* directly verifiable; for example, that under certain conditions the sun will emit radiation of a certain type, which is detectable on the earth. So it seems reasonable to try to test Lobachevskian geometry as a theory of physical space by testing a suitably chosen *theorem* rather than the postulates.

4. The Angle Sum of a Triangle as a Crucial Property

Which theorem should we choose? A natural choice seems to be the theorem on the sum of the angles of a triangle. This should be a crucial property for distinguishing between Euclidean geometry and Lobachevskian geometry as theories of physical space, since the first theory predicts a result of 180°, the second a result less than 180°.

So we have what seems to be a simple program: construct a physical model of a triangle, measure its angles and find their sum. Unfortunately, the process of physical testing is not as simple as it sounds. First we have the problem of the model.

We might take "point" to be a dot on the blackboard and "line segment" a ruled chalk streak joining two dots, similar to our interpretation in Section 2. Or we might refine this by the use of draftsman's instruments, taking "points" as tiny ink dots and "segments" as slender ruled ink streaks on high quality drawing paper. Or, as surveyors sometimes do, we might take "points" as stakes in the ground and "segments" as taut cords. Or as physicists and astronomers, we might take "points" as positions in space marked by "small" physical objects, and "segments" as beams of light projected from one physical object to another.

Then we must decide how physically to measure the angles of the triangle model. Presumably we would use a protractor in some form, such as a surveyor's transit which is merely a pair of protractors, one horizontal, one vertical, over which is mounted a telescope to facilitate observation.

After having chosen a model for the triangle and a means of measurement we measure the angles and compute their sum. The result might be $179.996°$. To check it you measure again and get $180.003°$. Shall we say that physical space is Lobachevskian by the first measurement and Riemannian by the second? It is well known to people who make physical measurements professionally that if a physical quantity like the force of gravity at a location on the earth's surface, or the distance from the earth to the moon, or the height of Mt. Everest, is measured many times, different values are obtained which cluster about a mean value. Thus if several models of triangles were constructed and their angle measure sums determined the values obtained might be: $179.996°$, $180.003°$, $180.000°$, $179.996°$, $179.994°$, $179.999°$. The arithmetic mean of these values, $179.998°$, would be taken to be the "probable value" of the angle sum of a triangle, and an indication of the variation or deviation of the values from this mean would be stated as: $179.998° \pm 0.003°$. Essentially, this means that to a suitable degree of probability, any measurement of the angle sum of a triangle will differ from $179.998°$ by less than $0.003°$ and so will lie in the range $179.995°$–$180.001°$.

On this basis the Euclidean prediction would be considered to be verified empirically, since the predicted value $180°$ is in the range obtained from the measurement. But we can just as well say that the Lobachevskian prediction is verified, since the prediction of a value less than $180°$ is also in the indicated range.

Although "theoretically" the angle sum of a triangle should be a crucial property for the discrimination between Euclidean and Lobachevskian geometries as theories of physical space, the actual process of physical interpretation and measurement is not sufficiently refined to permit the discrimination. This suggests two possible ways out of our difficulty: (1) refine the process of interpretation and measurement; (2) choose a different crucial property.

5. Refinement of the Physical Processes

It seems natural to believe that if we sharpen sufficiently our physical interpretation of triangle, and refine the processes of physical measurement, the difficulty may be resolved by the increased accuracy of the measured quantities. There are three possible outcomes. First, we may obtain a narrowed range for the angle sum which includes 180°, for example, 179.999° ± 0.002°. This still would not discriminate between the two theories. Secondly, the process of narrowing the range might exclude 180°, for example we might get 179.996° ± 0.002°. In this case the angle sum property does discriminate and Lobachevskian geometry is favored. Finally, we might get a range such as 180.004° ± 0.002° which excludes the values predicted by both the Euclidean and Lobachevskian theories, and favors Riemannian geometry. It seems clear that refinement of the physical processes involved in testing can only yield a decision at the expense of the Euclidean theory.

6. Analysis of the Angle Sum Property of Lobachevskian Geometry

It may seem that we made a mistake in choosing the angle sum property to discriminate between the geometric theories, and that our problem will be resolved by a better choice of crucial property. However, it is disturbing that the angle sum property, which is so basic and so pervasive in geometry, fails to discriminate. If this failed, others too may fail. Let us try to see the geometric grounds for its failure. This leads us to digress from consideration of empirical validity to a rather long discussion of the relation between Euclidean geometry and the non-Euclidean geometries as abstract theories. In the end we hope to clarify the problem of empirical validity.

The angle sum property failed because in Lobachevskian geometry it says merely that the angle sum of a triangle is less than 180°. How much less? It does not say. The answer lies in the theory of area (Ch. 4, Sec. 6), since the angle sum of a triangle determines its area. We see that for a triangle to have an angle sum close to 180°, its defect must be close to 0 and so its area measure must be small. On the other hand, if a triangle is "small" in the sense that its sides are small, its defect is small (Ch. 4, Exercises II, 6) and its angle sum is close to 180°. Thus "small" triangles have angle sums very close to 180°. "Large" triangles have angle sums "remote" from 180°.

This throws light on the failure of the angle sum property to discriminate. Triangles which yield the angle sums indicated in Section 4 are "small" from the standpoint of Lobachevskian theory. If very large triangles were measured we might obtain angle sums appreciably below 180°, so that Lobachevskian geometry would be confirmed, not Euclidean.

7. Lobachevskian Geometry in the Small as an Approximation to Euclidean

The discussion in Section 6 suggests that Lobachevskian and Euclidean geometries may be more closely related than their status as contradictory theories has led us to believe.

We know that in a neutral geometry the property that one triangle has an angle sum of 180° is equivalent to Euclid's parallel postulate and makes the neutral geometry Euclidean (Ch. 4, Th. 12, Cor. 3). It may not seem too farfetched to say that if a triangle has an angle sum "practically" equal to 180°, its properties are "practically" Euclidean. This is a rather vague idea and shall we give some examples of what we have in mind before sharpening it. Suppose a triangle in Lobachevskian geometry has an angle sum practically equal to 180°. Then any triangle contained in it has an angle sum practically equal to 180°, since it has smaller defect than the given triangle (Ch. 4, Exercises II, 5) and so has an angle sum closer to 180°. Any quadrilateral contained within the given triangle would have an angle sum practically equal to 360°. It follows that if a quadrilateral with three right angles is constructed inside the original triangle, its fourth angle would be practically a right angle, and the figure would be practically a rectangle. These examples, simple though they are, indicate that many Euclidean properties would hold to a high degree of approximation in the interior of the triangle.

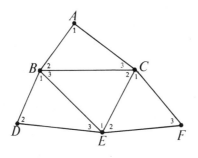

FIGURE 6.2

A somewhat deeper example is this: suppose in the interior of the given triangle we put together four congruent triangles, with angles corresponding as indicated in Figure 6.2, to form hexagon *ABDEFC*. This hexagon is practically a triangle since, for example, points *A, B, D* are practically in a line. Thus $\triangle ADF$ is practically similar to $\triangle ABC$. But the existence of noncongruent similar triangles is a typically Euclidean property (see discussion of Wallis' postulate Ch. 2, Sec. 6). Here again we see that the interior of the "small" Lobachevskian triangle has an approximately Euclidean character.

This example suggests the possibility of a theory of "approximate similarity" for small triangles, in which two small triangles whose corresponding angles are approximately equal would have corresponding sides approximately proportional. Further, it suggests a paraphrase of the familiar Euclidean proof of the Pythagorean theorem based on similar triangles, which would make plausible an "approximate" Pythagorean theorem for small Lobachevskian triangles, namely: For small right triangles the square of the hypotenuse is approximately equal to the sum of the squares of the legs. Stated precisely: *For a right triangle, the ratio of the square of the hypotenuse to the sum of the squares of the legs can be made as close as we please to unity, provided the sides of the triangle are made sufficiently small.* This result is correct. There is in Lobachevskian geometry an analogue of the Pythagorean formula for a right triangle, which approximates the Pythagorean relation more and more closely as the sides of the triangle approach zero.*

Summary. If in Lobachevskian geometry a triangle has an angle sum very close to 180° it behaves very much like a Euclidean triangle. (Slogan: If a triangle has an angle sum of almost 180°, it is almost Euclidean.) Somewhat more precisely, its properties are approximately Euclidean, the approximation getting better and better as the angle sum gets closer and closer to 180°. The difference of the angle sum from 180°, or the defect, is, so to speak, an indication of how "non-Euclidean" the triangle is. If in a portion of space all triangles had defects less than $1/10^6$ the geometry would be practically Euclidean.

Although we have not proved that Lobachevskian geometry in the small approximates Euclidean, we hope the examples have helped to make it plausible. Many more such examples can be found. It is difficult to formulate the principle precisely, especially if we do not restrict the type of geometric property to which it is applied. There is, however, a restricted example of it which can be made quite sharp. As we have remarked (Ch. 4, Sec. 17), there is a Lobachevskian theory of trigonometry, and there exist Lobachevskian analogues of the law of sines and the law of cosines for triangles. When the

* See S. Kulczycki, *Non-Euclidean Geometry*, Pergamon Press, New York, 1961, pp. 147–150.

sides of a triangle are permitted to approach zero, the Lobachevskian laws approach as limits the familiar corresponding Euclidean laws. Thus for a sufficiently small region in Lobachevskian geometry, Lobachevskian trigonometric formulas would yield practically the same results as the corresponding Euclidean ones. This is closely related to the existence, mentioned above, of a Lobachevskian analogue of the Pythagorean theorem which approximates it in the small.

8. Spherical Geometry in the Small as an Approximation to Euclidean

It is hard to convey an intuitive feeling for Lobachevskian geometry as an approximation to Euclidean geometry without a more detailed treatment of the subject than we have been able to include, since there is no simple representation of Lobachevskian geometry that is accessible to geometric intuition. The representation described in Chapter 5 is certainly not adequate. There is, however, an analogous situation which is easily grasped intuitively.

We observed in Section 1 that engineers treat small portions of the earth's curved surface as portions of a Euclidean plane. In other words, they use Euclidean plane geometry as an approximation to spherical geometry, applying, for example, the simpler formulas for ordinary triangles in a plane to spherical triangles. This idea, reversed, suggests that spherical geometry can be used to approximate Euclidean plane geometry. For instance, choose $\triangle ABC$ in a given plane. We can find a sphere through A, B, C with a radius so large that the spherical triangle with vertices A, B, C is a good approximation to $\triangle ABC$ of the plane (Figure 6.3). This can be done so that the sides and angles of spherical triangle ABC approximate as closely as we wish the corresponding sides and angles of $\triangle ABC$. (In the face of this, it is not

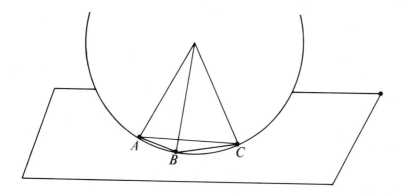

FIGURE 6.3

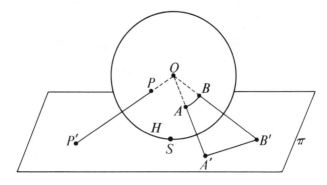

FIGURE 6.4

remarkable that the formulas of spherical trigonometry for spherical triangle *ABC* approximate closely the formulas of plane trigonometry for $\triangle ABC$.) Thus we can say roughly that by making the radius of the sphere large, we can approximate closely the geometry of the Euclidean figure $\triangle ABC$ by the geometry of the spherical triangle *ABC*; and that the approximation gets better and better as the radius of the sphere increases.

A similar relation holds true between a plane and a fixed sphere tangent to it, for suitably chosen *small* portions of the sphere. We present this relationship in connection with the problem of making a planar map of the earth's surface conceived as a sphere. Let us think of the sphere as a geographical globe and for the sake of convenience let us take the plane, denoted π, tangent to the globe at its south pole *S* (Figure 6.4). Let *O* be the center of the sphere. Then any point *P* of the sphere in the southern hemisphere *H* can be projected from *O* into a corresponding point *P'* on plane π by taking *P'* as the intersection of *OP* and π. Let us think of the process of projecting the points of *H* onto π as a way of making a map of *H* on π. We represent each point *P* of *H* by the corresponding point *P'* of π, and any curve (or region) in *H* will be represented by a corresponding curve (or region) in π, formed by the points in π which correspond to the points of the given curve (or region) in *H*. Consider *A*, *B* two points of *H*, and *A'*, *B'* their "projections" in π. In *H*, points *A*, *B* are joined by an arc $\overarc{AB}$ of a great circle. The points of $\overarc{AB}$ correspond to the points of the segment $\overline{A'B'}$ which joins *A'* and *B'* in π. That is, $\overarc{AB}$ in *H* corresponds to $\overline{A'B'}$ in π, or $\overarc{AB}$ is represented in the map by $\overline{A'B'}$. Of course the representation produces distortion: $\overarc{AB}$ is curved, $\overline{A'B'}$ is not; in general $\overarc{AB}$ and $\overline{A'B'}$ will have different lengths. In the neighborhood of the equator the distortion is very great, in the neighborhood of the pole *S*, quite small. To be more specific: if points *A*, *B* are very close to *S*, $\overarc{AB}$ will approximate $\overline{A'B'}$, in fact, as *A*, *B* approach *S* the ratio of the lengths of $\overarc{AB}$

and $\overline{A'B'}$ will approach unity. A similar statement holds for the spherical angle ABC determined by three points A, B, C of H and $\angle A'B'C'$ the angle in π determined by points A', B', C', the projections of A, B, C.

Thus a small portion of H in the neighborhood of the pole S closely approximates the corresponding portion of π. For example a small spherical triangle whose vertices A, B, C are close to S is a good approximation to the corresponding figure in π which is the Euclidean plane triangle $\triangle A'B'C'$ whose vertices are the projections of A, B, C. Furthermore, the approximation can be made as close as we please by taking A, B, C sufficiently near to S. Thus the small spherical triangle must in itself be approximately Euclidean: its angle sum must be approximately 180°; and a formula of spherical trigonometry applied to the triangle must give approximately the same result as the corresponding formula of plane trigonometry. In other words, the formulas of spherical trigonometry for a triangle on a sphere approach corresponding Euclidean plane trigonometric formulas as limiting forms when the sides of the triangle approach zero. This is well known in the formal study of spherical trigonometry.

Summary. A small portion of a sphere approximates a corresponding portion of a *suitably chosen* plane, so that the geometry of a small portion of a sphere approximates Euclidean geometry.

This conclusion is important in itself, and gives a concrete analogue of the approximation of Euclidean geometry by Lobachevskian.

9. Other Crucial Properties

Our discussion in Section 7 indicates that although Euclidean geometry and Lobachevskian geometry contradict each other, Lobachevskian properties for "small" portions of space closely approximate Euclidean properties. Is it remarkable then that many geometric properties fail, as does the angle sum property, to discriminate between the empirical validity of the two geometries? The existence of a rectangle or of noncongruent similar triangles, the Pythagorean relation, the equidistance of parallel lines, all fail as crucial properties, since all are approximately correct in sufficiently small portions of Lobachevskian space.

It may be felt that the difficulty can be resolved by performing measurements on large portions of physical space; but the basic issue is unchanged. If the measurements confirm the Euclidean theory, we can argue that a "large" portion of physical space corresponds to a "small" portion of abstract Lobachevskian space, and that the measurements are consistent with the Lobachevskian theory.

We do not assert that all Lobachevskian principles are confirmed empirically: take for example the existence of a triangle whose angle sum is less than 90° (Ch. 4, Exercises III, 7). On the other hand, the failure to find a physical model of such a triangle does not imply that physical space is necessarily non-Lobachevskian. It may be that if we found methods of testing sufficiently large models of triangles, we would discover some with sizable defects.

10. Conclusion

We are left with a sort of deadlock. Euclidean geometry, interpreted suitably, is confirmed to a high degree of approximation as a theory of physical space. Lobachevskian geometry in the small approximates Euclidean geometry. Consequently, a large number of its principles, though not all, may be considered confirmed to a high degree of approximation. It follows that engineers and scientists who use Euclidean geometry practically could use Lobachevskian geometry instead, by applying the appropriate principle to a "small" portion of space. Of course it is simpler to use Euclidean geometry since the formulas tend to be simpler.

This conclusion may be remote from the kind of resolution you expected to the question: Which theory is correct? We prefer sharp "black or white" answers and should try to attain them, but we would delude ourselves if we claimed to have them here. Indeed, it seems very doubtful that we could ever find a mode of physical interpretation which would confirm Euclidean geometry (or Lobachevskian for that matter) in any but an approximative sense. Euclidean geometry is not a perfect copy of physical space, but a good working approximation.

Similar considerations apply to the Riemannian geometries. The principle (Ch. 4, Sec. 16) that the angle sum of a triangle is greater than 180° is confirmed empirically by the data suggested in Section 4, since the measurement $179.998° \pm 0.003°$ covers values greater than 180°. This is not an accident, for in elliptic geometry, as in Lobachevskian, the angle sum of a "small" triangle is close to 180°. Furthermore, the elliptic geometries also approximate Euclidean geometry in the small. This follows from our discussion in Section 8 of the relation between spherical geometry and Euclidean plane geometry. For the double elliptic case the result is immediate, since double elliptic geometry is faithfully represented by spherical geometry which, in the small, approximates Euclidean. For the single elliptic case, observe that single elliptic geometry *is* represented faithfully by spherical geometry provided we restrict the representation suitably to a hemisphere. Thus for sufficiently small regions, single elliptic geometry is faithfully represented by spherical geometry and so, in the small, is approximately Euclidean. Hence our observations on the empirical validity of Lobachevskian geometry apply as well to elliptic geometry.

Introduction

We have made two principal points in Part I:

(1) The conventional treatment of Euclidean geometry in secondary school does not have a sound mathematical basis.

(2) There are several mutually contradictory geometrical disciplines which are equally valid as self-consistent logical systems and which are applicable, with certain restrictions, to the physical world.

Our object in this part of the book is to develop a logically sound treatment of geometry which will be sufficiently general to cover a broad family of geometrical systems, and which will do justice to Euclidean geometry as an important and central member of such a family.*

We propose to do this by laying down initially a small number of basic geometrical assumptions and by introducing gradually, in successive chapters, further assumptions until we have a complete basis for Euclidean geometry. Initially then our theory will have a broad scope, which gradually will be narrowed and finally will cover only Euclidean geometry.

The question of logical organization of our material must be clarified and settled before we proceed. If our treatment is merely to be general enough to cover Euclidean and Lobachevskian geometries, we must base our proofs

---

* Part II is mathematically complete in itself, although it does contain occasional references to Part I.

on properties formally and objectively stated, not on intuitive geometrical judgments or visual perceptions of diagrams. And the reason for this is not merely that an intuitive judgment or observation of a diagram is never acceptable as a valid reason in the formal logical development of a subject. More importantly it is that the development of non-Euclidean geometry has widened and liberated geometric intuition and perception of diagrams. There will never again be a unique "correct" geometric intuition. Geometric intuition and mathematical theory interact as they grow, each stimulating the other. A geometer has today, in contrast with the year 1800, many kinds of intuition: a Euclidean intuition which tells him that parallel lines are everywhere equidistant, a Lobachevskian intuition that the locus of points equidistant from a line is curved, and a Riemannian intuition that there are no parallel lines.

Our problem in setting up logical foundations for geometry is neither to reject intuition as a nonlogical element in mathematics, nor to sanctify it as a basis for mathematical ideas, but to refine, formalize and codify our basic intuitive judgments of a given branch of geometry in order to give the subject an autonomous logical structure.

Structure Chart for Part II

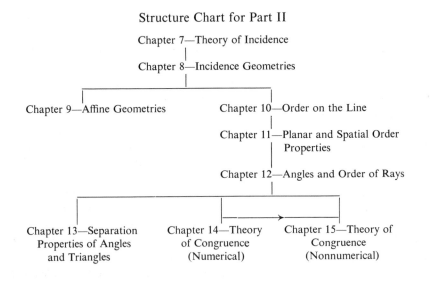

Chapter 7—Theory of Incidence

Chapter 8—Incidence Geometries

Chapter 9—Affine Geometries   Chapter 10—Order on the Line

Chapter 11—Planar and Spatial Order
Properties

Chapter 12—Angles and Order of Rays

Chapter 13—Separation   Chapter 14—Theory   Chapter 15—Theory of
Properties of Angles   of Congruence   Congruence
and Triangles   (Numerical)   (Nonnumerical)

Note. The arrow in the chart indicates that although Chapters 14 and 15 are logically independent, it is normally advisable to study them in numerical order.

CHAPTER 7

Theory of Incidence

In this chapter we begin our formal development of geometry by studying the simplest and most basic of geometrical relations, those of incidence, which may be considered to underlie the more complicated notions of order, congruence, similarity, et cetera. In later chapters we introduce, in terms of the foundation laid here, theories of parallelism, order, and congruence.

1. The Concept of Incidence

Incidence is a very elementary idea though the term is not commonly used in elementary texts. For example an incidence relation is given by the statement, "a point is on a line" or the equivalent forms "a point lies on a line," "a point is in a line," "a line passes through a point." Other statements which express incidence relations are: "a point is in or on a plane," "a line lies in a plane," or even "two lines meet." Thus incidence relations express the basic positional interconnections of points, lines and planes—how they are aligned with or united with each other.

We express the statement "a point is on a line" more formally by saying the point *is incident with* (or *aligned* or *united with*) the line, or equally the line *is incident with* the point, or the point and the line are *incident*. In exact analogy we express "a point is in a plane" formally as "a point *is incident with* a plane" or "a plane *is incident with* a point" or "a point and a plane are *incident*." And we use exactly similar language to express the incidence relation between line and plane, usually stated "a line lies in a plane."

Thus there are three basic incidence relations which interconnect pairs of the basic geometric entities: point, line, plane. Other incidence relations may be defined in terms of the basic ones. For example the familiar statement "two lines meet" may be expressed by "two lines are both incident with the same point." Other examples: "two lines are parallel" will be given by "two lines are both incident with the same plane, but there is no point incident with both"; and "two points determine a line" by "two points are both incident with some line, and there is only one line with which both points are incident."

The most elementary and basic properties of incidence which are involved (implicitly or explicitly) in a geometric theory are these: (1) The relation of incidence is symmetric, for example, if a point is incident with a line then the line is incident with the point. (2) If a point is incident with a line and the line is incident with a plane then the point is incident with the plane.

2. Incidence in Terms of the Set Concept

In school geometry it is not wholly clear how incidence is to be conceived. We tend to think of a line as a continuous path, so that a point is incident with a line if it overlaps or is physically attached to the line, somewhat like a bird on a telephone wire, or a bead on a string. For some purposes it is simpler to consider a line as *composed* of points, that is, to conceive a line as a *collection* or *set* of points. This concept is not very remote from the previous one, since if a line is conceived as a continuous figure of some sort, there is associated with a line a collection of points, namely the collection of all points that are incident with the line—and the properties of the line should correlate with corresponding properties of the associated collection of points.

So we shall conceive of lines, and planes also, as sets of points. This enables us to explain or define incidence relations in terms of the concept of set and so effects economy of treatment, since the idea of set is needed in geometry anyway. A set consists of its *members* or *elements*, as a circle of its points or a mathematics class of its students. So we will frequently use phrases of the form "*a* is a member (or element) of set *A*" or simply "*a* is in *A*." Then a point is incident with a line if the point is a *member* of the set of points which constitute the line, equivalently if the line contains the point as one of its *members*, or simply if the point is *in* the line. Similarly for the idea of a point incident with a plane. Finally, a line is incident with a plane, if each point of the line is a point of the plane, that is, if the line is *contained* in the plane as a *subset* of the plane. Consequently, we can dispense with the *term* incidence in studying the theory of incidence and simply use the corresponding set theoretic forms:

(1) *A point is in* or *belongs to a line;* equivalently: *A line contains a point* (*as a member*).

(2) *A point is in* or *belongs to a plane;* equivalently: *A plane contains a point* (*as a member*).

(3) *A line is contained in a plane* (*as a subset*)*;* equivalently: *A plane contains a line* (*as a subset*).

3. Postulates for Incidence

We now begin the *formal* development of geometry by stating postulates concerning incidence of points, lines, and planes, and deriving theorems from them. Our postulates, which are suggested by familiar incidence properties of points, lines and planes in Euclidean plane and solid geometry,* are as follows:

POSTULATE I1. (EXTENSION)   *A line is a set of points, containing at least two points.*

POSTULATE I2. (DETERMINATION)   *Two distinct points are contained in one and only one line.*

POSTULATE I3. (EXTENSION)   *A plane is a set of points, containing at least three points which do not belong to the same line.*

POSTULATE I4. (DETERMINATION)   *Three distinct points which do not belong to the same line are contained in one and only one plane.*

POSTULATE I5. (LINEARITY)   *If a plane contains two distinct points of a line, it contains the line* (*that is, it contains all points of the line*).

POSTULATE I6. (DIMENSIONALITY)   *If two planes have one point in common, they have a second point in common.*

4. Some Remarks on the Postulates

Our postulates, though very simple and familiar, merit a few words. I1 and I3 formalize the idea, expressed in Section 2, that lines and planes are to be composed of points. Do not be misled by the form in which I1 is stated into considering it a definition of line. (Compare, for example, the statement Paul Smith is a boy with red hair.) I1 states merely two properties of a line—that a line is a set of points and that it contains two distinct points.

---

* See and compare D. Hilbert, *The Foundations of Geometry*, Open Court, 1902, pp. 4–5, the Axioms of Connection.

It does not assert the converse, that anything with these properties is a line. These considerations apply as well to I3.

Further, note that I1 requires merely that a line contain two distinct points, not infinitely many as is sometimes assumed. Thus if we want to assert that a line contains infinitely many points (or even three!) we shall have to deduce this, if possible, from the postulates. Similarly, by I3, a plane contains three distinct points, possibly no more.

I2 is a precise form of the familiar statement, two points determine a line. It can be stated even more formally: *If a, b are distinct points, there exists a unique line L such that a and b are elements of L.* Similar remarks apply to I4 which concerns the "determination" of planes.

I5 says, roughly speaking, that a plane is *linear*, or line-like, or can be "generated" by lines. It is a familiar and essential principle in Euclidean solid geometry which serves to distinguish a plane from a sphere, cone, or any other surface.

I6, though easily grasped, plays a rather subtle role in the theory and deserves a paragraph to itself. In Euclidean solid geometry, it is usually formulated: If two planes in *space* have a point in common, they have a second point in common. The key term here is *space*, meaning of course three-dimensional space. Mathematicians have developed theories of four-dimensional space, and in such theories, the given property fails. That is, two planes in a four-dimensional space can have exactly one common point. This turns out to be possible if the planes are not contained in a common three-dimensional space. Actually, if two planes in a four-dimensional space *do* have two distinct common points, it can be proved that the two planes are contained in a common three-dimensional space. Thus I6 is not merely a property of planes, but also of the whole space in which they are immersed, and guarantees that our geometry is not four-dimensional, but is three-dimensional at most. I6 may then be considered a *postulate of dimensionality*, and in effect ensures that we will be studying nothing more complicated than plane or solid geometries. It does this very simply and cleverly, without involving us, at this early stage, in the complex notion of dimensionality of a space. You may expect to appreciate the point more fully when you study spaces of higher dimension.

Finally observe that in constructing a mathematical theory we must decide which terms are to be basic terms—these will be characterized by the postulates and will be used in defining other terms. A theory which had no basic terms would have to define *all* its terms, and either would involve an endless chain of definitions or would commit the logical sin of circularity. The *basic* or *primitive* or *undefined* geometrical terms in our treatment are "point," "line," and "plane." We make no attempt to define or clarify them in terms of simpler ideas.

Traditional texts sometimes attempt definitions of these terms. For example: A point is that which has position but no magnitude. This may be useful in giving an intuitive idea of how point is to be interpreted in a diagram. But it does not *function* as a definition in the development of the subject. It is not employed as a property of points to prove other properties of points. It does not characterize point in terms of more basic ideas, on the contrary, it is hard to see how to characterize "position" and "geometrical magnitude" without using the notion point. Such "definitions" play no role in the *logical* development of geometry, though they may have an incidental psychological or pedagogical value.*

In summary, points, lines, and planes literally are undefined and our study of geometry must be based on their *properties* as specified by the postulates not on their *substance* or *content*. In a formal treatment, what points, lines, and planes *are* is immaterial—the issue is how they *behave*. Thus in our development we may use any properties of points and lines that are stated in the postulates: for example that a line contains two distinct points, that if two distinct points of a line are in a plane then the line is contained in the plane, et cetera. We may not use properties of points and lines not stated in or not deducible from the postulates, no matter how intuitively evident these properties may seem. For example, we may not use that a line is endless, that its points are ordered, that it has no width or thickness.

5. A Note on Language

In stating the postulates we use "two" and "two distinct" in different senses. This rather strange usage is common in mathematical works. To explain and motivate it consider the problem of stating in English—without employing mathematical symbols—the closure law for subtraction in the real number system. You might say: *The difference of two real numbers is a real number.* But this is not quite complete. It omits the very important special case that the difference of a real number and itself is a real number. Thus to use English precisely we should say: *The difference of two real numbers, or of any real number and itself, is a real number.* This is rather involved and singles out the special case, which is really not intended in the original statement. By using algebraic symbolism we can state the principle simply: *If a is a real number and b is a real number then a − b is a real number.* The point is that we are asserting $a - b$ is a real number for *every* choice of a and b, whether distinct or not. To have a brief and simple way of saying this

---

* Such nonfunctional definitions are found in Euclid, for example: A point is that which has no part.

in English we adopt the convention that "two real numbers" shall be the English equivalent of the phrase "a real number a and a real number b." Thus "two real numbers" may refer to a *pair* of real numbers or to a real number and itself. Then the statement above can be phrased: *The difference of two real numbers is a real number*. This is a peculiar mode of expression—a singular use of the word "two"—but a very convenient form of "mathematical-English" for stating a general principle that involves two arbitrary choices of an entity. The situation gets even worse if there are three or more choices to be made for an entity since many more special cases are involved. Thus we use the phrase "three points" to refer to "a point a, a point b, a point c," without implying that a, b, c are distinct. If we wish to state that the entities considered are distinct we explictly indicate this by using phrases such as, "two distinct lines," "three distinct points," "at least two points," et cetera.

6. Elementary Theorems on Incidence

We start to develop the formal theory of incidence by proving a few elementary theorems. Our aim is to formalize geometrical properties completely—every theorem is to be deducible from the stated postulates. But we make no attempt to formalize the logical apparatus of deduction. We feel there are pedagogical advantages in dividing the difficulties by concentrating on the formalization of geometry and using logical principles informally.

A few words on notation. Systematically we use $a, b, c, \ldots$, with or without subscripts or superscripts, to denote points; similarly L, M, N lines; and P, Q, R planes. The equality sign denotes identity, for example, $a = b$ means a and b are the same point or are identical. Similarly, $a \neq b$ means a and b are not identical, or are distinct.

THEOREM 1.   Two distinct lines have at most one point in common.

Proof. Suppose lines L and M are distinct and have two distinct points a and b in common. By I2, $L = M$. This contradicts the hypothesis that L and M are distinct. Thus L and M have at most one common point.

To state Theorem 2 we need two definitions. The first is suggested by Postulate I2; the second introduces a very convenient way of expressing that points belong to the same line.

DEFINITION.   If a and b are distinct points we use the symbol ab to denote the unique line containing a and b, and call it the *line determined by a and b*. We also say line ab *joins a and b*. (If a and b are the same point the symbol ab is not defined.)

DEFINITION.   Points $a_1, a_2, \ldots, a_n$ are said to *colline*, or be *collinear*, if there is a line which contains all of them. Similarly, we define figures (sets of points) $S_1, S_2, \ldots, S_n$ to *colline* or to be *collinear* if there is a line which contains all of them.

THEOREM 2.   If point a is not in line bc, then a, b, c are distinct and noncollinear.

Proof.   By definition of bc, $b \neq c$. Suppose $a = b$. Then, since b is in bc, so is a. This contradicts the hypothesis. The supposition $a = b$ is therefore false, so that $a \neq b$. Similarly, $a \neq c$. Thus a, b, c are distinct.

Now suppose a, b, c colline. By definition, there is a line L containing a, b, and c. Since L contains b and c, and $b \neq c$, $L = bc$ by I2. But a is in L; hence a is in bc, contradicting the hypothesis. The supposition is therefore false, and a, b, c are noncollinear.

THEOREM 3.   A line and a point not in it are contained in one and only one plane.

Proof.   Let a be a point not in line L. We show first there is at least one plane containing L and a. By I1 there are two distinct points, say b and c, in L. Thus $L = bc$. Since a is not in L, a is not in bc. Hence by Theorem 2 a, b, c are distinct and noncollinear. By I4 a, b, c are contained in a plane P. Since b and c are in P, by I5 P contains bc or L.

Suppose plane P' contains L and a. Then P' contains b and c, since it contains L. Thus P' contains a, b, and c. Hence by I4, $P' = P$, and P is the only plane containing L and a.

The theorem suggests the following definition.

DEFINITION.   Suppose a does not belong to L. Then the unique plane containing L and a is called the *plane determined by L and a*, or the *plane of L and a*, and is denoted La. Similarly, if a, b, c are distinct and noncollinear the unique plane containing a, b, c is called the *plane determined by a, b, c*

and is denoted abc. (La is not defined if a is in L. Similarly, abc is not defined if a, b, c are not distinct or are collinear.)

The concept of parallelism is an important topic in the theory of incidence:

DEFINITION.   Two lines, L and M, are *parallel* (written $L \parallel M$) if they are contained in the same plane and have no common point.

In defining parallel lines we do not assume tacitly that parallel lines exist. We merely introduce the brief phrase "L, M are parallel" to stand for the longer phrase "L, M are contained in the same plane and have no common point." When we have an important notion that often recurs, we introduce a convenient abbreviation for it.

COROLLARY.   If $L \parallel M$, there exists a unique plane containing L and M.

PROOF.   By definition of parallel lines there is a plane P containing L and M. Suppose plane P' contains L and M. Let a be in M. Then each of P' and P contains L and a, so that $P' = P$ by Theorem 3.

DEFINITION.   If S and T are figures (that is, sets of points) which have a common point, we say S and T *intersect* or *meet*. Note that in this relation S and T may have many points in common or may even coincide, that is $S = T$ may occur. If S and T have a point a in common, we say S and T *intersect* or *meet in a*. The set of points common to S and T is called the *intersection* of S and T.

THEOREM 4.   If two distinct lines intersect, they are contained in one and only one plane.

Proof.   Let L, M be distinct lines that intersect. We show first that there is at least one plane containing L and M. Note there is a point a common to L and M. By I1, there is a point b in M such that $b \neq a$. Moreover b is not in L, otherwise $L = M$ contrary to hypothesis. By Theorem 3 there is a plane P containing L and b. Since P contains L, it contains a. By I5, P contains M, and so P contains L and M.

Suppose plane P' contains L and M. Since P' contains M it contains b. Thus P' contains L and b, and so by Theorem 3 $P' = P$. Thus P is the only plane containing L and M.

THEOREM 5.   If two distinct planes intersect, their intersection is a line.

Proof.   Let planes *P*, *Q* be distinct and intersect, that is have a point *a* in common. By I6, *P* and *Q* have a second point *b* in common. Thus by I5, *ab* is contained in *P* and in *Q*. Thus each point in *ab* is common to *P* and *Q*.

We complete the proof by showing the intersection of *P* and *Q* is *ab*. To do this we must show, according to the definition of the term intersection, that *ab* is the set of points which are common to *P* and *Q*. That is (1) each point in *ab* is common to *P* and *Q*, and (2) each point common to *P* and *Q* is in *ab*. We have already established (1). To prove (2), let *c* be any point common to *P* and *Q*, and suppose *c* is not in *ab*. Then by Theorem 3, *ab* and *c* are contained in only one plane. But *ab* and *c* are contained in *P* and in *Q*. Hence *P* = *Q*, contradicting the hypothesis. Thus our supposition that *c* is not in *ab* is false, and *c* is in *ab*. Thus (2) is verified, and the proof is complete.

COROLLARY.   If a line is contained in each of two distinct planes, it is their intersection.

7. The Role of Intuition

As you proved the preceding theorems you may have made diagrams of the familiar type, picturing point as dot, line as streak, and plane as sheet. Or you may have formed such images in your mind. In the process you may have made intuitive judgments about points, lines, and planes based on such diagrams or mental images. Does intuition have a valid role in mathematics, particularly in geometry? Geometric intuition, as we have seen in the preceding chapters, may be fallible. For twenty centuries mathematicians entertained as an intuitive certainty a belief about Euclid's parallel postulate that the development of non-Euclidean geometry proved false.

If geometric intuition always yielded correct results, it would still not be a valid instrument of proof. For in a mathematical science, we aim to deduce our theorems from objectively stated postulates, rather than validate them by their coherence with the subjective combination of experience and preconception that we call geometric intuition. A principle is not validated as a theorem until we have constructed a proof of it, no matter how great our intuitive certainty of its correctness.

This is not to say that intuition has *no* role in the total mathematical process. It often suggests the choice of postulates in formalizing a branch of mathematics, and many theorems (and proofs) are discovered in a flash of intuitive insight.* In the process of learning mathematics, logical under-standing and intuitive insight always complement each other. In short, intuition is an important and essential factor in the process of mathematical *learning* and mathematical *discovery*; but no intuitive judgment, no matter how great its subjective certainty, is valid *justification* for a mathematical statement.

In the face of this, shall we banish diagrams from geometry or images from our minds? Clearly not. A diagram is often helpful when proving a theorem. It indicates, by a sort of shorthand, what we are given and what we are trying to conclude. Consequently, it often suggests modes of procedure to carry us toward the conclusion. It often helps us to perceive a pattern in the proof not easily discerned in its formal presentation. This is all to the good. We must, however, guard against justifying mathematical assertions on visual or empirical evidence. No matter how many dots appear on a streak in the diagram, Postulate I1 controls: *A line contains at least two points.*

8. Parallel Planes and Parallel Lines

We begin with a definition, familiar in Euclidean solid geometry.

DEFINITION.   Two planes, P and Q, are said to be *parallel* (written $P \parallel Q$) if they have no common point.

We now prove a simple and important relation between parallel planes and parallel lines.

THEOREM 6.   If planes P and Q are parallel, and plane R intersects P and Q, then the intersections of R with P and Q are parallel lines.

Proof.   In order to use Theorem 5 to show that the intersections are lines, it is necessary to prove the intersecting planes distinct. Suppose $R = P$. Then the hypothesis R intersects Q, implies P intersects Q, contrary to $P \parallel Q$. The supposition is therefore false, and we conclude $R \neq P$. Similarly $R \neq Q$.

By Theorem 5, therefore, the intersection of R and P is a line L, and the intersection of R and Q is a line M. L and M are contained in the same plane, namely R. Furthermore L and M do not meet, for if they had a point a in

---

* See, for example, H. Poincaré, *Science and Method*, Dover (reprint), pp. 46–63, and J. Hadamard, *The Psychology of Invention in the Mathematical Field*, Dover (reprint).

common, a would be in P (since L is contained in P) and in Q (since M is contained in Q). Thus P and Q would meet contrary to the hypothesis $P \parallel Q$. Thus L and M are in the same plane and do not meet. By definition $L \parallel M$.

DEFINITION. Lines $L_1, L_2, \ldots, L_n$ are said to *concur*, or to be *concurrent*, if they all contain a common point. Figures $S_1, S_2, \ldots, S_n$ are said to *coplane* or to be *coplanar*, if there is a plane containing all of them.

THEOREM 7. If three lines are coplanar in pairs, but not all coplane, then either the three lines concur, or else they are parallel in pairs.

Proof. Let L, M, N be such lines; let L, M be contained in P; M, N in Q; and L, N in R. We show P, Q, R distinct. Suppose $P = Q$. Then L, M, N coplane, contrary to hypothesis. Thus $P \neq Q$. Similarly, $Q \neq R$ and $P \neq R$. It follows that the planes intersect in pairs in the lines as indicated in the table:

Planes	Line of Intersection
P, Q	M
Q, R	N
P, R	L

Now suppose two of the lines meet, say L, M meet in a. Since a is in L, we see from the table that a is in P and in R. Similarly, since a is in M, it must be in P and in Q. Thus a is in Q and in R, and by the table, a is in N. Thus if two of L, M, N meet, the three lines concur.

Finally, suppose no two of L, M, N meet. Since they are coplanar in pairs, we have immediately that they are parallel in pairs.

COROLLARY. If $L \parallel M$ and a is not in the plane of L and M, then there exists a unique line N containing a such that $N \parallel L$ and $N \parallel M$.

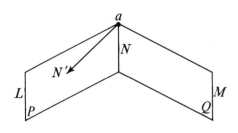

FIGURE 7.1

Proof. There is a plane P containing L and a, and a plane Q containing M and a, by Theorem 3. Clearly, $P \neq Q$, since a is not in the plane of L and M. Let N be the intersection of P and Q. By the theorem, $N \parallel L$ and $N \parallel M$.

To prove uniqueness, suppose N' contains a, $N' \parallel L$ and $N' \parallel M$. Then N' and L are contained in a plane R. Necessarily R contains L and a, so that $R = P$. Thus N' is contained in P. Similarly, N' is contained in Q. Thus $N' = N$, the intersection of P and Q.

NOTE: The diagram accompanying this corollary is intended as an aid in grasping the proof, not as part of the proof.

9. Existence Theorems

In constructing geometrical proofs we often make statements of this sort: Given a line L contained in plane P, select a point which is in P and not in L. Surely we cannot literally *select* a point when we are reasoning abstractly. Points are not on display like oranges on a fruitstand. We take the statement to mean that at least one point having the given property is available—then we can proceed to derive further properties of such a point. Thus, in our abstract theory of geometry, we must justify the *existence* of a point with the given property, that is, we must prove an *existence theorem*.

In the given instance the existence theorem follows easily from Postulate I3:

THEOREM 8.   In a plane P, if line L is given, there is a point not in L.

Proof. By I3 there are in P three noncollinear points a, b, c. Hence at least one of a, b, c is not in L.

THEOREM 9.   Each plane contains three distinct nonconcurrent lines.

Proof. By I3, any plane P contains three distinct noncollinear points a, b, c. By I5, P contains ab, bc, ac. $ab \neq bc$, otherwise c is in ab, and so a, b, c colline, contrary to above. Similarly, $ab \neq ac$ and $bc \neq ac$, so that ab, bc, ac are distinct lines. Since ab, bc, ac intersect in pairs in the distinct points a, b, c they cannot concur.

COROLLARY 1.   In a plane P, if point a is given there is a line not containing a.

Proof.   By the theorem, there are in *P* three nonconcurrent lines. Hence at least one of these does not contain *a*.

COROLLARY 2.   In a plane *P*, any point *a* is contained in at least two lines.

Proof.   By Corollary 1, *P* contains a line, say *bc*, which does not contain *a*. Then *ab* and *ac* are distinct lines containing *a*.

The important notion of skew lines is familiar in Euclidean solid geometry —it forms a natural complement to the idea of coplanar lines.

DEFINITION.   If two distinct lines are not coplanar we say they are *skew* and each is *skew* to the other.

Finally we have

THEOREM 10.   Suppose there are four points *a*, *b*, *c*, *d* that are distinct, noncollinear, and noncoplanar. Then
    (i) given a plane, there is a point not in it;
    (ii) given a line, there is a line skew to it;
    (iii) given a point, there is a plane not containing it;
    (iv) there are at least six lines and at least four planes.

Proof.   (i) Let plane *P* be given. Since *a*, *b*, *c*, *d* are noncoplanar, at least one of them is not in *P*.
    (ii) Let line *L* be given. Since *a*, *b*, *c*, *d* are noncollinear, at least one of them is not in *L*. Let *p* be a point not in *L*. Consider plane *Lp*. By (i) there is a point *q* not in *Lp*. It follows that line *pq* is skew to *L*.
    (iii) Let point *r* be given. Since *a*, *b*, *c*, *d* are distinct, there exists a point *s* distinct from *r*. Consider line *rs*. By (ii) there is a line *M* skew to *rs*. It follows that plane *Ms* cannot contain *r*.
    (iv) To be proved by the reader.

E X E R C I S E S   I

1. Prove: If each of two intersecting lines is parallel to a third line, then the three lines are coplanar.
2. Prove: If a plane contains two distinct lines, each parallel to the same line not contained in the plane, then the two lines in the plane are parallel.

3. Let L_1, L_2, L_3 be three lines, $L_1 \parallel L_3$, $L_2 \parallel L_3$, and let M intersect L_1, L_2, and L_3. Prove all four lines are coplanar.

4. Let L_1, L_2, ..., L_n be lines that concur in a, and suppose line M intersects L_1, L_2, ..., L_n, and does not contain a. Prove L_1, ..., L_n, M are coplanar.

DEFINITION.   A line and a plane are said to be *parallel* (and each is said to be *parallel* to the other) if they have no common point.

5. Prove: If a line is parallel to a plane, it is parallel to the intersection with the plane of any plane containing it.

6. Given a point in the first of two parallel planes and a line contained in the second, prove there is a unique line contained in the first plane which contains the given point and is parallel to the given line.

7. Prove: If a line is parallel to a plane, it is parallel to a unique line which is contained in the plane and contains a given point of the plane.

8. Prove: A line is parallel to no line contained in a plane if it meets the plane in just one point.

9. Prove: If a line is parallel to a line contained in a plane, it is contained in the plane or it is parallel to the plane.

10. Suppose two distinct planes intersect in L, and M is contained in the first, is distinct from L and meets L. Prove that no line contained in the second plane is parallel to M.

11. Suppose P, Q are planes such that $L \parallel M$, where L is contained in P but not in Q, and M is contained in Q but not in P. Prove that $P \parallel Q$ or L and M are parallel to the intersection of P and Q.

12. Prove: If each line contained in one plane is parallel to some line contained in a distinct plane, then the planes are parallel.

13. Let P, Q be distinct planes and a a point of P. Suppose that all lines containing a and contained in P are parallel to lines of Q. Prove $P \parallel Q$. Show the result holds if we assume the condition for all lines except possibly one.

14. Prove: If two distinct planes intersect, then any line contained in one that is parallel to a line contained in the other coincides with or is parallel to their intersection.

15. Prove: If for each point p of a plane there are two distinct lines which contain p, are contained in the plane, and are parallel to lines of a different plane, then the planes are parallel.

16. Prove: If point a is not in plane bcd then a, b, c, d are distinct, noncollinear and noncoplanar. (Compare Th. 2.)

17. Prove: If one line has a skew line, then every line has a skew line.

18. Let a be a point in one of two distinct intersecting planes, but not in their intersection L. Let b be a point in the other but not in L. Prove that ab and L are skew.

19. Suppose L and M are skew, a is not in L or M, plane La meets M and plane Ma meets L. Prove that there is one and only one line which contains a and intersects L and M.

E X E R C I S E S   1 1

1. Prove: If a plane contains two parallel lines, then it contains at least four points.
2. If a plane contains exactly three points, prove:
   (a) each of its lines contains exactly two points;
   (b) it contains no parallel lines.
3. If a plane contains exactly four points, and all of its lines have the same number of points, prove:
   (a) each of its lines contains exactly two points;
   (b) it satisfies the Euclidean parallel postulate (Playfair form).
4. Suppose in a plane each line L has a parallel line M such that every point in the plane is in L or M. Prove that each line contained in the plane has exactly two points, and that the plane has exactly four points.
5. Prove: If each plane contains exactly three points, then there are no parallel lines.
6. Prove: If each line contains exactly two points, and there are no parallel lines, then each plane contains exactly three points.
7. Prove: If each line contains exactly two points and the Euclidean parallel postulate (in the Playfair form) holds, then each plane contains exactly four points.
8. Prove: If each line contains exactly two points, and the Lobachevskian parallel postulate holds, then each plane contains at least five points.
9. Suppose each line contains exactly two points and each plane exactly five points. Prove that if p is not in L there are exactly two lines containing p that are parallel to L.
10. If each line contains at least three points, prove:
    (a) each point of a plane is contained in at least three lines of the plane;
    (b) each plane contains at least seven points;
    (c) each plane contains at least seven lines.
†11. Prove: If each line contains exactly three points, and there are no parallel lines, then each plane contains exactly seven points and exactly seven lines.

12. Prove: If each line contains exactly three points and each plane exactly seven points, then there are no parallel lines.

13. Suppose that each line contains at least three points and that there exist four distinct points which are noncollinear and noncoplanar. Prove:
    (a) there are two distinct intersecting lines both skew to a given line;
    (b) there exist three mutually skew lines.

†14. If each line contains exactly three points, and Euclid's parallel postulate (in the Playfair form) holds, prove:
    (a) each plane contains exactly nine points;
    (b) each point of a plane is in exactly four lines that are contained in the plane;
    (c) each plane contains exactly twelve lines.

15. Prove: If every line has a unique skew line, there are exactly two points in each line and exactly three points in each plane.

†16. Suppose Euclid's parallel postulate (in the Playfair form) holds, and that there exist four distinct points that are noncollinear and non-coplanar. Prove:
    (a) each plane contains at least four points and at least six lines;
    (b) there are at least eight points;
    (c) there are at least twenty-eight lines;
    (d) each line is contained in at least three planes;
    (e) there are at least three lines parallel to a given line;
    (f) there are at least fourteen planes.

†17. If we add to the hypothesis of Exercise 16 that each line contains exactly two points, then each of the conclusions of Exercise 16 holds, with the words "at least" changed to "exactly."

E X E R C I S E S   I I I

(Suggested by Desargues' Theorem)

1. Let a, b, c, a', b', c' be distinct noncoplanar points such that a, b, c are noncollinear and a', b', c' are noncollinear. Suppose ab and $a'b'$ meet in p; bc and $b'c'$ in q; ac and $a'c'$ in r. Prove that p, q, r colline.

2. Let a, b, c, a', b', c' be distinct noncoplanar points such that a, b, c are noncollinear and a', b', c' are noncollinear. Suppose ab and $a'b'$ meet in p; bc and $b'c'$ meet in q; but $ac \parallel a'c'$. Prove pq is parallel to both ac and $a'c'$ or it coincides with one of them.

3. Let a, b, c, a', b', c' be distinct noncoplanar points such that a, b, c are noncollinear and a', b', c' are noncollinear. Suppose ab and $a'b'$ meet;

bc and $b'c'$ meet; ac and $a'c'$ meet. Prove that aa', bb', cc' concur or are parallel in pairs. (Compare Exercise 1.)

4. Let a, b, c, a', b', c' be distinct noncoplanar points such that a, b, c are noncollinear and a', b', c' are noncollinear. Suppose ab and $a'b'$ meet; bc and $b'c'$ meet; but $ac \parallel a'c'$. Prove aa', bb', cc' concur or are parallel in pairs. (Compare Exercise 2.)

5. Let a, b, c, a', b', c' be distinct noncoplanar points such that a, b, c are noncollinear and a', b', c' are noncollinear. Suppose $ab \parallel a'b'$, $bc \parallel b'c'$, $ac \parallel a'c'$. Prove aa', bb', cc' concur or are parallel in pairs.

6. Let a, b, c, a', b', c' be distinct noncoplanar points such that a, b, c are noncollinear and a', b', c' are noncollinear. Suppose aa', bb', cc' concur. Prove:

   (a) ab and $a'b'$ are coplanar; similarly for bc and $b'c'$; and for ac and $a'c'$.

   (b) Suppose two of these pairs consist of intersecting lines. Then the third pair are parallel or the intersections of the three pairs are collinear.

7. What can be inferred in Exercise 6 if the concurrence of aa', bb', cc' is replaced by the condition that they are parallel in pairs?

CHAPTER 8

Incidence Geometries—Models
of the Theory of Incidence

In this chapter we give concrete meaning to the Theory of Incidence studied abstractly in Chapter 7. We exhibit models of the theory, that is, specific systems which exemplify the theory, which satisfy its postulates. Several important ideas such as isomorphism and automorphism of geometric systems and the use of models to prove consistency and independence fit naturally into the context.

1. The Abstract Nature of the Theory of Incidence

In Chapter 7 we deduced theorems of incidence on the basis of Postulates I1–I6. We insisted that our treatment be abstract—that no specific content be assigned to the terms point, line, plane. Nor was content to be implicitly assumed by conceiving the basic terms intuitively as dots, streaks and sheets or by a *sub rosa* agreement that we really are studying Euclidean geometry and could represent our theory by the familiar diagrams of school geometry—only we must not say so.

There is a very striking procedure to exhibit the abstract nature of geometric reasoning which Professor C. J. Keyser (1862–1947) employed in his lectures. If point, line, plane are abstract terms, devoid of significance, and if our proofs are logically correct, then the proofs depend on the *form* of Postulates I1–I6, not on the choice of the particular words point, line, plane, for the basic terms. Thus if we replace point, line, plane, by three other terms the logical structure of our theory will be unchanged. So let us replace point, line, plane, by the meaningless vocables, hoig, boig, loig. Then Postulate I1 is restated as: *A boig is a set of hoigs, containing at least two hoigs.* Theorem 5 becomes: *If two distinct loigs have a hoig in common then the set of all their common hoigs is a boig.* This is not a trivial game but can be a critical test of whether in proving the theorems of Chapter 7 we actually justified the steps by the postulates rather than by reference to intuitive ideas or surreptitious glances at diagrams. You might find it an interesting and challenging exercise to restate Postulates I1–I6 in terms of hoig, boig, loig and write out a proof of a theorem or two in terms of the restated postulates.

2. Interpretation of the Theory of Incidence

This use of meaningless terms for point, line, plane may seem a sterile intellectual exercise—an indication of the absurdity into which one is led by too much insistence on abstract logic.

On the contrary, it helps to point up the advantages of an abstract formulation of a mathematical theory. For if the basic terms are literally meaningless, literally devoid of content, the possibility is opened of assigning them content in new and unforseen ways. To clarify this observe that our postulates and theorems are not statements in the ordinary sense but abstract "statement forms," so-called *propositional functions* or *open sentences*. They become statements, true or false, when suitable content is assigned to the basic terms point, line, plane. To make the theory "significant," or apply it "concretely," we merely *interpret* or assign meaning to the abstract terms point, line, plane, in such a way that Postulates I1–I6 become true statements. If this is done, we argue that the theorems (of course with the same interpretations for point, line, plane) also become true statements.

The theory of incidence is surely applicable to Euclidean and to Lobachevskian geometry, since Postulates I1–I6 are valid in both. Let us try to find an interpretation of the theory different from the classical ones. We must find specific meaning for the terms point, line, plane such that I1–I6 become true statements. Suppose we try to do this in the simplest way we can. I1 offers a clue. To satisfy I1, a line must consist of points and contain at least two points. Surely the simplest way to satisfy I1 is to require a line to consist of two distinct points—that is to be a *couple* of points. Similarly, to satisfy I3 in the simplest way we require a plane to consist of three distinct points— that is to be a *triple* of points. So we have interpretations of line and plane in terms of point, but it is not clear how to interpret point. Without trying to resolve this at the moment, let us find the interpretations of I1–I6 by assigning the meaning couple of points, to "line," and triple of points to "plane."

The results appear in the following table:

Postulate	Interpretation
I1. *A line is a set of points, containing at least two points.*	I1. *A couple of points is a set of points, containing at least two points.*
I2. *Two distinct points are contained in one and only one line.*	I2. *Two distinct points are contained in one and only one couple of points.*
I3. *A plane is a set of points, containing at least three points which do not belong to the same line.*	I3. *A triple of points is a set of points, containing at least three points which do not belong to the same couple of points.*
I4. *Three distinct points which do not belong to the same line are contained in one and only one plane.*	I4. *Three distinct points which do not belong to the same couple of points are contained in one and only one triple of points.*
I5. *If a plane contains two distinct points of a line, it contains the line.*	I5. *If a triple of points contains two distinct points of a couple of points, it contains the couple.*
I6. *If two planes have one point in common, they have a second point in common.*	I6. *If two triples of points have one point in common, they have a second point in common.*

Upon examining the table we see that the interpretations of I1–I5 are true, regardless of the meaning assigned to point. I6, however, is in doubt since we have made no decision about how different triples of points are to be related, nor even about how many triples there should be. We can satisfy I6 very simply by requiring the number of points to be three, so that there will be only one triple of points, that is only one plane. Then I6 is obviously satisfied, since any triple of points has three distinct points in common with itself, and certainly two. Observe that the use of the phrase "two planes" in I6 does not require distinctness (Ch. 7, Sec. 5), and that I6 would be valid when applied to any plane and itself, merely if a plane were to contain two distinct points.

Summary. Postulates I1–I6 are satisfied if we choose any set of three distinct elements; interpret point to mean any of the three elements; line, any couple of the elements; and plane, the set of all three elements. If the elements of the set are *a*, *b*, *c* then our system has points *a*, *b*, *c*; lines (*a*, *b*), (*b*, *c*), (*a*, *c*); and a single plane (*a*, *b*, *c*).

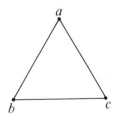

FIGURE 8.1

It is interesting that the system can be represented by the simple diagram of Figure 8.1. Here the dots labeled a, b, c denote points. The pairs of dots (a, b), (b, c), (a, c) denote lines. The connecting streaks are inserted only for suggestiveness and have no formal significance for the system represented.

We have thus constructed a miniature geometrical system consisting of three distinct points, three distinct lines, and one plane complete with its own pictorial representation. This system shares with Euclidean and Lobachevskain geometry the distinction of satisfying Postulates I1–I6.

A system which satisfies a set of postulates is called a *model* (or *interpretation* or *concrete application*) of the set of postulates, or of the theory which is based on the set of postulates. Thus the miniature geometric system above is a *model* of the Theory of Incidence, and we label it M1.

Now we have reached our main object of study in this chapter: models of the Theory of Incidence. Formally expressed, a model of the Theory of Incidence is a system (S_1, S_2, S_3) comprising three specific sets S_1, S_2, S_3 whose elements are called points, lines, and planes respectively, which satisfies Postulates I1–I6. To specify such a model, we must give explicit meaning to the sets S_1, S_2, S_3; or equivalently to the terms point, line, and plane. For example for model M1, S_1 the set of points is composed of a, b, c. S_2 the set of lines is composed of (a, b), (b, c), (a, c). Finally S_3 is composed of the single plane (a, b, c).

Models of the theory of incidence are of sufficient importance to deserve a name of their own.

DEFINITION.   Any system (S_1, S_2, S_3) comprising three sets S_1, S_2, S_3 whose elements are called points, lines, and planes, respectively, which satisfies Postulates I1–I6 is called an *incidence geometry*.* An incidence geometry is said to be *planar* or *two-dimensional* if it contains only one plane; it is *three-dimensional* if it contains more than one plane.

3. Incidence Geometries

We present in this section a list of incidence geometries, models of the Theory of Incidence of Chapter 7, which form the basis of discussion in this chapter and will occasionally be referred to in later chapters.

As you proceed, the advantages of the abstract formulation of the Theory of Incidence should become quite evident. The first and most obvious of these

---

* Saul Gorn, *On Incidence Geometry*, Bulletin of the American Mathematical Society, Vol. 46, pp. 158–167, 1940. [In this paper the concept of Incidence Geometry was introduced in a form, more general than that presented here, which permits application to n-dimensional geometries.]

is economy. For the postulates imply the theorems formally and abstractly, independently of the content of the basic terms. Consequently, whenever the postulates are satisfied the theorems necessarily are satisfied. In other words, *the theorems are valid for all models*. Therefore, instead of having to prove the incidence theorems of Chapter 7 for Euclidean geometry, and again for Lobachevskian geometry, and possibly a third time for single elliptic geometry—we merely show that each of these systems satisfies Postulates I1–I6, and we infer that the incidence theorems hold for all three systems. In effect the proofs of Chapter 7 are done once and applied three times. This saves effort and time; paper and ink; thought.

Secondly, in addition to economy we have richness. For the theory applies to *all* models, to *all* incidence geometries, not just to the ones we have in mind when we set up the theory, but to *undiscovered* ones also. Indeed the existence of an abstractly formulated theory leads us to search for and discover new models, as we discovered the simple model M1 of the Theory of Incidence. The abstract formulation of a theory need not lead to sterility—on the contrary it can be very fruitful in enlarging our knowledge of the concrete systems which form the basic subject matter of mathematics and give significance to its abstractions. Other subtler advantages appear in further study. For example, concrete systems are often better understood when viewed and compared in the light of an abstract theory.

It is unfortunate that abstract theories and concrete knowledge are so often thought to be antithetic. The interplay of the abstract and concrete is the life of mathematics. We have indicated that an abstract theory can foster the enrichment of concrete knowledge. The reverse can happen in several ways. The examination and comparison of concrete systems may suggest new theories, for example the Theory of Fields in algebra, or the Theory of Incidence in geometry; new concepts, for example isomorphism (Sec. 5 below); and new modes of proof, for example proofs of independence (Sec. 9 below).

The models are shown in the following table:

Model	Point	Line	Plane	Diagram
M1	a, b, c (three distinct elements)	(a, b), (b, c), (a, c)	(a, b, c)	
M2	a, b, c, d (four distinct elements)	(a, b), (a, c), (a, d), (b, c), (b, d), (c, d)	(a, b, c), (a, b, d), (a, c, d), (b, c, d)	
M3	a, b, c, d (four distinct elements)	(a, b), (a, c), (a, d), (b, c), (b, d), (c, d)	(a, b, c, d)	
M4	a, b, c, d, e, f, g, h (eight distinct elements)	(a, b), (a, c), (a, d), (a, e), (a, f), (a, g), (a, h), (b, c), (b, d), (b, e), (b, f), (b, g), (c, d), (c, e), (c, f), (c, g), (c, h), (d, e), (d, f), (d, g), (d, h), (e, f), (e, g), (e, h), (f, g), (f, h), (g, h)	(a, b, c, d), (e, f, g, h), (a, b, f, e), (d, c, g, h), (a, b, g, h), (e, f, c, d), (a, c, g, e), (b, d, h, f), (a, d, h, e), (b, c, g, f), (a, d, g, f), (b, c, h, e), (a, c, f, h), (b, d, e, g)	

Model	Point	Line	Plane	Diagram
M5	1, 2, 3, 4, 5, 6, 7 (seven distinct elements)	The columns in the table: 1 2 3 4 5 6 7 2 3 4 5 6 7 1 4 5 6 7 1 2 3	(1, 2, 3, 4, 5, 6, 7)	
M6	Dots labeled a, b, c, d, e, f, g in the diagram	Triples of dots lying on a streak in the diagram: (a, b, d), (b, c, e), (c, d, f), (d, e, g), (e, f, a), (f, g, b), (g, a, c)	The set of dots in the diagram (a, b, c, d, e, f, g)	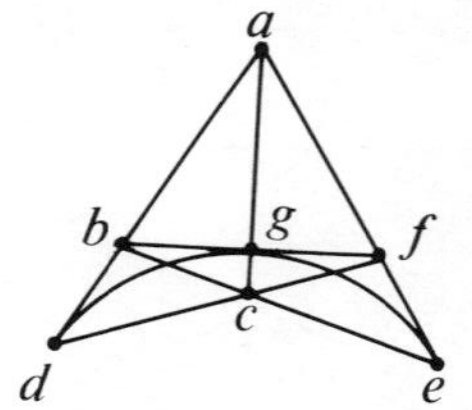
M7	Same description as M6, but with different diagram			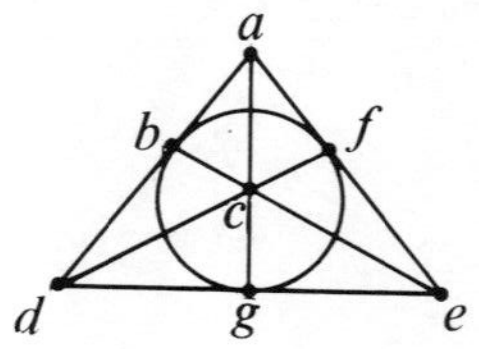
M8	1, 2, 3, 4, 5, 6, 7, 8, 9 (nine distinct elements)	(1, 2, 3), (4, 5, 6), (7, 8, 9) (1, 4, 7), (2, 5, 8), (3, 6, 9) (1, 5, 9), (2, 6, 7), (3, 8, 4) (1, 8, 6), (2, 4, 9), (3, 5, 7)	(1, 2, 3, 4, 5, 6, 7, 8, 9)	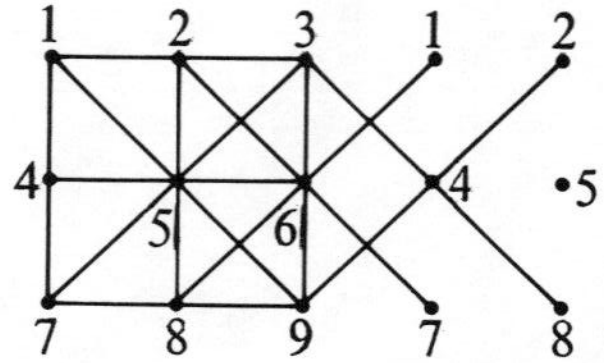

M9	Each point inside a given Euclidean circle	Each "open" chord of the circle (that is, the chord minus its end-points)	The interior of the circle
M10	Each point inside a given Euclidean sphere	Each "open" chord of the sphere	The interior of each circle lying on the sphere
M11	Each point inside a given Euclidean triangle, $\triangle abc$	Each "open" segment joining two points lying in different sides of the triangle	The interior of $\triangle abc$
M12	Each point inside a given Euclidean tetrahedron $abcd$	Each "open" segment joining two points lying in different faces of $abcd$	The interior of each cross-section of $abcd$ by a plane

Model	Point	Line	Plane	Diagram
M13	Each point in a Euclidean half-plane H; that is each point of a Euclidean plane which is on a given side of a given line of the plane	The intersection of H with any Euclidean line which contains two distinct points of H	The half-plane H	
M14	Similar to M13, using a Euclidean half-space, which is the set of points of space on a given side of a given plane			
M15	Each point of a given Euclidean Plane P	Each line of P	Plane P	
M16	Each point of Euclidean 3-space S	Each line of S	Each plane of S	
M17	Similar to M15, using a Lobachevskian plane			
M18	Similar to M16, using Lobachevskian 3-space			
M19	Each ordered pair (x, y) of real numbers	The set of (x, y) which satisfy $ax + by + c = 0$, a linear equation in the real number system	The set of all (x, y)	

M20	Each ordered triple (x, y, z) of real numbers	The set of (x, y, z) which satisfy a pair of nonequivalent, consistent linear equations $ax + by + cz + d = 0,$ $ex + fy + gz + h = 0,$ in the real number system	The set of (x, y, z) which satisfy a linear equation $ax + by + cz + d = 0,$ in the real number system

M21   Similar to M19 but replacing the real number system by the system of integers modulo 2, which consists of the two elements 0, 1 combined according to the rules:

$$0 + 0 = 0, 0 + 1 = 1 + 0 = 1, 1 + 1 = 0;$$
$$0 \times 0 = 0 \times 1 = 1 \times 0 = 0, 1 \times 1 = 1.$$

There are four points and six lines.

M22	Each ordered triple (x, y, z) of real numbers except $(0, 0)$ with the agreement that (x, y, z) and (kx, ky, kz) are to be considered the same point for every real $k \neq 0$	The set of "points" (x, y, z) which satisfy $ax + by + cz = 0,$ a *homogeneous* linear equation in the real number system	The set of all "points" (x, y, z)
M23	Each point of a given Euclidean plane P, excluding a given point o	Each line L of plane P, excluding point o if L contains o	The set of points of P excluding point o

EXERCISES I

1. Verify that M1, M2, M3, M4, M5, M6 and M7 are incidence geometries; that is, they satisfy Postulates I1–I6.
2. Convince yourself by intuitive geometric reasoning that M9, M10, M11, M12, M13, M14, M23 are incidence geometries.
3. Determine which of the models are finite, (that is, contain a finite number of points), and which infinite.
4. Examine the models for parallelism properties. List those that satisfy the Euclidean parallel postulate (Playfair form), those that satisfy the Lobachevskian, and those that satisfy the Riemannian parallel postulate. Are there any models that satisfy none of these?
5. Which models are two-dimensional; which three-dimensional? Do you find cases where a three-dimensional model "contains" a two-dimensional model already listed?
6. List all possible "points" (x, y) in M21, and all possible linear equations. List all solutions (x, y) for each linear equation. Make a table showing the "lines" as composed of "points." Compare M3 and M21.
7. Consider M19 with the real number system replaced by the system of integers modulo 3, which consists of the elements 0, 1, 2 with the following addition and multiplication tables:

+	0	1	2
0	0	1	2
1	1	2	0
2	2	0	1

×	0	1	2
0	0	0	0
1	0	1	2
2	0	2	1

List all possible "points" (x, y), and all possible linear equations. List all solutions (x, y) for each linear equation. Make a table showing each "line" as composed of three "points." Verify that the system formed is an incidence geometry. Compare it with M8.
8. Consider the following system: Point, line, and plane are as defined in M22, except that the numbers of the system are the integers modulo 2 (as defined in M21). List all possible "points" (x, y, z), and all possible homogeneous linear equations $ax + by + cz = 0$. List all solutions (x, y, z) of each equation. Verify that this system is an incidence geometry. Compare it with M7.
9. Show that models M2 and M10 satisfy Theorems 3 and 5 of Chapter 7.
10. Consider the following system:
   Point: a, b, c, d, e (five distinct elements)

Line: $(a, b), (a, c), (a, d), (a, e), (b, c), (b, d), (b, e), (c, d), (c, e), (d, e).$
Plane: $(a, b, c), (a, b, d), (a, b, e), (a, c, d), (a, c, e), (a, d, e), (b, c, d),$
$(b, c, e), (b, d, e), (c, d, e).$

Which of Postulates I1-I6 does it satisfy? Which does it falsify? How is this system related to M2?

11. Consider the following system: Point and line as in Exercise 10. Plane is the set (a, b, c, d, e); that is, there is only one plane. Show that this system is an incidence geometry, and that it satisfies the Lobachevskian parallel postulate.

12. Consider the following system:

Point: Any point of a Euclidean sphere S.

Line: Any great circle of the Euclidean sphere S; that is, a circle on the sphere whose radius is the radius of the sphere.

Plane: The sphere S.

Which of Postulates I1–I6 does the system satisfy? Which does it falsify?

13. Consider the following system (S_1, S_2, S_3): Let S be a Euclidean sphere. An element of S_1 is a pair of diametrically opposite points of S, and S_1 is the set of all such pairs. Thus a "point" in this system is a diametral pair of Euclidean points. An element of S_2 is the set of all such diametral pairs in a great circle of S. Thus a "line" in this system is a great circle of S, conceived as composed of diametral pairs. S_3 is the set of all diametral pairs of S, that is, there is only one plane in the system, namely S conceived as composed of diametral pairs.

Verify that this system is an incidence geometry. Which parallel postulate does it satisfy?

14. If a planar incidence geometry satisfies the Lobachevskian parallel postulate, prove that it contains at least five points, and that there are at least three lines parallel to a given line. Compare the system in Exercise 11.

15. M6 and M7 appear to be "similar" or to have the" same structure." Do you observe other examples of models which have the "same structure?"

16. Prove that an incidence geometry is three-dimensional if and only if it contains four distinct points which are noncollinear and noncoplanar.

17. Prove: In a three-dimensional incidence geometry there are four distinct points, six distinct lines, and four distinct planes.

18. In M4, how many lines are parallel to ab? Skew to ab. Do you get the same number for ac? For ag?

19. In any three-dimensional incidence geometry, show that each plane determines a planar incidence geometry, that is, it and its points and lines form a planar incidence geometry.

20. Show how to construct all planar incidence geometries with exactly two points in each line. Are they all finite?

4. Discussion of the Algebraic Model M19

We refer to an ordered pair (x, y) of real numbers as a *dyad*, and to the set of solutions (x, y) of a real linear equation as a *linear system*, for the sake of conciseness and to point up the importance of these entities in our discussion. Thus the phrase "a point is in a line" becomes in this system, "a dyad is an element of a linear system." The notions of equality (or identity) for points and for lines appear in the Postulates I1–I6 and so must be clarified in the system. Equality of dyads is simply identity for ordered pairs, that is $(x_1, y_1) = (x_2, y_2)$ means $x_1 = x_2$ and $y_1 = y_2$. Equality of two linear systems means simply that the linear systems are identical as sets of dyads, that is, they comprise the same set of dyads.* For example, the linear systems determined by $2x - 3y + 7 = 0$ and $4x - 6y + 14 = 0$ are identical. This illustrates a general criterion: The linear systems determined by

$$a_1x + b_1y + c_1 = 0, \qquad a_2x + b_2y + c_2 = 0$$

are identical if and only if their coefficients are *proportional*, that is there exists a real number k such that

$$a_1 = ka_2, \qquad b_1 = kb_2, \qquad c_1 = kc_2, \qquad k \neq 0.$$

We proceed to verify that M19 is an incidence geometry.

Interpretation of Postulate I1.   A linear system is a set of dyads, containing at least two dyads.

Let linear system L be determined by the linear equation $ax + by + c = 0$. If $a \neq 0$, then $(-c/a, 0)$ and $(-(b + c)/a, 1)$ are easily seen to be dyads of L. If $a = 0$ then by definition of a linear equation, $b \neq 0$, and $(0, -c/b)$, $(1, -c/b)$ are dyads of L.

Interpretation of Postulate I2.   Two distinct dyads are contained in one and only one linear system.

Let distinct dyads (x_1, y_1), (x_2, y_2) be given. Suppose there is a linear system L, determined by linear equation $ax + by + c = 0$, which contains the two dyads. Then

$$ax_1 + by_1 + c = 0,$$
$$ax_2 + by_2 + c = 0.$$

We determine how a, b, c are related by these equations. Subtraction to eliminate c yields

$$a(x_1 - x_2) + b(y_1 - y_2) = 0.$$

---

* Stated formally, sets A, B are identical if each element of A is an element of B and each element of B is an element of A.

Suppose $y_1 - y_2 \neq 0$, so that we may solve for b. Then

(1) $$b = \frac{-a(x_1 - x_2)}{y_1 - y_2}.$$

We now express c in terms of a.

$$c = -ax_1 - by_1$$

$$= -ax_1 + \frac{a(x_1 - x_2)}{y_1 - y_2} y_1$$

$$= \frac{a(x_1 y_2 - x_2 y_1)}{y_1 - y_2}.$$

Substitution for b and c in $ax + by + c = 0$ yields

(2) $$ax - \frac{a(x_1 - x_2)y}{y_1 - y_2} + \frac{a(x_1 y_2 - x_2 y_1)}{y_1 - y_2} = 0.$$

This equation determines the linear system L. Observe that $a \neq 0$, otherwise by Equation (1) $b = 0$, contrary to the definition of a linear equation. Hence we may multiply the terms of Equation (2) by $(y_1 - y_2)/a$, getting the equation

(3) $$(y_1 - y_2)x + (x_2 - x_1)y + (x_1 y_2 - x_2 y_1) = 0,$$

which also determines L.

Since the coefficients of Equation (3) are determined by the given numbers x_1, y_1, x_2, y_2, we see that there is only one possible linear system containing the dyads (x_1, y_1) and (x_2, y_2), and that it is determined by Equation (3). To see that the linear system determined by Equation (3) actually does contain the two given dyads, substitute (x_1, y_1) and (x_2, y_2) in Equation (3).

This discussion was based on the supposition $y_1 - y_2 \neq 0$. Suppose $y_1 - y_2 = 0$; then $y_1 = y_2$. Clearly $x_1 \neq x_2$, and $x_1 - x_2 \neq 0$. An argument exactly similar to that above yields the same linear Equation (3) and the same conclusion.

The verifications of I3, I4, I5, and I6 are left as exercises. Note that in this model I4, I5, and I6 are obviously satisfied since there is only one plane.

EXERCISES II

1. Complete the verification that M19 is an incidence geometry by showing:
   (i) The linear systems determined by

   $$a_1 x + b_1 y + c_1 = 0, \qquad a_2 x + b_2 y + c_2 = 0$$

   are identical if and only if there exists a real number k such that

   $$a_1 = ka_2, \qquad b_1 = kb_2, \qquad c_1 = kc_2, \qquad k \neq 0.$$

    (ii) I2 holds if we assume $y_1 - y_2 = 0$.

    (iii) I3, I4, I5, and I6 hold.

2. Prove, by algebraic reasoning, that M22 is an incidence geometry.

5. The Concept of Isomorphism

You may have noticed that sometimes two of our models seem to have the same structure, for example, M6 and M7 or M3 and M21. Roughly speaking, this means that the models can be "matched," point with point, line with line, plane with plane, in such a way that if two objects (point, line, or plane) of one model are incident, then the corresponding objects of the other also are incident.

To formalize this notion it is desirable first to clarify the essential idea of one-to-one correspondence.

DEFINITION.   A *one-to-one correspondence* between a set S and a set S' is a correspondence $a \to a'$ (read "*a* corresponds to *a'*") between the elements in S and the elements in S' such that each element a in S has a unique correspondent a' in S', and each element b in S' is the correspondent, under the correspondence, of a unique element c in S; that is $b = c'$.

For example, the correspondence $x \to 2x$ is a one-to-one correspondence between the set of all positive integers and the set of all even positive integers. Note that there exists a one-to-one correspondence between two finite sets if and only if they have the same number of elements.

For simplicity we first define isomorphism for *planar* incidence geometries.

DEFINITION.   Two *planar* incidence geometries G, G' are said to be *iso-morphic* (or to have the *same structure*) if there exist two one-to-one correspondences $p \to p'$, $L \to L'$, between the points of G and the points of G', and the lines of G and the lines of G', which have the following property: Suppose

$$p \to p', \qquad L \to L'.$$

Then p is in L if and only if p' is in L'. We may describe this by saying: *the correspondences preserve incidence relations of points and lines.* *

---

  * This use of the "preserve" terminology is the simplest and most convenient for our purposes. A more common usage would restrict the italicized phrase to mean that p is in L implies p' is in L'.

EXAMPLE.   Models M6 and M7 are isomorphic. To see this, let points of M6 and M7 correspond if they are denoted by the same letter, and lines correspond if their points are denoted by the same letters. At first glance this may seem trivial—it merely means that the systems can be described by the same labels. But it is a remarkable fact about M6 and M7 that they can be labeled in exactly the same way; if you were given the unlabeled diagrams you might never have guessed that they are isomorphic. Finally note that our reference to "labeling" is only an intuitive way of indicating the existence of the appropriate one-to-one correspondences. If taken too literally it may suggest that M6 and M7 are the same system, which is not so.

Let G, G' be planar incidence geometries which are isomorphic. Suppose there are exactly three points a, b, c in line L of G. Then the corresponding points a', b', c' are in the corresponding line L' of G'. Since a, b, c are distinct so are a', b' c'. Thus L' has at least three points. Suppose L' contained a fourth point say d'. Then d, its correspondent in G, would be in L, and L would contain a fourth point. Thus L' has exactly three points. This type of argument enables us to show that G and G' agree in their incidence properties —any incidence property of one is mirrored in the other. Thus if three lines of G concur, their correspondents in G' also concur. If two lines of G are parallel their correspondents in G' also are parallel. If G satisfies the Lobachevskian parallel postulate so does G'.

To show G and G' isomorphic it is not necessary to establish explicitly both correspondences required by the definition. For suppose G and G' are isomorphic; consider the one-to-one correspondence, $p \to p'$, between the points of G and the points of G'. It has the following property: Suppose

$$a \to a', \qquad b \to b', \qquad c \to c'.$$

Then a, b, c are collinear (in G) if and only if a', b', c' are collinear (in G'). This may be described by saying: *the correspondence preserves collinearity*. Now suppose G, G' are planar incidence geometries and there exists a one-to-one correspondence between the points of G and the points of G' which satisfies the above condition. Then we can construct a one-to-one correspondence between the lines of G and the lines of G' as follows: If L is a line of G, the correspondents of the points of L will form a line L' of G'. (Why?) Form the correspondence $L \to L'$. This is a one-to-one correspondence between the lines of G and G', and the two correspondences $p \to p'$, $L \to L'$ satisfy the conditions of the definition of isomorphism. We summarize the discussion in

THEOREM 1.   Two planar incidence geometries G, G' are isomorphic if and only if there exists a one-to-one correspondence between the points of G and the points of G' which preserves collinearity.

This is often the most useful criterion for proving isomorphism.

We now formulate the general definition of isomorphism for incidence geometries.

DEFINITION.   Two (not necessarily planar) incidence geometries G and G' are said to be *isomorphic* if there exist three one-to-one correspondences $p \rightarrow p'$, $L \rightarrow L'$, $P \rightarrow P'$, between the points, lines, planes of G and the points, lines, planes of G' which have the following properties: Suppose

$$p \rightarrow p', \qquad L \rightarrow L', \qquad P \rightarrow P'.$$

Then

(i) p is in L if and only if p' is in L';

(ii) p is in P if and only if p' is in P';

(iii) L is contained in P if and only if L' is contained in P'.

This may be stated: *the correspondences preserve incidence relations of points, lines, and planes.*

Observe that the general definition is equivalent to the preceding one for isomorphism of *planar* incidence geometries if G and G' are planar.

There is a generalization of Theorem 1 which applies to arbitrary incidence geometries. To state it we explain the phrase *preservation of coplanarity.*

Let $p \rightarrow p'$ be a one-to-one correspondence between the points of incidence geometry G and the points of incidence geometry G' such that if

$$a \rightarrow a', \qquad b \rightarrow b', \qquad c \rightarrow c', \qquad d \rightarrow d',$$

then a, b, c, d are coplanar (in G) if and only if a', b', c', d' are coplanar (in G'). Then we say: *the correspondence preserves coplanarity.*

THEOREM 2.   Two incidence geometries G, G' are isomorphic if and only if there exists a one-to-one correspondence between the points of G and the points of G' which preserves collinearity and coplanarity.

The proof is analogous to that of Theorem 1.

E X E R C I S E S   I I I

1. Prove that M5 and M6 are isomorphic. What do you infer about the relation of M5 and M7?

2. Prove that M3 and M21 are isomorphic.

3. Prove that M8 and the model of Exercises I, 7 are isomorphic.

4. Show that there exists a one-to-one correspondence between the points of M2 and the points of M3 which preserves collinearity. Are M2 and M3 isomorphic?

5. Consider M6, M7. Let points a, b, c, d, e, f, g of M6 correspond respectively to b, c, d, e, f, g, a of M7. Show that this correspondence preserves collinearity and effects an isomorphism of M6 and M7.

6. Show that each plane of M4 with its points and lines forms an incidence geometry isomorphic to M3.

7. Prove: If two incidence geometries are isomorphic and one is three-dimensional so is the other; if one satisfies the Lobachevskian parallel postulate so does the other.

8. Prove that M9 and M15 are not isomorphic.

9. Prove that M15 and M23 are not isomorphic.

10. Prove that M9 and M11 are not isomorphic.

11. Prove: If a one-to-one correspondence between the points of one incidence geometry and the points of another preserves collinearity, it also preserves noncollinearity.

12. Show that isomorphism of incidence geometries is an equivalence relation: that is, if G, G', and G'' are incidence geometries, then (a) G is isomorphic to itself; (b) if G is isomorphic to G', then G' is isomorphic to G; and (c) if G is isomorphic to G', and G' is isomorphic to G'', then G is isomorphic to G''.

6. Isomorphic and Automorphic Correspondences

Let G, G' be isomorphic incidence geometries. Theorems 1 and 2 indicate the importance of one-to-one correspondences between the points of G and the points of G' which preserve collinearity and coplanarity. A correspondence of this type is not merely useful in establishing that G and G' are isomorphic, but it completely determines an isomorphism relation between G and G', since it determines how the lines of G correspond to the lines of G', the planes to the planes. Such a correspondence is, so to speak, the "core" of an isomorphism relation between G and G'. Thus we introduce the following definition.

DEFINITION. Given incidence geometries G, G' let $p \to p'$ be a one-to-one correspondence between the points of G and the points of G' which preserves collinearity and coplanarity. Then $p \to p'$ is called an *isomorphic correspondence* or an *isomorphism* between G and G'. If $G = G'$, that is, G and G' are the same incidence geometry, we employ the corresponding terms *automorphic correspondence* or *automorphism* of G.

The existence of automorphisms of an incidence geometry G is an indication of the "symmetry of structure" of G. For example if two points (or two lines) correspond under an automorphism of G they play similar roles in the structure of G. Further, suppose G is so rich in automorphisms that any two of its lines can be made to correspond under some automorphism of G—then G has, in a certain sense, a homogeneous structure, for every line behaves like every other line.

Automorphisms are examples of *collineations*, that is correspondences between two geometrical systems which associate collinear points of the first with collinear points of the second. They may be thought of as analogous to motions, such as translations or rotations—not that they preserve distance (no such notion is required in an incidence geometry) but that, like motions, they are collineations.

7. Some Additional Models

M24.   Let E denote Euclidean 3-space, and let a be a point in E. S_1 is the set of all lines of E which contain a. Thus a "point" of the model is a line of E which contains a. S_2 is the set of all pencils of lines with vertex a. (A pencil of lines with vertex a is the set of all lines which contain a and are contained in a plane P, such that a is in P.) Thus a "line" of the model is a certain set of coplanar lines in E. S_3 is the set whose sole element is the set of all lines in E which contain a, and the model has only one "plane."

M25.   Let H be an open Euclidean hemisphere, that is, the set of all points of a hemisphere that are not in its bounding circle. Let c be the center of H. The intersection of H with any plane that contains c is an open

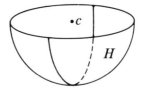

FIGURE 8.2

great semi-circle that is, a semi-circle without its endpoints. S_1 is the set of all points of H. Thus a "point" of the model is a point of H. S_2 is the set of all open great semi-circles contained in H. Thus a "line" of the model is a semi-circle. S_3 is the set whose sole element is H, so that H is the only "plane" in the model.

M26. In Euclidean 3-space, we are given a sphere S and a point b of S. S_1 is the set of all points of S except b. Let C be any circle which is contained in S and which contains b. The "punctured" circle C', that is, the set of all points of C other than b, is an element of S_2, and S_2 is the set of all such "punctured" circles C'. Finally S_3 is the set whose sole element is the "punctured" sphere S', that is, the set of all points of S other than b.

M27. Point: Each ordered triple of real numbers (x, y, z) satisfying

$$x^2 + y^2 + z^2 = 1, \qquad z < 0.$$

     Line: The set of ordered triples (x, y, z) which satisfy simultaneously $ax + by + cz = 0$, (a homogeneous linear equation in the real number system), $x^2 + y^2 + z^2 = 1$ and $z < 0$.

    Plane: The set of all ordered triples of real numbers (x, y, z) satisfying

$$x^2 + y^2 + z^2 = 1, \qquad z < 0.$$

E X E R C I S E S   I V

1. Prove that M24 is an incidence geometry. Which parallel postulate does it satisfy?
2. Prove that M25 is an incidence geometry. Which parallel postulate does it satisfy?
†3. Prove that M26 is an incidence geometry. Which parallel postulate does it satisfy?

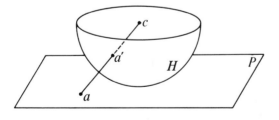

FIGURE 8.3

4. Prove that M15 and M25 are isomorphic. [*Hint.* (Figure 8.3) Let the Euclidean plane P be placed tangent to the open Euclidean hemisphere H, and parallel to the plane containing the circle which bounds H. Let c be the center of H. Corresponding to each point a in P the line ca meets

H in a point a'. Show that the correspondence $a \rightarrow a'$ is an isomorphism between the geometry of P and the geometry of H.]

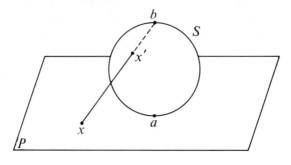

FIGURE 8.4

5. Prove that M15 and M26 are isomorphic. [*Hint.* (Figure 8.4) Let the sphere S be placed tangent to the Euclidean plane P at point a which is diametrically opposite to b. For each point x in P the line bx meets S', the "punctured" sphere, in a unique point x'. Show that the correspondence $x \rightarrow x'$ is an isomorphism between the geometry of P and the geometry of S'. This is an interesting and important type of correspondence called a *stereographic projection* of plane on sphere.]
6. Prove that M27 is an incidence geometry. Which parallel postulate does it satisfy?
7. Show that M27 and M25 are isomorphic. (Use solid analytic geometry.)
8. Show that M27 and M19 are isomorphic. [To find an isomorphic correspondence, interpret (x, y, z) as the coordinates of a point in Euclidean 3-space on the hemisphere described by $x^2 + y^2 + z^2 = 1$, $z < 0$. Let line L contain the points (x, y, z) and $(0, 0, 0)$, and let L meet the plane $z = -1$ in $(x', y', -1)$. Then point (x, y, z) is "projected" from $(0, 0, 0)$, the center of the hemisphere, into point $(x', y', -1)$ on plane $z = -1$. (Compare the correspondence in Exercise 4.) It follows easily that $x' = -(x/z)$, $y' = -(y/z)$. This suggests that $(x, y, z) \rightarrow (-x/z, -y/z)$ is an isomorphic correspondence between M27 and M19.]
9. Show that M25 is not isomorphic to the incidence geometry described in Exercises I, 13.

8. The Use of Models to Prove Consistency

We can now appreciate better the method used in Chapter 5 to prove Lobachevskian geometry consistent. Our procedure was to construct a "representation" of Lobachevskian geometry in the interior of a Euclidean circle which rendered faithfully in Euclidean terms the properties of the Lobachevskian plane. Particularly the postulates of Lobachevskian geometry were represented by Euclidean propositions about O-points, O-chords, O-segments, et cetera. Thus the representation must be a *model* of Lobachevskian geometry, conceived as an abstract theory, in which point, line, segment, and so forth, are *interpreted* as O-point, O-chord, O-segment, et cetera. Consequently, the model (representation) must satisfy the theorems of Lobachevskian geometry. Thus every theorem of Lobachevskian geometry, when interpreted concretely in the model, becomes a Euclidean proposition. We conclude that if Euclidean geometry is consistent then Lobachevskian geometry must be consistent.

The same type of argument is available to prove that any mathematical theory based on a set of postulates is consistent. Consider as an example the Theory of Incidence. Since M1 is a model of this theory it satisfies all the propositions (postulates and theorems) of the theory. Thus if the Theory of Incidence were inconsistent, contradictory assertions about M1 would be true. Hence the Theory of Incidence is consistent provided M1 is free from contradiction, or *consistent*. It is practically impossible to doubt that a simple, finite system like M1 is consistent and so it is practically impossible to doubt the consistency of the Theory of Incidence.

Summary. If a model of an abstract theory can be found, the theory is consistent provided the model is consistent. This is a *relative* test of consistency but is practically indubitable if the model is finite. Often the model is constructed in terms of the notions of another mathematical theory and we have a relative test of consistency for one theory in terms of another, of the type exhibited in Chapter 5.

9. The Use of Models to Prove Independence

In studying a branch of mathematics you may encounter an interesting and distressing kind of problem. You conjecture a theorem and have trouble proving it. You may suppose your conjecture wrong and try to prove the opposite theorem. Suppose you do not succeed. You may wonder whether the theorem can not be deduced from the postulates of the theory, or whether you have not been clever enough to find a proof. (This difficulty often arises in mathematical research.)

If the theory is formulated abstractly there is a procedure for coping with the problem. Consider a specific example. Suppose in the Theory of Incidence we conjecture the property that every line contains at least three points. If this is a theorem, that is if it is deducible from Postulates I1–I6, then it must be valid for *every* model of the Theory of Incidence. But we can easily find a model which falsifies the given property, namely M1. Consequently, the given property can not be deduced from Postulates I1–I6, or as we say, it is *independent* of Postulates I1–I6. Thus to prove in a given abstract theory that a statement is independent of the postulates it is sufficient to find a model, a so-called *independence model*, which falsifies the statement. Of course if we fail to find a model which falsifies the statement, that is if all the models we know satisfy the statement, we can not conclude that the statement is not independent of the postulates. Often the construction of the desired model is very difficult and requires a good deal of ingenuity.

Critique. In our argument we assumed that the property, every line contains at least three points, is deducible from I1–I6 and concluded that every line of M1 contains at least three points. Since this contradicts a known property of M1, that each of its lines consists of two points, we inferred that our assumption is false. Clearly we assumed implicitly that M1 is consistent—that two contradictory assertions about M1 could not both be true. As we observed in Section 8, it is practically impossible to doubt the consistency of a system like M1. We must however make the point that the use of a model to prove independence assumes implicitly the consistency of the model and so is a relative, not absolute, test for independence. The method is nonetheless very important and very useful.

EXERCISES V

1. Show that each of the following statements is independent of the postulates of incidence by exhibiting appropriate models:

 (a) Any two lines contain the same number of points.
 (b) There are two lines which have different numbers of points.
 (c) Every line has a skew line.
 (d) No line has a skew line.
 (e) There exists a line which has more than two points.
 (f) There exists a line which has precisely two points.
 (g) The Euclidean parallel postulate (Playfair form) holds.
 (h) The Lobachevskian parallel postulate holds.
 (i) One of the three classical parallel postulates holds.

2. Can you prove in an incidence geometry: if there is one line skew to a given line then there is another?

3. Can you prove in an incidence geometry: if two of three noncoplanar lines are parallel to the third then they are parallel to each other?

4. In an incidence geometry, suppose there is a line L and a point a not in L such that there exists a line which contains a and is parallel to L. Can you prove that for any line M and any point b not in M, there is a line which contains b and is parallel to M?

5. In an incidence geometry, suppose there is a line L and a point a not in line L such that every line which contains a meets L. Can you prove there is another line M and point b not in M with the same property?

6. Prove that I6 is independent of I1–I5.

7. Prove that I2 is independent of I1, I3–I6.

10. Existence of Points, Lines, Planes

What can we say about the existence of points, lines, planes? Our postulates tell us that if two distinct points exist, then a line exists—or that if a plane exists, then three distinct points exist. But they do not tell us unconditionally that a point or a line or a plane exists. It is customary to assume an *Existence Postulate* which asserts the existence of four distinct points that are non-collinear and noncoplanar.* We have not included this postulate in the Theory of Incidence since it is unduly restrictive: it would exclude planar models. We have not even assumed the existence of three distinct non-collinear points (or equivalently of a plane) since we have no particular need for it, and wish to keep our list of postulates as short as possible. This means we cannot prove that a plane exists, or a line, or a point. Thus our Theory of Incidence is completely hypothetical as regards existence. If in a theorem we want to assert the existence of a line, or a plane, or of noncoplanar points, the condition must be assumed in the hypothesis of the theorem or deduced from it.

If we can not deduce the theorem that a plane exists, it would seem likely that there exists a model with no planes. For example, a system S that has just four points a, b, c, d; a single line (a, b, c, d); and no plane; which might be described as *one-dimensional*. Is S an incidence geometry? By testing the postulates we see that S satisfies I1 and I2. I3 requires that a plane be a set of points and contain three distinct noncollinear points. Since there are no planes in system S, it is customary to say that S satisfies I3 *vacuously*. This is not unreasonable since in effect I3 imposes a restriction on entities called planes; if there are no such entities in a system, I3 imposes no restriction. To put this in another form, we ask how could a system fail to satisfy I3, and so falsify I3. The answer is that the system must contain a plane which

---

* See D. Hilbert, *Foundations of Geometry*, Open Court, 1902, p. 5, Axioms of Connection, (7).

either (1) is not a set of points, or (2) does not contain three distinct non-collinear points. Since S has no plane at all, it can not falsify I3.

S satisfies I4, I5, I6 vacuously also. For example, I4 involves the hypothesis that three points are noncollinear, which never holds in S.

A more formal explanation of vacuous satisfaction can be based on Bertrand Russell's definition of implication.* According to this the statement, "p implies q," or, "if p then q," is taken to mean, "not p or q." Suppose p false. Then "not p" is true and "not p or q" holds, regardless of the truth or falsity of q. Thus in a sense, when p is false "p implies q" holds "vacuously," that is, without imposing any restriction on q.

In conclusion we note that there are incidence geometries more "degenerate" than system S. For example an incidence geometry with a single point and no lines or planes, or one which is devoid of points, lines, and planes.

Exercise 1.   Show that four distinct noncoplanar points may be collinear. Show that four noncoplanar points (or three noncollinear points) need not be distinct.

Exercise 2.   Show that the following statement is independent of the postulates of incidence: There exist four distinct points that are noncollinear and noncoplanar.

11. Two Types of Incidence Geometry

Two important types of incidence geometry which have received much attention and have elaborate theories of their own are affine geometry and projective geometry. They are characterized as satisfying appropriate parallel postulates.

DEFINITION.   An *affine geometry* is an incidence geometry which satisfies the Euclidean parallel postulate in the Playfair form. A *projective geometry* is an incidence geometry in which there are no parallel lines or equivalently any two coplanar lines meet.

---

* See B. Russell, *Introduction to Mathematical Philosophy*, Allen and Unwin, 1920, p. 147.

Chapter 9 is devoted to affine geometry. We make no formal study of projective geometry despite its importance, since we feel a brief treatment would not do it justice and a lengthy one would divert us from one of our principal objectives, the formalization of Euclidean geometry.

E X E R C I S E S   V I

1. Which of the models M1–M27 are affine geometries and which are projective geometries? (Compare Exercises I, 4 above.)
2. Prove that in a projective geometry there are no parallel planes, and that no line is parallel to a plane. (That is, every line and every plane meet.)
3. Show that M1 is "essentially" the only planar projective geometry with exactly two points in each line; that is, any two such geometries are isomorphic.
4. Show that M3 is "essentially" the only planar affine geometry with exactly two points in each line.
5. Show that M7 is "essentially" the only planar projective geometry with exactly three points in each line.
6. Show that M8 is "essentially" the only planar affine geometry with exactly three points in each line.

E X E R C I S E S   V I I   (Automorphisms)

1. Show that there are exactly six automorphisms of M1, twenty-four of M2, and twenty-four of M3.

DEFINITION. An automorphism holds or leaves a point *fixed* if it associates the point with itself. It holds a line L (or plane P) fixed if each point of L (or P) is associated with some point of L (or P), not necessarily itself.

2. How many automorphisms of M1 hold a point fixed? How many hold exactly one point fixed? Exactly two points? No points?
3. How many automorphisms of M2 hold no points fixed? Exactly one point? Exactly two points?
4. Find an automorphism of M3 that is like a translation in the sense that each point corresponds to a point distinct from itself, and each line to itself or to a parallel line. Are all automorphisms of M3 of this type? If not, how many are there of this type?

5. Find an automorphism of M6 that is like a rotation in the sense that one and only one point is fixed? How many such automorphisms of M6 are there?
6. Find an automorphism of M3 that is like a rotation (as described in Exercise 5), or prove there is none.
7. Find an automorphism of M6 that is like a translation (as described in Exercise 4), or prove there is none.
8. Find an automorphism of M3 that holds some line fixed. How many automorphisms of M3 hold a particular line fixed? How many hold at least one line fixed?
9. Find an automorphism of M6 that holds some line fixed. How many automorphisms of M6 hold a particular line fixed? Is it the same number for each line? How many hold at least one line fixed?
10. Find all automorphisms of M2 that hold plane abc fixed. How many automorphisms of M2 hold at least one plane fixed? How many hold no plane fixed?
11. Show that each of the following is an automorphism of M21:

    (a) $(x, y) \rightarrow (x + 1, y)$;
    (b) $(x, y) \rightarrow (x, y + 1)$;
    (c) $(x, y) \rightarrow (y, x)$;
    (d) $(x, y) \rightarrow (x + 1, y + 1)$;
    (e) $(x, y) \rightarrow (y + 1, x + 1)$.

    Can you find any others?
12. In M7 let p_1, p_2, p_3 be three distinct noncollinear points, and let q_1, q_2, q_3 be three distinct noncollinear points. Show that there is exactly one automorphism of M7 that makes p_1, p_2, p_3 correspond to q_1, q_2, q_3 respectively. Will the conclusion follow if the points are collinear?
13. Prove that M7 has exactly 168 automorphisms. How many of them hold no points fixed? Exactly one point? Exactly two points? Three collinear points? Three noncollinear points? (See Exercise 12.)
14. Given lines L, L' of M4. Show there is an automorphism of M4 which makes L correspond to L'. Do the "diagonal" lines in M4 behave differently from the other lines? Is there an automorphism of M4 that makes a correspond to b and b to a?

E X E R C I S E S   V I I I   (Miscellaneous)

1. Prove that the following system is an incidence geometry:
   Point: 1, 2, 3, 4, 5, 6, 7, 8, 9, 10, 11, 12, 13 (thirteen distinct elements)
   Line: Each column in the table

1	2	3	4	5	6	7	8	9	10	11	12	13
2	3	4	5	6	7	8	9	10	11	12	13	1
4	5	6	7	8	9	10	11	12	13	1	2	3
10	11	12	13	1	2	3	4	5	6	7	8	9

Plane: (1, 2, 3, 4, 5, 6, 7, 8, 9, 10, 11, 12, 13).

2. Verify that the geometry in Exercise 1 is projective.

3. Delete all the points of one line in Exercise 1, say 1, 2, 4, 10, and consider the following system:

Point: Each point that remains—3, 5, 6, 7, 8, 9, 11, 12, 13.

Line: Each "deleted line," namely each column in the table

.	3	.	5	6	7	8	9	.	11	12	13
3	.	5	6	7	8	9	.	11	12	13	.
5	6	7	8	9	.	11	12	13	.	.	3
11	12	13	.	.	3	.	5	6	7	8	9

Plane: (3, 5, 6, 7, 8, 9, 11, 12, 13).

Show that this system is an incidence geometry.

4. Show that the incidence geometry in Exercise 3 is affine.

5. Show that the incidence geometry in Exercise 3 is isomorphic to M8.

6. Show that by deleting the points of a line in M5, the remaining points and "deleted lines" form an incidence geometry (see Exercise 3). Show that although M5 is projective, the "sub-geometry" obtained by deleting the points of a line is affine, and is isomorphic to M3.

7. Show that the following system is an incidence geometry:

Point and plane are the same as in M19. Line is interpreted as "lineal system" which is defined as follows.

Let (x_0, y_0), (x_1, y_1) be distinct fixed dyads. Then the set of all dyads (x, y), which are expressible in the form

(1)
$$x = x_0 + (x_1 - x_0)t$$
$$y = y_0 + (y_1 - y_0)t$$

for some real value of t, is called a *lineal system*. Thus as t ranges over the real number system the dyad (x, y) as determined by Equations (1) ranges over a lineal system.

[*Hint.* To show that two distinct dyads belong to one and only one lineal system, observe that (x_0, y_0) and (x_1, y_1) belong to the lineal system determined by Equations (1). Moreover if (x_0, y_0) and (x_1, y_1) also belong to the lineal system determined by

(2)
$$x = x_2 + (x_3 - x_2)t$$
$$y = y_2 + (y_3 - y_2)t,$$

where (x_2, y_2), (x_3, y_3) are distinct dyads, then the two lineal systems must be identical, that is, a dyad satisfies (1) for some value of t if and only if it satisfies (2) for some value of t.]

8. Show that the following system is an incidence geometry:
   Point and plane are the same as in M20, but line is defined as the set of all ordered triples of real numbers (x, y, z) which are expressible in the form

$$x = x_0 + (x_1 - x_0)t$$
$$y = y_0 + (y_1 - y_0)t$$
$$z = z_0 + (z_1 - z_0)t$$

   for some real value of t, where (x_0, y_0, z_0), (x_1, y_1, z_1) are distinct ordered triples of real numbers.

9. Show that the incidence geometry described in Exercise 7 is the same as M19. (The points and plane are given to be the same. Show that the interpretations of line are the same, that is, the notions of *linear system* and *lineal system* are equivalent; or points colline in one system if and only if they colline in the other.)

†10. Show that the incidence geometry described in Exercise 8 is the same as M20, thus verifying that M20 is an incidence geometry.

11. Show that the following is an incidence geometry: Point, line, and plane are the same as in M19, with the added restriction that each dyad (x, y) satisfies $x^2 + y^2 < 1$. (Compare M9.)

12. Show that each of the following is an automorphism of the incidence geometry of Exercise 11:

   (a) $(x, y) \rightarrow (x, -y)$
   (b) $(x, y) \rightarrow (-x, -y)$
   (c) $(x, y) \rightarrow (y, x)$
   (d) $(x, y) \rightarrow (-y, x)$
   (e) $(x, y) \rightarrow (x', y')$ where $x' = \dfrac{\sqrt{2}}{2}(x - y)$, $y' = \dfrac{\sqrt{2}}{2}(x + y)$.

   Are these also automorphisms of M19 when (x, y) is an arbitrary dyad?

13. Show that each of the following is an automorphism of M19:

   (a) $(x, y) \rightarrow (x + a, y + b)$ where a, b are fixed real numbers.
   (b) $(x, y) \rightarrow (ax, by)$ where a, b are fixed nonzero real numbers.
   (c) $(x, y) \rightarrow (x + y, x - y)$.

   Are these also automorphisms of the incidence geometry of Exercise 11 when (x, y) satisfies $x^2 + y^2 < 1$?

†14. Prove that the following is an automorphism of the incidence geometry of Exercise 11:

$(x, y) \to (x', y')$ where $x' = \dfrac{(a^2 + 1)x - 2a}{2ax - (a^2 + 1)}$, $y' = \dfrac{(a^2 - 1)y}{2ax - (a^2 + 1)}$ and a is a fixed real number.

15. Prove that the following system is an incidence geometry: Point, line, and plane exactly as in M19, except that the words "real number" are replaced throughout by "rational number."

16. Prove that the following system is an incidence geometry: Point, line, and plane are exactly as in M19, except that the words "real number" are replaced throughout by "number of the form $a + b\sqrt{2}$, where a and b are rational numbers."

17. The same as Exercise 15 but replace M19 by M20 or M22.

18. The same as Exercise 16 but replace M19 by M20 or M22.

DEFINITION.   A set F of complex numbers is called a *number field* if for any two numbers a, b in F, the numbers $a + b$, $a - b$, ab also are in F; and a/b is in F, provided $b \neq 0$.

19. Prove that the following system is an incidence geometry: Point, line, and plane are the same as in M19, except that the numbers involved are all chosen from a given number field.

20. In M19, show that the dyads (x_0, y_0), (x_1, y_1), (x_2, y_2) belong to the same linear system if and only if

$$\begin{vmatrix} x_0 & y_0 & 1 \\ x_1 & y_1 & 1 \\ x_2 & y_2 & 1 \end{vmatrix} = 0.$$

21. Let a cartesian coordinate system be set up in a Euclidean plane P. Let A be an arbitrary point of P and (x, y) its coordinates in the given coordinate system. Show that $A \to (x, y)$ is an isomorphism of incidence geometries M15 and M19.

22. Consider the following system suggested by Poincaré's model of Lobachevskian plane geometry:

Point: Each point in the interior of a given Euclidean circle S.

Line: The intersection with the interior of S of a circle or line which is orthogonal* to S at its intersections with S.

Plane: The interior of S.

* A line and a circle are *orthogonal* at a point of intersection A if the line is perpendicular to the tangent line to the circle at A. Similarly, two circles are *orthogonal* at intersection A if the tangents to the circles at A are perpendicular.

Show that this system is an incidence geometry.

[*Hint.*   To verify I2 let A, B be distinct points in the interior of S, let O be the center of S and r the radius of S. Let A' be the point on the ray $\overrightarrow{OA}$ such that the product of the distances $OA \cdot OA'$ is r^2. If A, A', B are noncollinear verify that the circle containing them is orthogonal to S. If A, A', B are collinear the line containing them is a diameter and clearly orthogonal to S. Conversely, show that any circle which contains A and B and is orthogonal to S meets OA in a point A' such that $OA \cdot OA' = r^2$.]

23. Verify that the model of Exercise 22 satisfies the Lobachevskian parallel postulate.

24. Project: Study the theory of inversion transformations with respect to a circle in Euclidean geometry, and use it to learn about automorphisms of the model of Exercise 22.

25. If two incidence geometries are isomorphic, and there are exactly five lines in the first show there are exactly five lines in the second. If two lines in the first geometry are skew, show the lines corresponding to them under the isomorphism in the other geometry also are skew.

26. Find a planar incidence geometry in which each line has a unique parallel line. Is there such a three-dimensional incidence geometry?

27. Can you construct an incidence geometry in which some lines have exactly two points and some more than two?

Theory of Affine Geometries

It is evident from our discussion in Chapter 8, that there are many interesting and important geometric systems. It is doubtful that they can all be encompassed by one theory: some are finite, some infinite; they differ in dimensionality and parallelism properties; in some, the points of a line are ordered in a natural way; some involve metrical ideas like distance. Thus there will be *many* geometrical theories, each characterized by suitable postulates, and each applicable to a set of models which satisfy its postulates. In effect we set up different theories to classify and illumine the concrete models, and to learn—by the processes of deductive reasoning—more about them.

This chapter is devoted to the theory of affine geometries: those incidence geometries which satisfy the Euclidean parallel postulate in Playfair's form (Ch. 2, Sec. 2). The theory is applicable to many of the models of Chapter 8 and in particular to Euclidean geometry, the first and most famous affine geometry.

1. Introduction

An incidence geometry is called an *affine geometry* if it satisfies the following postulate:

POSTULATE E.   *If point a is not in line L there exists one and only one line M such that M contains a and M ∥ L.*

The study of affine geometries involves the familiar *qualitative* properties of parallel lines and planes of Euclidean plane and solid geometry. The methods of proof are more restrictive than those in high school geometry where, for example, parallelism properties may be proved by means of congruence or perpendicularity properties. Such methods are not available here, since we assume only the postulates of an incidence geometry and Postulate E. The theorems of Chapter 7 may of course be used, since they hold in any incidence geometry, whether or not it is an affine geometry. We have tried to unify the treatment of parallelism by introducing the idea of transversality for two lines, a line and a plane, and two planes.

2. The "On" Language for Incidence

We shall often in this chapter be concerned simultaneously with incidence relations involving points, lines, and planes, and it is very helpful to have a brief and convenient notation for such incidence relations. We adopt the convention of Veblen and Young* that any incidence relation (Ch. 7, Sec. 1) involving points, lines, planes be expressed by the word "on." Thus "*a* is on *L*" or "*L* is on *a*" means "point *a* is in line *L*" or "line *L* contains point *a* (as an element)." Similarly for "*a* is on *P*" or "*P* is on *a*." Finally "*L* is on *P*" or "*P* is on *L*" means "line *L* is contained in plane *P* (as a subset)" or "plane *P* contains line *L* (as a subset)."

This form of expression is not a reversion to our somewhat vague use of the term "on" in Part I of the book. Rather it is a technical mathematical use of the word "on," deliberately adopted to facilitate expression and thought. The air of strangeness quickly disappears with use. This phraseology almost seems too simple, since the same word is used to express an incidence relation, say *a* is on *L*, and the reverse relation, *L* is on *a*. But this causes no confusion since our notational agreements (Ch. 7, Sec. 6) tell us that *a* denotes a point and *L* a line. In fact a main advantage of the "on" phraseology is that it permits the automatic reversal of incidence statements.

3. Parallelism of Lines

We begin by showing that Postulate E, the parallel postulate, is valid in a plane. This follows readily from Postulate E.

* See O. Veblen and J. W. Young, *Projective Geometry*, vol. 1, Ginn and Company, Boston, 1910, p. 15.

COROLLARY TO POSTULATE E.   If *a* is not on *L*, *a* is on *P*, and *L* is on *P*, then there exists one and only one line *M* such that *M* is on *a*, *M* ∥ *L* and *M* is on *P*.

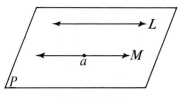

FIGURE 9.1

Proof.   By Postulate E, there exists a unique line *M* such that *M* is on *a* and *M* ∥ *L*. By definition of parallel lines *L* and *M* coplane, say on *Q*. Since *a* is on *M*, *a* is on *Q*. By Theorem 3 of Chapter 7 there is a unique plane on *a*, *L*. By hypothesis *a*, *L* are on *P*. Thus *Q* = *P*, so that *M* is on *P*.

THEOREM 1.   Two distinct lines which are parallel to the same line are parallel to each other.

Restatement.   If *L* ∥ *M*, *N* ∥ *M* and *L* ≠ *N*, then *L* ∥ *N*.

Proof.   Case 1. *L, M, N* coplane.

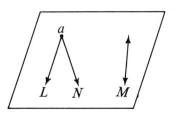

FIGURE 9.2

Suppose *L* meets *N*. Let *a* be on *L* and on *N* (Figure 9.2). Then *a* is not on *M*, since *L* ∥ *M*. By hypothesis *L* ≠ *N*. Thus there exist two distinct lines *L* and *N*, each on *a* and each parallel to *M*, contradicting Postulate E. The supposition *L* meets *N* is therefore false. Since *L, N* coplane, by the definition of parallel lines, *L* ∥ *N*.

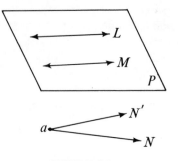

FIGURE 9.3

Case 2.   *L, M, N* do not coplane.

Since $L \parallel M$, by the Corollary to Theorem 3 of Chapter 7 there exists a unique plane *P* on *L, M* (Figure 9.3). Since *L, M, N* do not coplane, there is a point *a* on *N* which is not on *P*. By the Corollary to Theorem 7 of Chapter 7 there exists a unique line N' on *a* such that $N' \parallel L$ and $N' \parallel M$. Since *a* is on *N* and $N \parallel M$, by Postulate E $N' = N$. Thus $L \parallel N$.

COROLLARY.   If $L \parallel M$ and $M \parallel N$, then $L = N$ or $L \parallel N$.

The conclusion of the corollary suggests that parallelism and coincidence may be related notions. This is not very farfetched for when we say two lines have the same direction, they are either parallel or coincident. A familiar principle of (Euclidean) analytic geometry asserts that if the slope of line *L* is the same as the slope of line *M* then $L \parallel M$ or $L = M$. These considerations suggest the following definition.

DEFINITION.   If *L, M* have the property that $L \parallel M$ or $L = M$ we say *L, M* have the *same direction*, or are *codirectional*, or *L* is *codirectional* to *M*, written *L* cod *M*.

Remark.   Codirectionality of lines may be considered a generalization of parallelism since, in addition to parallelism, it covers coincidence which is a sort of "degenerate" case of parallelism. In certain situations it is more convenient to study codirectionality than parallelism since its formal properties are more regular. In particular codirectionality of lines is an *equivalence relation*. That is, for any lines *L, M, N* the following statements hold:

(i) *L* cod *L*.

(ii) If *L* cod *M*, then *M* cod *L*.

(iii) If *L* cod *M* and *M* cod *N*, then *L* cod *N*.

Observe that (i) and (iii) do not hold for the relation parallelism of lines, and that (iii) is a bit neater than the corollary above. Postulate E is also somewhat simplified if we replace parallelism by codirectionality. It may be stated: *Given a and L, there is a unique line M such that M is on a and M cod L.* (We do not need to assume that *a* is not on *L*.)

Since equivalence relations often occur in geometry we introduce the notion formally.

DEFINITION.   Let S be a set and R a relation involving two elements of S. We write aRb to denote that relation R holds for the elements a, b of S in the stated order. Then R is an *equivalence relation in* S if the following properties hold, where a, b, c denote arbitrary elements of S.
   (i) (Reflexiveness) aRa.
   (ii) (Symmetry) If aRb then bRa.
   (iii) (Transitiveness) If aRb and bRc then aRc.
For example, congruence or similarity is an equivalence relation in the set of all triangles.

4. Transversality of Lines

If lines L, M are coplanar they must satisfy one of three possible relations: (1) $L \parallel M$ or (2) $L = M$ or (3) $L \not\parallel M$ and $L \neq M$. In the third case L and M meet and are distinct. This is an important relation between two lines, is very useful in the study of parallelism, and also deserves a name. Thus we introduce the following definition.

DEFINITION.   We say L is *transverse* to M, or L is a *transversal* of M, or L and M are *transverse*, written L **tr** M, if L intersects M and $L \neq M$.

Then it is easy to see that the following statements hold.

*L **tr** M if and only if the intersection of L and M is a point.*
*For any coplanar lines L, M either (i) $L \parallel M$, (ii) L **tr** M, or (iii) $L = M$.*

The basic connection between parallelism and transversality is given in

THEOREM 2.   In a plane a line transverse to one of two parallel lines is transverse to the other.

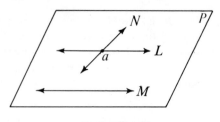

FIGURE 9.4

Restatement.   If L, M, N are on P, $L \parallel M$, and N **tr** L, then N **tr** M.

Proof.   Let a be on L, N and suppose N is not transverse to M. Then either $N = M$ or $N \parallel M$. $N \neq M$, otherwise a is common to two parallel lines. N is not parallel to M, otherwise there are two distinct lines, L and N, each of which is on a, and each of which is parallel to M, contradicting Postulate E. Our supposition is therefore false, and N **tr** M.

5. Transversality of Lines and Planes

We extend the notions of parallelism and transversality to a line and a plane, and use the latter to derive parallelism properties of two lines, two planes, and a line and a plane.

DEFINITION.   If line L and plane P have no common point, we say L is *parallel* to P or P is *parallel* to L, and write $L \parallel P$ or $P \parallel L$. We say L is *transverse* to P, or P is *transverse* to L, written L **tr** P, or P **tr** L, if the intersection of L and P is a point.

It is immediate that for any L, P either (i) $L \parallel P$; (ii) L **tr** P, or (iii) L is on P.

THEOREM 3.   A plane transverse to one of two parallel lines is transverse to the other.

Proof.   Let $L \parallel M$, P **tr** L. We show P **tr** M. Let a be the point of intersection of L and P, and let Q be the plane containing L, M. Then $Q \neq P$; otherwise L is on Q implies L is on P, contradicting hypothesis P **tr** L. Also, a is on L and L is on Q implies a is on Q. Thus a is common to P

and Q; hence P intersects Q. The intersection of P and Q is thus a line N, which is on a. $N \neq L$, otherwise L is on P, contradicting the hypothesis. Thus N **tr** L by definition. Since L, M, N are on Q, by Theorem 2 N **tr** M. Let b be the point of intersection of N and M. Then b is on P, and b is common to P and M.

Suppose there is another point c common to P and M. By Postulate 15, M is on P. Since M is also on Q, it follows that $M = N$. This contradicts N **tr** M. The supposition is therefore false, and b is the only point common to M and P. By definition P **tr** M.

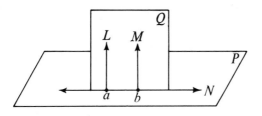

FIGURE 9.5

COROLLARY 1. If $L \parallel M$ and P is not transverse to L, then P is not transverse to M.

Proof. Assume P **tr** M. Then by the theorem P **tr** L, contradicting the hypothesis.

COROLLARY 2. If $L \parallel M$ and $P \parallel L$, then $P \parallel M$ or P is on M.

Proof. $P \parallel L$ implies P is not transverse to L. By Corollary 1, P is not transverse to M. The only possibilities are therefore $P \parallel M$ or P is on M.

COROLLARY 3. If $L \parallel M$ and P is on L, then P is on M or $P \parallel M$.

Proof. Since P is on L, P is not transverse to L. By Corollary 1, P is not transverse to M. Thus P is on M or $P \parallel M$.

COROLLARY 4. (RESTATEMENT OF COROLLARY 3) If a plane is on one of two parallel lines and not on the other, it is parallel to the other. Equivalently, if a line is parallel to a line which is on a plane, and is not itself on the plane, then it is parallel to the plane.

THEOREM 4.   A line transverse to one of two parallel planes is transverse to the other.

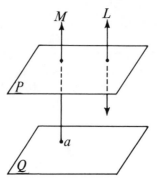

FIGURE 9.6

Proof. Let $P \parallel Q$, L **tr** P. Let a be a point on Q which is not on L. By Postulate E there exists a unique line M on a such that $M \parallel L$. By Theorem 3, M **tr** P. M is not parallel to Q, since M intersects Q in a. M is not on Q, since M intersects P and $P \parallel Q$. Thus M **tr** Q. Since $M \parallel L$, we infer L **tr** Q by Theorem 3.

Alternate Proof.   We sketch a proof which is longer but independent of Theorem 3. Using notation above let $R = La$. Then R intersects P, Q in lines N, N' and $N \parallel N'$. Show L **tr** N. Infer L **tr** N'. Show L **tr** Q.

COROLLARY 1.   If $P \parallel Q$ and L is not transverse to P, then L is not transverse to Q.

Proof.   Suppose L **tr** Q. Then by the theorem, L **tr** P, contradicting hypothesis.

COROLLARY 2.   If $P \parallel Q$ and $L \parallel P$, then $L \parallel Q$ or L is on Q.

Proof. $L \parallel P$ implies L is not transverse to P. Thus by Corollary 1, L is not transverse to Q. The only remaining possibilities are $L \parallel Q$ or L is on Q.

Observe that Theorem 4 and its corollaries are thus far analogous to Theorem 3 and its corollaries. If the pattern were to continue, we would have here a third corollary of this form: If $P \parallel Q$ and L is on P, then $L \parallel Q$. But this is clear by definition, and requires no further proof.

6. Transversality of Planes

We define transversality of planes in a natural way and relate it to parallelism of planes.

DEFINITION. P is *transverse* to Q, or P and Q are *transverse*, written P **tr** Q, if the intersection of P and Q is a line.

It is immediate that for any planes P, Q either (i) $P \parallel Q$; (ii) P **tr** Q; or (iii) $P = Q$.

THEOREM 5.   A plane transverse to one of two parallel planes is transverse to the other.

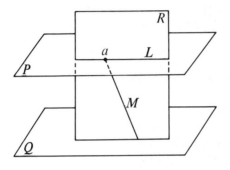

FIGURE 9.7

Proof.   Let $P \parallel Q$, R **tr** P. Let L be the intersection of R and P, and let a be on L. By Corollary 2 to Theorem 9 of Chapter 7 there is on R a line M such that M is on a and $M \neq L$. Thus M is not parallel to P. Moreover M is not on P, since the only line common to R and P is L. Hence M **tr** P. Thus by Theorem 4, M **tr** Q, so that M intersects Q. Thus R intersects Q, and R is not parallel to Q. Clearly $R \neq Q$, since R meets P and $P \parallel Q$. Hence the only remaining possibility is R **tr** Q.

COROLLARY 1.   If $P \parallel Q$ and R is not transverse to P, then R is not transverse to Q.

COROLLARY 2.   If $P \parallel Q$ and $Q \parallel R$, then $P = R$ or $P \parallel R$.

Proof. $P \parallel Q$ implies P is not transverse to Q. By the preceding corollary, P is not transverse to R. The only possibilities are $P = R$ or $P \parallel R$.

COROLLARY 3.   Two transverse planes are not both parallel to the same plane.

Proof. Let $R \text{ tr } P$, and suppose both $P \parallel Q$ and $R \parallel Q$. By the theorem, $P \parallel Q$ and $R \text{ tr } P$ implies $R \text{ tr } Q$, which contradicts the supposition.

Remark. In view of Corollary 2 above, it would be natural to define *codirectionality* for planes as we did for lines in Section 3:

DEFINITION.   If P, Q have the property that $P \parallel Q$ or $P = Q$, we say P and Q are *codirectional*, written $P \text{ cod } Q$.

The relation of codirectionality for planes is an equivalence relation, as it was for lines, and may be considered a generalization of parallelism of planes.

7. Conditions for Parallelism of Planes

We have not yet proved the existence of parallel planes. To do so we need a condition sufficient to ensure parallelism of planes. It is easy to see that if two planes are parallel, each line on the first is parallel to the second plane. This suggests a converse question: How many lines on one plane must be parallel to a second plane to guarantee the parallelism of the two planes? An answer is given in

THEOREM 6.   If two distinct intersecting lines on P are parallel to Q, then P is parallel to Q.

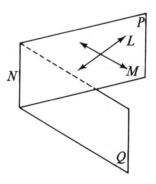

FIGURE 9.8

Proof. Let L, M be distinct intersecting lines on P which are parallel to Q. Then $P \neq Q$. Suppose P **tr** Q. Let N be the intersection of P and Q. Then L and N have no common point, since N is on Q and $L \parallel Q$. Since L, N are on P, $L \parallel N$. Similarly $M \parallel N$. This contradicts Postulate E. The supposition P **tr** Q is therefore false. The remaining possibility, $P \parallel Q$, must therefore hold.

The preceding theorem reduces the problem of determining parallel planes to that of determining lines parallel to a plane. But the latter was settled in Corollary 4 to Theorem 3. This suggests

THEOREM 7. Let a be on P and not on Q. Let two distinct lines on P which are on a be parallel to lines on Q. Then $P \parallel Q$.

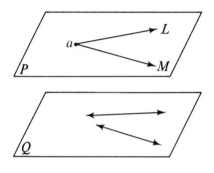

FIGURE 9.9

Proof. Let the lines on P be L and M. Then L is parallel to a line on Q. But L is not on Q. Hence by Corollary 4 to Theorem 3, $L \parallel Q$. Similarly $M \parallel Q$. Thus by Theorem 6, $P \parallel Q$.

We conclude with an analogue of Postulate E for planes.

THEOREM 8. If a is not on Q, there exists one and only one plane which is on a and parallel to Q.

Proof. (Existence) Let L', M' be distinct intersecting lines on Q. Since a is not on Q, a is not on L' or on M'. Thus by Postulate E there is a line L on a and a line M on a such that $L \parallel L'$, $M \parallel M'$. If $L = M$, then $L' \parallel L$, $M' \parallel L$, contrary to Postulate E. Thus $L \neq M$. By Theorem 4 of Chapter 7 there is a plane P on L and M. By Theorem 7, $P \parallel Q$.

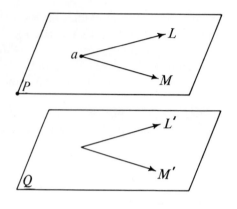

FIGURE 9.10

(Uniqueness)   Let P' be on a, $P' \parallel Q$, and suppose $P' \neq P$. Then P' **tr** P. Since this contradicts Corollary 3 to Theorem 5, the supposition is false and $P' = P$, proving uniqueness.

E X E R C I S E S   I

1. In Theorem 6, could we discard the term "intersecting"? Would it be sufficient in Theorem 6 to require that just one line of P be parallel to Q?
2. If $L \parallel P$, is L parallel to each line on P? Explain.
3. What conclusions can be drawn about planes which are parallel to the same line? Justify your answer.
4. What conclusions can be drawn about lines which are parallel to the same plane? Justify your answer.
5. If a is not on P, prove that all lines on a which are parallel to P are on a plane which is parallel to P.
6. If a line is parallel to a plane, prove that any line parallel to the line, which intersects the plane, is on the plane.
7. If $L \parallel P$, Q is on L, and Q intersects P, prove L is parallel to the intersection of P and Q.
8. If P **tr** Q, $L \parallel P$ and $L \parallel Q$, prove L is parallel to the intersection of P and Q.
9. Prove: In a plane, if $L \neq M$, and every line transverse to L is transverse to M, then $L \parallel M$.
10. Prove in a three-dimensional geometry: If $L \neq M$ and every plane transverse to L is transverse to M, then $L \parallel M$. [*Hint.* Suppose L is not

parallel to M, and consider the cases L, M do and L, M do not coplane.]

11. Prove: If $P \neq Q$ and every line transverse to P is transverse to Q, then $P \parallel Q$.

12. Prove: If $P \neq Q$ and every plane transverse to P is transverse to Q, then $P \parallel Q$.

13. If a line is transverse to a plane, prove that any plane on the line is transverse to the plane.

14. If a line is parallel to a plane, prove that any plane transverse to the line is transverse to the plane.

15. If L **tr** P and M **tr** L, does it follow that M **tr** P? Explain.

16. Prove: P **tr** Q if and only if P is transverse to a line of Q. (See Exercise 13.)

17. If L is skew to M, and M is skew to N, is L necessarily skew to N? What are the possible relations between L and N?

18. Prove: If L, M are skew, there exists a unique plane on L which is parallel to M.

19. Prove: Any two skew lines are contained in a unique pair of parallel planes.

20. Prove: If L, M are skew and a is any point, there is a unique plane on a which is "codirectional" to L and M. (A plane is *codirectional* to a line if it is parallel to the line or contains it.)

21. Let L, M be skew; let P be the unique plane on L which is parallel to M, and Q the unique plane on M which is parallel to L. (See Exercise 18.) Let a be any point not on P or on Q. Prove there is a unique line on a which is transverse to both L, M.

E X E R C I S E S   I I

1. Which of the models M1–M23 are affine geometries?

2. Prove: Any plane of a three-dimensional affine geometry "forms" an affine plane geometry.

3. If a geometry is affine, can it contain a "subgeometry" which is projective? (A subgeometry of an incidence geometry G is a system of points, lines, and planes, each point of which is a point of G and each line (plane) of which is a subset of a line (plane) of G.) (See Ch. 8, Exercises VIII, 6.)

4. Prove that M19 is affine.

5. Prove that we obtain an affine geometry if (as in Ch. 8, Exercises VIII, 15, 16, 19) we replace "real numbers" in M19 by
   (i) rational numbers;

(ii) complex numbers;

(iii) numbers chosen from a given number field.

6. If two planes are transverse, prove there is a line on one that is not parallel to any line on the other. Can you prove that there are three such lines? Five?

7. Prove: Any two parallel planes contain a pair of skew lines. Show that at least twenty-four pairs of skew lines are contained in any two parallel planes. How many can you prove there are in a pair of transverse planes?

8. Can we deduce from the postulates of affine geometry the theorem that a line contains more than two points?

9. Can we deduce that in a three-dimensional incidence geometry there are three mutually skew lines? Can we deduce this in a three-dimensional affine geometry?

10. Prove that $(x, y) \to (x', y')$ where $x' = ax + by + c$, $y' = dx + ey + f$, $\begin{vmatrix} a & b \\ d & e \end{vmatrix} \neq 0$, is an automorphism of M19.

11. Prove that M19 has an automorphism of the type described in Exercise 10, which "maps" any three given noncollinear points onto $(0, 0)$, $(1, 0)$, $(0, 1)$.

†12. Prove that the only automorphism of M19 which maps each of $(0, 1)$, $(1, 0)$, $(0, 0)$ on itself is the identity automorphism $(x, y) \to (x, y)$. [*Hint.* If the automorphism is $x' = f(x, y)$, $y' = g(x, y)$, show that the lines $x = 0$, $y = 0$, $x = y$, and $x + y = 1$ are fixed; use the fact that the "images" or correspondents of parallel lines are parallel, so that lines which are parallel to a fixed line map onto lines that are parallel to that line. Then show $f(x, y) = f(x, 0) = F(x)$; $g(x, y) = g(0, y) = G(y)$; $f(x, x) = g(x, x)$, so that $F(x) = G(x)$. Then $F(x + y) = F(x) + F(y)$, $F(xy) = F(x)F(y)$, and finally (difficult) that $F(x) = x$.]

13. Assume that each line of a geometry (affine) contains exactly n points (n a positive integer) and prove:

  (i) Each point on a plane is on exactly $n + 1$ lines of the plane.

  (ii) Each plane contains exactly n^2 points.

  (iii) Each plane contains exactly $n^2 + n$ lines.

†14. Assume that each line of a three-dimensional (affine) geometry contains exactly n points and prove:

  (i) There are exactly n^3 points.

  (ii) Each point is on exactly $n^2 + n + 1$ lines.

  (iii) There are exactly $n^2(n^2 + n + 1)$ lines.

  (iv) There are exactly $n^3 + n^2 + n$ planes.

15. Prove that any three dimensional (affine) geometry with exactly two points on each line is isomorphic to M4.

E X E R C I S E S   III   (Parallel Congruence)

In the following exercises assume that the (affine) geometry is three-dimensional.

1. Let aa', bb', cc' be parallel in pairs and not all coplanar; suppose $ab \parallel a'b'$, $bc \parallel b'c'$. Prove $ac \parallel a'c'$. (Compare Ch. 7, Exercises III, 5.)

2. Prove: If c is not on ab, thereexists a point d such that $ab \parallel cd$ and $ac \parallel bd$.

3. Prove: (Form of Desargues Theorem) If aa', bb', cc' are parallel in pairs, and $ab \parallel a'b'$, $bc \parallel b'c'$, then $ac \parallel a'c'$.

DEFINITION.   We say a, b is *parallel congruent* to c, d (written $a, b \equiv c, d$) provided one of the following conditions holds:

(1) if $ab \neq cd$ then $ab \parallel cd$ and $ac \parallel bd$;

(2) if $ab = cd$ then there exist p, q such that a, b and c, d bear the relation (1) to p, q; that is $ab \parallel pq$, $ap \parallel bq$ and $cd \parallel pq$, $cp \parallel dq$.

NOTE: If $a, b \equiv c, d$ then $a \neq b$, $c \neq d$. We shall consider $a, b \equiv c, d$ to be a relation between two ordered pairs of distinct points, denoted simply a, b and c, d.

4. Prove: If $a \neq b$ then $a, b \equiv a, b$.

5. Prove: If $a, b \equiv c, d$ then $c, d \equiv a, b$.

6. Prove: If $a, b \equiv c, d$ then $b, a \equiv d, c$.

7. Prove that from $a, b \equiv c, d$ we can *not* deduce $a, b \neq d, c$. [*Hint*.   Find a suitable model.]

8. Prove: If c is on ab and $a, b \equiv c, d$ then d is on ab.

9. Prove: If $a, b \equiv a', b'$ and $b, c \equiv b', c'$ and a, b, c colline, then a', b', c' colline.

10. Prove: If points a, b, c are given with $a \neq b$, then there exists a point d such that $a, b \equiv c, d$. (Compare Exercise 2.)

11. Prove: If $a, b \equiv c, d$ and $c, d \equiv e, f$ then $a, b \equiv e, f$. (Note there are several cases.)

NOTE: Exercises 4, 5, and 11 show that parallel congruence is an equivalence relation in the set of all ordered pairs of distinct points.

12. Prove: If points a, b, c are given with $a \neq b$, then there exists a unique point d such that $a, b \equiv c, d$. (Compare Exercise 10.)

13. Prove: If $a, b \equiv a', b'$ and $b, c \equiv b', c'$ then $a, c \equiv a', c'$, provided $a \neq c$. [*Hint.* Apply Exercise 3.]

14. Prove: If $a, b \equiv c, d$ and $a \neq c$, then $a, c \equiv b, d$. [*Hint.* First consider case $ab \neq cd$. If $ab = cd$ use first case and Exercise 13.]

15. Prove: If $a, b \equiv b', a'$ and $b, c \equiv c', b'$ then $a, c \equiv c', a'$, provided $a \neq c$. [*Hint.* Let $c, d \equiv b', a'$.]

DEFINITION.   If $a, m \equiv m, b$ and $a \neq b$ we say m is a *midpoint* of the ordered pair a, b.

16. Prove: If m is a midpoint of a, b then
    (i) m is a midpoint of b, a;
    (ii) m, a, b colline.

17. Prove: If a, d and b, c have a common midpoint then $a, b \equiv c, d$, provided $a \neq b$.

18. Suppose $a, b \equiv c, d$ and ad meets bc just in point m. Prove that m is a common midpoint of a, d and b, c. [*Hint.* Use the indirect method.]

NOTE: The last two exercises say, in a sense, that if the diagonals of a quadrilateral bisect each other the quadrilateral is a parallelogram; and that if the diagonals of a parallelogram intersect, they bisect each other.

Query.   In an affine geometry, must the diagonals of a parallelogram intersect?

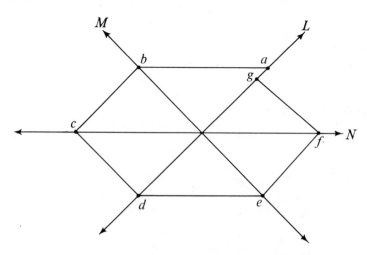

FIGURE 9.11

19. (Hexagonal Closure) Let L, M, N be distinct, coplanar, and concurrent (Figure 9.11). Let a be on L and b on M such that $ab \parallel N$. Let c be on N such that $bc \parallel L$, d on L such that $cd \parallel M$; e on M such that $de \parallel N$; f on N such that $ef \parallel L$; and g on L such that $fg \parallel M$. Prove $g = a$.

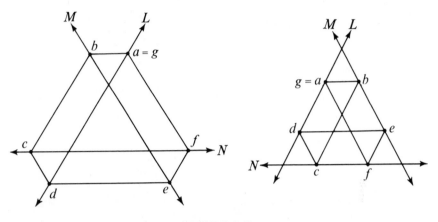

FIGURE 9.12

20. (Hexagonal Closure) The same as Exercise 19, except that L, M, N intersect in pairs, and are nonconcurrent (Figure 9.12).

CHAPTER 10

Theory of Order on the Line

We formalized the basic theory of incidence in Chapter 7, and we now come to grips with the problem of putting the theory of order in geometry on a firm foundation. The need to do this became apparent in Chapter 1, when we saw that notions of order were not formalized in the conventional treatment of Euclidean geometry but just assumed implicitly from diagrams. This chapter is devoted to the study of the order relations of points of a line.

1. The Concept of Order

Order is one of the most basic and pervasive of mathematical ideas. We encounter it in algebraic form when we learn to count, in geometric form when we observe that an object is *to the left of* another, or is *between* two others, or is on the *opposite side* of a path from another. As we saw in Chapters 7, 8, 9, geometric theories of incidence can be developed which are interesting and reasonably complex, but geometry is measurably enriched by the introduction of the order concept. It is obviously needed for the study of the relative position of points on a line, but is essential also for the definition and study of many important "nonlinear" ideas. Without it (or an equivalent) we could not clarify ideas of direction, separation, and interiority —without it we could not even define triangle.

How can order be introduced formally into our theory of incidence? There are only two ways of studying a concept in a mathematical theory: (1) to define it in terms of other basic or primitive notions; (2) to take *it* as a basic notion and characterize it by suitably chosen postulates. It seems very difficult if not impossible to define the idea of order in terms of point, line, plane, the basic notions of the theory of incidence—so we shall adopt the second procedure.

There are two well-known theories of order called the theory of *precedence* and the the theory of *betweenness*. In the first, the elements of a set are "ordered" by specifying a two-term (or binary) relation called *precedence*, for example, "to the left of" in the set of points of a line, or "greater than" for the set of rational numbers. In the second theory a three-term (or ternary) relation called *betweenness* is specified in a set, for example, "betweenness for points of a line" or "betweenness for real numbers." Of course in each theory suitable postulates are assumed.

Formally stated, the theory of precedence involves a two-term relation $a < b$ (read *a precedes b*, not *a is less than b*) and a basic set S of elements $a, b, c, \ldots$, which satisfy the following postulates:

P1. $a < a$ *is always false.*

P2. $a < b, b < c$ *imply* $a < c$.

P3. *If* a, b *are distinct then one of the relations* $a < b, b < a$ *holds.*

In algebra, theories of order are invariably based on precedence rather than betweenness because of its simplicity. In geometry, it seems more natural to take betweenness as fundamental. For there is no unique precedence relation for points on a given line; it is just as natural to order them by the relation "to the left of" as by the inverse relation "to the right of." In fact there is no intrinsic geometric method for distinguishing between these relations. (Lines do not come with little built-in arrows pointing left or right.) Further since there are many lines in a geometry, it would be necessary to choose precedence relations on all of them, and there would be no natural way to tie these relations together. The basing of the theory of order in geometry on betweenness avoids these difficulties and seems intuitively natural in any case.

2. Postulates for Betweenness

There are many systems of postulates for betweenness—ours is chosen to be reasonably simple, not too hard to remember, and to facilitate generalization to the study of order in the plane and space. We consider a general incidence geometry satisfying I1–I6, and introduce an additional basic notion, "between," indicated by the symbol (abc) which is read: points a, b, c are in the *order abc* or *b is between a* and *c*. (Postulate E of Chapter 9 is not assumed.) We assume that the relation "between" satisfies the following postulates:

B1. (SYMMETRY PROPERTY)   (*abc*) *implies* (*cba*).

B2. (ANTICYCLIC PROPERTY)   (*abc*) *implies the falsity of* (*bca*).

B3. (LINEAR COHERENCE)   *a, b, c are distinct and collinear if and only if* (*abc*), (*bca*) *or* (*cab*).

B4. (SEPARATION PROPERTY)   *Let p colline with, and be distinct from a, b, c. Then* (*apb*) *implies* (*bpc*) *or* (*apc*) *but not both.*

B5. (EXISTENCE)   *If* $a \neq b$ *there exist x, y, z such that* (*xab*), (*ayb*), (*abz*).

These postulates merit a few remarks. Note that they are suggested by familiar diagrams of school geometry and are easily verified pictorially. Observe that B1 is a simple symmetry property, which asserts that we can symmetrically permute the elements in the relation (*abc*) without destroying its validity. B2 asserts that we do destroy the validity of (*abc*) if we apply the cyclic permutation which replaces *a, b, c* by *b, c, a*.

Postulate B3 relates, in a natural way, the basic idea *between* to the basic ideas *point* and *line* of the theory of incidence. Without some such property we would have two separate, noninteracting theories—one for incidence, one for betweenness. B3 is easily remembered since the order relations involved are cyclic permutations of (*abc*).

B3 has two simple and useful consequences:

B3.1.   (*abc*) *implies a, b, c are distinct and collinear.*

B3.2.   *If a, b, c are distinct and collinear then* (*abc*), (*bca*) *or* (*cab*).

Actually B3 is equivalent to B3.1 and B3.2 together and is a convenient formulation of these two properties.

B4 is a linear or one-dimensional form of Pasch's Postulate (Ch. 11, Sec. 1) which was formulated originally as a property of triangles. It may be considered a kind of weak *separation postulate*. To see this, read (*abc*) as *b separates a* from *c*. Then the conclusion of B4 says: *If p separates a from b it must separate a or b from c, but not both.* Intuitively speaking, *c* must be on the side of *p* opposite to *a* or to *b*, but not both. Note that in B4 no assumption is made about the distinctness of *a, b, c;* and that it is valid, for example, if $b = c$.

B5 is introduced to ensure the existence of points which arise in our discussions. It prevents the theory from becoming existentially trivial.

3. Elementary Properties of Betweenness

In this brief section we study the order properties of three points.

Among the most important of these are B3.1 and B3.2 of Section 2. From B3.1 (in view of the incidence postulates) we easily derive the following principles:

(i) (*abc*) *implies* $ab = bc = ac$.

(ii) (*abc*) *implies that ab contains c, bc contains a, ac contains b*.

Given a betweenness relation, say (*abc*), we naturally ask which betweenness relations follow from it. B1 and B2 give a partial answer. The complete answer is found in

THEOREM 1.   (*abc*) implies (*cba*), and (*abc*) implies the falsity of (*bca*), (*bac*), (*acb*), and (*cab*).

Proof.   (*abc*) implies (*cba*) by B1. (*abc*), (*cba*) respectively imply the falsity of (*bca*), (*bac*) by B2. Suppose (*acb*); then by B1 (*bca*), which is false. Hence (*acb*) must be false. A similar argument shows (*cab*) false.

COROLLARY.   (*abc*) if and only if (*cba*). That is (*abc*) and (*cba*) are equivalent.

This completes, in essence, the theory of order for three points. Later (Sec. 15 below) we discuss order theory for four points. We continue now by using the theory of order to define and study segments and rays.

4. Segments

The simplest and most important geometric figure, after line, would seem to be segment, which is easily defined in terms of order:

DEFINITION.   If $a \neq b$, the set of all points x such that (*axb*) is called *segment ab*, denoted $\overline{ab}$. a and b are called *endpoints* of segment ab, which is said to *join a and b*.

Note that a segment as defined is simply a set of points. We cannot yet measure segments or even compare them as smaller or larger.

The most elementary properties of segments appear in

THEOREM 2.   If $a \neq b$ then

(i) $\overline{ab} = \overline{ba}$;

(ii) $\overline{ab}$ is a subset of ab;

(iii) a, b are not elements of $\overline{ab}$;

(iv) $\overline{ab}$ is a nonempty set.

Proof. (i) asserts that $\overline{ab}$ and $\overline{ba}$ are identical sets. This means they comprise the same elements. Precisely stated, each element of $\overline{ab}$ is an element of $\overline{ba}$, and conversely each element of $\overline{ba}$ is an element of $\overline{ab}$. In symbols we must show that if x is in $\overline{ab}$ it is in $\overline{ba}$ and conversely. By definition of $\overline{ab}$, x is in $\overline{ab}$ only if (axb). Similarly x is in $\overline{ba}$ only if (bxa). Thus we must show that (axb) implies (bxa) and conversely. That is (axb) and (bxa) are equivalent. But this holds by the Corollary to Theorem 1. Thus we conclude $\overline{ab} = \overline{ba}$.

(ii) We must show that each element of $\overline{ab}$ is an element of ab. That is, x is in $\overline{ab}$ implies x is in ab. Now x is in $\overline{ab}$ means (axb). This implies, as we observed in Section 3, that x is in ab, and the proof is complete.

(iii) Suppose a is an element of $\overline{ab}$. Then by definition of $\overline{ab}$ we have (aab), contradicting B3.1. Thus a is not an element of $\overline{ab}$. Similarly for b.

(iv) This means that $\overline{ab}$ has at least one element. Since $a \neq b$ there exists a point x such that (axb) by B5. By definition x is in $\overline{ab}$, and the proof is complete.

Remark.   Property (iii) asserts that a segment as we defined it, does not contain its endpoints. Your study of school geometry may have suggested that a segment *does* contain its endpoints. There is no contradiction: we have here two related concepts that may be called *open segment* and *closed segment*. We find it more convenient to study open segments, since it seems easier to convert an open into a closed segment by adjoining endpoints, than the reverse by deleting endpoints. In general, "open" geometric figures are in a sense easier to study than "closed" ones since they are more regular; none of their points are boundary points.

5. Rays or Half-Lines

The notion ray (or half-line) appears implicitly in Euclid (and in school geometry) in the form of a side of an angle. It may be described intuitively as the path traced by a point which starts from a given point and moves endlessly in a given direction.* If the starting point is a, and b is a point in the

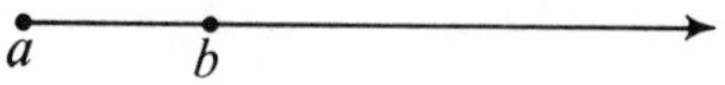

FIGURE 10.1

---

* The term direction is used ambiguously in mathematics. In an "ordered" geometry it refers to the two senses in which a line can be traversed. In an affine geometry two lines have the same direction if they are parallel or coincident (Ch. 9, Sec. 3).

given direction from a, then the ray will consist of all points between a and b, together with b, and all points "beyond" b "relative to" a (Fig. 10.1).

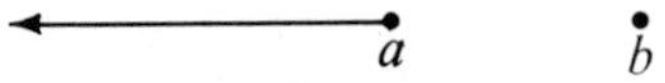

FIGURE 10.2

There is another mode of construction. To describe it let a still be the starting point, but suppose the given direction is opposite to the preceding one, that is, it is from a directly opposite to point b (Fig. 10.2). Then the ray will consist of all points "beyond" a "relative to" b. Both modes of construction (or definition) are important and are needed in the development. The first seems intuitively more natural, but involves three separate components; the second, involving only one is logically simpler. So we base the definition of ray on the second construction—as we shall see, the first falls nicely into place.

DEFINITION.   If $a \neq b$, the set of all points x such that (xab) is called a *ray* or *half-line* and is denoted a/b, read "a over b." Sometimes a/b is called the *extension* or *prolongation* of $\overline{ab}$ beyond a. Point a is said to be an *endpoint* of ray a/b.

Remark.   Note that ray is defined solely in terms of point and betweenness. The intuitive description in terms of "direction" yielded, on analysis, a formal definition in which "direction" does not appear. However the intuitive idea of direction remains part of the substructure of our geometric knowledge: it helps us to understand and assimilate the properties of rays and might even suggest new ones.

Our precise definition of ray was motivated by the intuitive idea of direction. It is interesting that, having formalized the concept of ray, we can use it to make direction precise. For example we might define "b and c are *in the same direction from a*" to mean "b and c belong to the same ray with endpoint a." The more difficult idea that the direction (or sense) from a to b is the same as that from c to d (where a, b, c, d are collinear points) can also be defined in terms of the ray concept. (See Exercises III at the end of the chapter.)

We have for rays a partial analogue of Theorem 2 for segments.

THEOREM 3.   If $a \neq b$ then
  (i) a/b, b/a are subsets of ab;

(ii) a, b are not elements of a/b;

(iii) a/b is a nonempty set.

Proof. (i) We prove a/b is a subset of ab by showing that every element of a/b is an element of ab. Let x be an element of a/b. By definition of a/b we have (xab). This implies, as we saw in Section 3, x is in ab. Thus a/b is a subset of ab. Similarly for b/a.

(ii) Proceed as in Theorem 2, assume a (or b) is in a/b and get a contradiction.

(iii) Apply B5 as in Theorem 2.

Note, in view of (ii), that a ray like a segment is an "open" figure—it does not contain its endpoint.

This completes our introduction to the theory of order on a line. After having derived the elementary properties of betweenness, we introduced the basic "linear" figures, segment and ray, and studied their simplest properties. To facilitate a deeper study of the theory of order we digress to present the elements of the theory of sets.

6. Concepts and Terminology of Set Theory

The study of geometry is different from that of algebra since it effectively involves the concept of set from the very beginning. We postulate lines and planes to be *sets* of points; we study *intersection* properties of lines and planes; we define segments and rays to be certain *sets* of points and prove that they are nonempty *subsets* of lines. As we continue we become more and more involved with set relations and set operations, and it is essential to formalize them and adopt a convenient notation for them.

A set is a collection, totality, aggregate, or class of objects, which are called its elements. If a set is finite it can be defined by listing its elements, for example the set consisting of the endpoints and the midpoint of a given segment, or the set consisting of the numbers 1, 3, 997, and 13. In general a set is defined by specifying a "defining property," that is, a property which is to hold for each element of the set and for no other element. Thus we defined segment ab, when $a \neq b$, by the defining property (axb); that is point x is to be an element of set $\overline{ab}$ if and only if it is between a and b. In school geometry any locus problem involves such a definition, for example, the set of points each of which is equidistant from two given points.

Sometimes a property is specified which is satisfied by no element, for example, the set of Euclidean triangles which have two obtuse angles. In this case we say the property defines the *null* or *empty* set, characterized as having no elements. The null set is denoted $\varnothing$.

If set A is a subset of set B we write $A \subset B$ or $B \supset A$. Formally phrased, $A \subset B$ means that each element of A is an element of B. This does not exclude $A = B$. Clearly the *Reflexive Property* $A \subset A$ holds for every set A. (Compare divisibility in arithmetic, where each integer is a divisor of itself.) Note that the *Transitive Law* holds for containment of sets: $A \subset B$, $B \subset C$ imply $A \subset C$.

In dealing with elements and sets simultaneously, for example, points and geometric figures, it is desirable to distinguish them by suitable choice of notation. Thus we use $a, b, c, \ldots$ to denote elements and $A, B, C, \ldots$ to denote sets. It is essential to have a compact notation to indicate that x is an element of set A. The conventional notation is $x \in A$ (read x is *in* A or x *belongs to* A). It is often convenient to invert $x \in A$, that is to state A contains x as an element, but there seems to be no attractive notation for this. Sometimes $x \in A$ is inverted as $A \ni x$, which means that we have introduced four "containment" symbols: $\subset, \supset, \in, \ni$. So we take a different path and express, x is an element of A, simply by $x \subset A$ or $A \supset x$. There is a slight but definite objection to this usage, since we are employing the symbol $\subset$ in two senses: in $X \subset A$ it means *is a subset of* and in $x \subset A$ it means *is an element of*. These are related but *distinct* ideas; although a line is a *subset* of a plane we hardly could say that a line is an *element* of a plane. However no ambiguity arises in our treatment, since our "alphabets" distinguish elements from sets, and we have the advantage of using one convenient symbol for two related ideas.

Remark. A more formal way of viewing our slightly unorthodox usage of $\subset$ is this: Let (x) denote the set whose only element is x. Then $x \in A$ is logically equivalent to $(x) \subset A$. So our use of $\subset$ for element containment is essentially an agreement to omit parentheses in a context which precludes ambiguity.

This use of $\subset$ provides a simple and convenient notation for dealing with certain properties of sets. Thus $A \subset B$ may be stated: If $x \subset A$ then $x \subset B$. Further consider set equality or identity, expressed by $A = B$. This says: if $x \subset A$ then $x \subset B$ and conversely, that is, $x \subset A$ if and only if $x \subset B$. It follows that $A = B$ is equivalent to $A \subset B$ and $B \subset A$.

In geometry we often combine "smaller" sets to form "larger" ones or break down larger sets into smaller ones. This suggests the basic operation of forming unions of sets.

DEFINITION.   If A, B are sets, $A \cup B$, called the *set union* of A and B, is the set which contains each element of A, each element of B, and no other element.

$A \cup B$ is thus characterized by the property: $x \subset A \cup B$ if and only if $x \subset A$ or $x \subset B$ (where both possibilities may occur). Thus $x \subset A \cup B$ if and only if x is an element of at least one of A, B. And $x \nsubseteq A \cup B$ if and only if $x \nsubseteq A$ and $x \nsubseteq B$. Note that $A \cup B$ may be defined as consisting of all elements of A together with those elements of B which are not in A.

If A is a set and b an element $A \cup b$ (or $b \cup A$) denotes the set formed by adjoining b to the elements of A.* Note that if b is in A, $A \cup b = A$. More generally we have the *Absorption Law*: If $B \subset A$ then $A \cup B = A$. Thus the *Idempotent Law*, $A \cup A = A$, is valid.

Since $A \cup B$ is a set we can form the union of it and a set C, expressed $(A \cup B) \cup C$. Similarly $A \cup (B \cup C)$ is formed. $(A \cup B) \cup C$ contains each element of A, of B, and of C, and no other element. Likewise for $A \cup (B \cup C)$. Thus $(A \cup B) \cup C = A \cup (B \cup C)$, and we write $A \cup B \cup C$ for either expression, and refer to it as the *union* of sets A, B, C. Observe that $x \subset A \cup B \cup C$ if and only if x is in at least one of the sets A, B, C. And $x \nsubseteq A \cup B \cup C$ if and only if $x \nsubseteq A$, $x \nsubseteq B$, and $x \nsubseteq C$. These ideas generalize immediately for several sets.

We can treat equations somewhat as in ordinary algebra, for example, given $A = B$ we can infer $A \cup X = B \cup X$. But the converse does not hold. For example,

$$\overline{ab} \cup ab = ab = ab \cup ab,$$

but $\overline{ab} \neq ab$. To state a correct converse we introduce the idea of disjointness of sets which we need in any case. Two sets A, B are *disjoint* or A is *disjoint to B*, if A and B have no common element; in general sets $A_1, \ldots, A_n$ are *disjoint* if no two have a common element. Then we can assert the

Cancellation Principle.   If $A \cup X = B \cup X$ then $A = B$, provided X is disjoint to A and to B.

Proof.   To prove $A = B$ we show $x \subset A$ implies $x \subset B$ and the converse. Suppose $x \subset A$. Then $x \subset A \cup X$ so that $x \subset B \cup X$. Thus $x \subset B$ or $x \subset X$. But $x \nsubseteq X$ since A and X are disjoint. Hence $x \subset B$. That is, $x \subset A$ implies $x \subset B$. The converse is shown similarly, and we conclude $A = B$.

* More formally $A \cup b$ stands for $A \cup (b)$ where (b) is the set whose only element is b. Compare the Remark above.

7. Decomposition of a Line Determined by Two of its Points

We are now prepared to justify some of the order properties, tacitly assumed in Euclid, that were discussed in Chapter 1. First we show that two points of a line effect a breakdown into a segment and two rays.

THEOREM 4.   If $a \neq b$ then

$$ab = a/b \cup a \cup \overline{ab} \cup b \cup b/a,$$

and the terms to the right of the equal sign are disjoint.

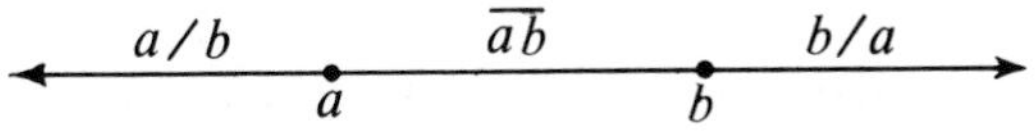

$$\xleftarrow{\hspace{2cm}} \overset{\displaystyle a/b}{} \quad \overset{\displaystyle \overline{ab}}{\underset{\displaystyle a}{\bullet}} \quad \overset{\displaystyle b/a}{\underset{\displaystyle b}{\bullet}} \xrightarrow{\hspace{2cm}}$$

FIGURE 10.3

Proof.   Let S denote the set $a/b \cup a \cup \overline{ab} \cup b \cup b/a$. We prove $S = ab$ by showing $S \subset ab$ and conversely $ab \subset S$. By Theorems 2 and 3, each of $\overline{ab}$, a/b, b/a is a subset of ab. Since $a, b \subset ab$, it follows that $S \subset ab$.

Now we show $ab \subset S$. Let $x \subset ab$. If $x = a$ or $x = b$, then x is in S. If $x \neq a, b$ then by B3.2 (abx), (bxa), or (xab). First suppose (abx). Then B1 implies (xba) and $x \subset b/a$ by definition of ray. Hence $x \subset S$. Next suppose (bxa). Then $x \subset \overline{ba} = \overline{ab}$, and so $x \subset S$. Finally if (xab) then $x \subset a/b$; hence $x \subset S$. Thus each point of ab is in S, or $ab \subset S$. We conclude $S = ab$.

By hypothesis $a \neq b$. By Theorems 2 and 3, $a \not\subset \overline{ab}$, a/b, b/a and $b \not\subset \overline{ab}$, a/b, b/a. Suppose $\overline{ab}$ and a/b not disjoint. Let x be a common point. Then (axb) and (xab), contradicting Theorem 1. Therefore $\overline{ab}$ and a/b are disjoint. Similarly, for $\overline{ab}$ and b/a. Finally if x is common to a/b, b/a then (xab) and (xba), contradicting Theorem 1. Hence a/b, b/a are disjoint and the proof is complete.

8. Determination of Rays

Ray a/b is determined by specifying its endpoint a and a second point b, but b is not in ray a/b (Th. 3). The relation between the ray concept and the intuitive idea of direction suggests that a/b can also be determined by specifying endpoint a and one of its points, say c—since we feel that c lies in a unique direction from a. This is settled in Corollary 1 below. Corollary 1 quickly yields a definition (and notation) for the new way of determining a ray, which is then related to the original formulation in Corollaries 3, 4, 6, 7. The key to the discussion is Theorem 5, which may be stated:

If two rays with the same endpoint have a common point, they must coincide.

THEOREM 5.   If p/a meets p/b then $p/a = p/b$.

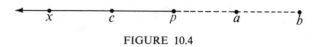

FIGURE 10.4

Proof.   First we show (apb) false by using B4. Let $c \subset p/a$, p/b. Then (cpa) and (cpb), so that $p \neq c$, a, b and p collines with c, a, b. Thus by B4, (cpa) implies (cpb) or (apb) but not both. Since (cpb), it follows that (apb) is false.

Now we show $p/a \subset p/b$. Let $x \subset p/a$; then (xpa). We show (xpb) using B4. Since $p \neq x$, a, b and p collines with x, a, b, by B4 (xpb) or (apb). Since (apb) is false, (xpb) follows, so that $x \subset p/b$. Thus $p/a \subset p/b$. If we interchange a and b in this argument, we get $p/b \subset p/a$. Thus $p/a = p/b$.

COROLLARY 1.   If $p \neq a$ there is one and only one ray with endpoint p which contains a.

Proof.   Since $p \neq a$, by B5 there is a point x such that (apx). Thus $a \subset p/x$, and p/x is a ray with the desired properties. Since any ray with endpoint p is of the form p/y, suppose p/y contains a. Then p/x meets p/y, and by the theorem, $p/x = p/y$.

This result settles the problem of determination of rays. It may be stated: *A ray is determined by specifying an endpoint and one of its points.* (Of course the two points must be distinct.) We may express it in "global" terms as follows: The set of all points, excluding a given point p, is "simply covered" by the family of rays with endpoint p.

(a)   (b)

FIGURE 10.5

Our introduction of the ray concept (Sec. 5) involved two informal constructions which may be described as follows:

(A) Start from point a and move endlessly in the direction given by point b (Fig. 10.5 (a)).

(B) Start from a and move endlessly in the direction opposite to b (Fig. 10.5 (b)).

Construction (B) yielded the definition of ray in the form a/b. Now Corollary 1 enables us to formalize (A).

DEFINITION.   If $p \neq a$ the unique ray with endpoint p which contains a is denoted $\overrightarrow{pa}$, read "ray pa" or "pa arrow."

COROLLARY 2.   Let R be a ray with endpoint p. Then $a \subset R$ implies $R = \overrightarrow{pa}$.

Proof. By Theorem 3, $p \neq a$. R is a ray with endpoint p which contains a. By Corollary 1, R is the only such ray. Thus by definition R is $\overrightarrow{pa}$.

This can be expressed more compactly by using a specific form for R, for example:

If $a \subset p/x$ then $p/x = \overrightarrow{pa}$.

COROLLARY 3.   Any ray p/x with endpoint p is expressible in the form $\overrightarrow{pa}$.

Proof. By Theorem 3 p/x is not empty and contains a point a. Thus $p/x = \overrightarrow{pa}$.

COROLLARY 4. (apb) implies $\overrightarrow{pa} = p/b$, $\overrightarrow{pb} = p/a$.

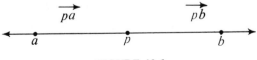

FIGURE 10.6

Proof. (apb) implies $a \subset p/b$. By Corollary 2, $p/b = \overrightarrow{pa}$. By B1 (apb) implies (bpa) and the above argument yields $p/a = \overrightarrow{pb}$.

This result is very useful: it enables us to convert rays from the "arrow" form to the "fractional" form or vice versa as needed. Roughly it is comparable to the algebraic principle $a/b = ab^{-1}$ which converts quotients into products. We use it to prove

COROLLARY 5. $\overrightarrow{ab} \subset ab$, provided $a \neq b$.

Proof.   Let c satisfy (cab). By Corollary 4 and Theorem 3

$$\overrightarrow{ab} = a/c \subset ac = ab.$$

The relation of rays a/b and $\overrightarrow{ab}$ to line ab is indicated in the diagram (Fig. 10.7).

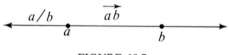

FIGURE 10.7

Corollary 4 has two additional consequences which relate the "arrow" and "fractional" forms of rays.

COROLLARY 6.   If $\overrightarrow{pa} = p/b$ then $\overrightarrow{pb} = p/a$.

Proof. $a \subset \overrightarrow{pa} = p/b$. Thus (apb) and Corollary 4 implies $\overrightarrow{pb} = p/a$.

Speaking intuitively, Corollary 6 says that if the direction of a from p is opposite to that of b, then the direction of b from p is opposite to that of a. Expressed formally this is a sort of "transposition" principle since we merely interchange a and b in the relation $\overrightarrow{pa} = p/b$ to get $\overrightarrow{pb} = p/a$.

COROLLARY 7. $\overrightarrow{pa} = \overrightarrow{pb}$ if and only if $p/a = p/b$.

Proof.   Let $\overrightarrow{pa} = \overrightarrow{pb} = p/x$. By Corollary 6, $\overrightarrow{px} = p/a$ and $\overrightarrow{px} = p/b$. Thus $p/a = p/b$.

Conversely, let $p/a = p/b = \overrightarrow{py}$. By Corollary 6, $\overrightarrow{pa} = p/y$ and $\overrightarrow{pb} = p/y$ so that $\overrightarrow{pa} = \overrightarrow{pb}$.

9. Opposite Rays

The notion of opposite rays is suggested by the intuitive idea of opposite directions from a point. It appears in school geometry in the form of the sides of a straight angle. It is as important in our study of geometry as "opposite" numbers, for example, -5 and 5 in algebra. Our definition is suggested by an analysis of a familiar diagram (Fig. 10.8) in terms of the betweenness concept.

FIGURE 10.8

DEFINITION.   Rays R, R' are *opposite* if they have a common endpoint p, and p is between each point of R and each point of R'.

THEOREM 6.   Let rays R, R' have a common endpoint p. Let there be a point a in R, and a point b in R' such that (apb). Then p is between each point of R and each point of R', so that R and R' are opposite.

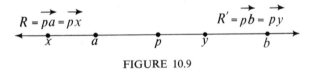

FIGURE 10.9

Proof.   Let $x \subset R$, $y \subset R'$; we show (xpy). Since $a \subset R$, $b \subset R'$ we have

(1) $$R = \overrightarrow{px} = \overrightarrow{pa},$$

(2) $$R' = \overrightarrow{py} = \overrightarrow{pb}.$$

We shall "eliminate" a, b in (1), (2) to obtain a relation involving x, y, p. (apb) implies (Th. 5, Cor. 4)

(3) $$\overrightarrow{pa} = p/b.$$

(1), (3) yield

(4) $$\overrightarrow{px} = p/b,$$

and we have eliminated a. Equation (4) implies (Th. 5, Cor. 6)

(5) $$\overrightarrow{pb} = p/x.$$

Equations (2) and (5) imply $\overrightarrow{py} = p/x$ and b is eliminated. Thus $y \subset p/x$ so that (ypx) and (xpy). To complete the proof note that R, R' are opposite by definition.

COROLLARY 1.   Rays $\overrightarrow{pa}$ and p/a are opposite and p is between any two points of the respective rays.

Proof.   Let $x \subset p/a$. Then (xpa). Since $x \subset p/a$ and $a \subset \overrightarrow{pa}$ the result is immediate by the theorem.

COROLLARY 2.   Suppose (apb). Then p/a and p/b are opposite and p is between any two points of the respective rays.

Proof. (apb) implies $a \subset p/b$ and $b \subset p/a$. Hence the theorem applies.

COROLLARY 3.   Any pair of opposite rays are disjoint.

Proof.   Suppose R, R' are opposite rays and point x is common to R, R'. Then by definition of opposite rays (xpx) for common endpoint p, contrary to B3.1.

COROLLARY 4.   No ray is opposite to itself.

Proof.   By Corollary 3, since no ray is an empty set.

The last corollary may be unfamiliar and sound a bit bizarre, but it is not trivial. This kind of "irreflexive" property occurs in many situations: In algebra no number is less than itself; in geometry, no line is perpendicular to itself; and in vertebrate zoology, no creature is its own parent.

10. The Concept of Separation

Separation is one of the most important geometric ideas and is deeply rooted in geometric intuition. As an example consider the familiar intuitive statement, *a point p of line L separates L into two parts or sides S, S'*. This sounds like the statement, *a knife separates a long thin loaf of bread into two halves.* But the similarity in the main is auditory. For points and lines are abstractions and we can not define geometric separation as a physical process, though it may be suggested by or related to a physical process. Moreover, point p is in line L, the knife, we hope, is not in the bread. The point does not do anything to the line, as the knife the bread—geometric separation is not a process at all, its essence must be found in the fixed interrelations of the four objects: p, L, S, S' (Figure 10.10).

FIGURE 10.10

How are p, L, S, S' interrelated? Surely the simplest relation among these will be that L is constituted by S, S' and p; we do not want to lose part of L in the analysis of the separation concept. Thus we require

(a) $L = S \cup S' \cup p$.

Moreover we suppose S, $S' \neq \varnothing$. If the knife slipped and missed the bread, we hardly would say the bread was separated but one of the components was empty.

Next we require

(b) S, S', p are disjoint.

For the "separator" p should not belong to the things S, S' it separates, and the latter to be "separated" by p should not overlap.

Conditions (a) and (b) are not sufficient to ensure separation, for they fail to indicate that p has a different role from S and S'. Point p is to be a "separator" or "barrier" for S and S': we take this to mean

(c) p is between each point of S and each point of S'.

Conditions (a), (b), (c) might seem sufficient to characterize the idea of separation, and they would make a reasonable and useful definition. However, if we want to use the phrase "p separates L into S, S'" as sharply as we can, it is desirable to exclude the possibility that p might separate S or S' into smaller sets. We want S, S' to be "ultimate" components of L, that is, they are not to be "separated" by p. Thus we require finally

(d) p is not between two points of S, or of S'.

We take (a), (b), (c), (d) to characterize the idea that point p separates line L into the nonempty sets S, S'. However our analysis is valid in other situations; it would apply, for example, to the separation of a segment into two segments by one of its points. Thus we are led to introduce the following.

DEFINITION.   We say point p *separates* a set of points A *into* the nonempty sets S and S' if the following conditions are satisfied:

(i) $A = S \cup S' \cup p$;
(ii) p is between each point of S and each point of S';
(iii) p is not between two points of S, or of S';
(iv) S, S', p are disjoint.

Condition (iv) is intimately related to (i) and is placed last simply because in practice it is often more easily justified after the others have been established.

11. Separation of a Line by One of its Points

In formulating the separation theorem for point p and line L, it is clear that the separation sets will be rays, but there is no simple or natural way to specify the rays in terms of p and L. Thus we take L to be in the form ab and assume that point p satisfies (apb). Then we can identify the separation sets as the rays p/a, p/b. Thus we state

THEOREM 7. (LINE SEPARATION)   Suppose (apb). Then p separates ab into p/a and p/b.

Proof.   Note p/a, p/b are not empty (Theorem 3). We have to show:

(i) $ab = p/a \cup p \cup p/b$;
(ii) p is between each point of p/a and each point of p/b;
(iii) p is not between two points of p/a, or between two points of p/b;
(iv) $p/a, p/b, p$ are disjoint.

Proof of (i).   Let $S = p/a \cup p \cup p/b$. We prove $ab = S$ by showing $S \subset ab$ and $ab \subset S$. (apb) implies $ab = pa = pb$. By Theorem 3 $p/a \subset pa, p/b \subset pb$. Thus $p/a, p/b \subset ab$. Since $p \subset pa = ab$ we have

$$p/a \cup p \cup p/b \subset ab$$

or $S \subset ab$.

Conversely suppose $x \subset ab$. If $x = p$ certainly $x \subset S$. If $x \neq p$ then (apb) implies $p \neq a, b, x$ and p collines with a, b, x. Thus by B4 (apb) implies (apx) or (bpx). By B1 (xpa) or (xpb), so that $x \subset p/a$ or $x \subset p/b$. In either case $x \subset S$. Thus $ab \subset S$, and so $ab = S$.

Proof of (ii).   By Corollary 2 of Theorem 6, (apb) implies p is between each point of p/a and each point of p/b.

Proof of (iii).   Suppose p between points x, y of p/a. By Theorem 6 p/a is opposite p/a, contradicting Corollary 4 of Theorem 6. Similarly for p/b.

Proof of (iv). $p \not\subset p/a$, p/b by Theorem 3. By (ii) p/a, p/b are opposite, hence disjoint by Corollary 3 of Theorem 6.

COROLLARY 1. (LINE SEPARATION) The theorem holds if we replace p/a, p/b by $\overrightarrow{pa}$, $\overrightarrow{pb}$.

Proof. (apb) implies $p/a = \overrightarrow{pb}$, $p/b = \overrightarrow{pa}$ by Corollary 4 to Theorem 5.

It is conceivable that a point p of line L may separate L into two sets which are not rays—or p may separate L into rays in two distinct ways. This can not happen as we show in

COROLLARY 2.   (UNIQUENESS OF SEPARATION)   Let $p \subset L$. Then p separates L uniquely into two sets; and these sets are rays with common endpoint p.

Proof.   Suppose p separates L into S, S'. By definition (Section 10)

$$L = S \cup S' \cup p,$$

where the right-hand terms are disjoint and S, $S' \neq \varnothing$. Let $a \subset S$, $b \subset S'$. By definition (apb). Thus p separates L into $\overrightarrow{pa}$, $\overrightarrow{pb}$ by Corollary 1. We show $S = \overrightarrow{pa}$, $S' = \overrightarrow{pb}$. Let $x \subset S$. Then $b \subset S'$ implies (xpb). Hence $x \not\subset \overrightarrow{pb}$, for p is not between two points of $\overrightarrow{pb}$. Since $x \neq p$ we infer $x \subset \overrightarrow{pa}$. Thus $S \subset \overrightarrow{pa}$. Symmetrically $\overrightarrow{pa} \subset S$, and we infer $S = \overrightarrow{pa}$. By interchanging a, b in the argument we get $S' = \overrightarrow{pb}$. Thus S, S' are uniquely determined and are rays with endpoint p. Finally to show that p does separate L into two sets, note that (cpd) holds for some pair of points c, d of L, and apply the theorem or Corollary 1.

We have an important decomposition of a line:

COROLLARY 3. (LINE DECOMPOSITION)   If $a \neq b$

$$ab = a/b \cup a \cup \overrightarrow{ab},$$

and a/b, a, $\overrightarrow{ab}$ are disjoint.

Proof. By B5 there is a point x such that (xab). Thus the theorem implies

$$ab = xb = a/b \cup a \cup a/x,$$

where a/b, a, a/x are disjoint. The result follows since (xab) implies $a/x = \overrightarrow{ab}$ by Corollary 4 to Theorem 5.

A similar decomposition for a ray:

COROLLARY 4. (RAY DECOMPOSITION)   If $a \neq b$

$$\overrightarrow{ab} = \overline{ab} \cup b \cup b/a,$$

and $\overline{ab}$, b, b/a are disjoint.

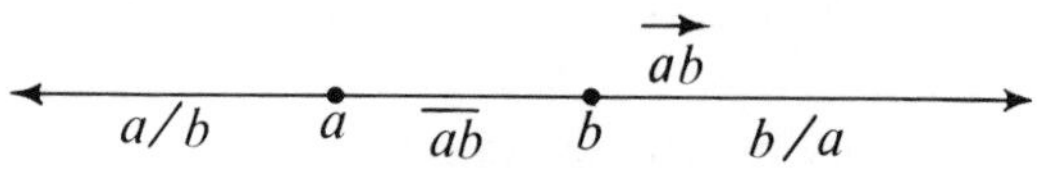

FIGURE 10.11

Proof. Corollary 3 and Theorem 4 imply

$$ab = a/b \cup a \cup \overrightarrow{ab},$$

$$ab = a/b \cup a \cup \overline{ab} \cup b \cup b/a.$$

By equating we have

(1) $$a/b \cup a \cup \overrightarrow{ab} = a/b \cup a \cup \overline{ab} \cup b \cup b/a.$$

Since the terms on the left in (1) are disjoint, as are the terms on the right, we may delete the common terms in (1) by the Cancellation Principle of Section 6, and obtain

$$\overrightarrow{ab} = \overline{ab} \cup b \cup b/a,$$

where the terms on the right are still disjoint.

The intuitive significance of this important formula is that $\overrightarrow{ab}$ is traversed by moving from a to b, passing through b, and continuing endlessly away from a. It is sometimes used as a definition of $\overrightarrow{ab}$. Such a tripartite definition is a bit awkward, but is adopted in order to study rays of the form $\overrightarrow{pa}$ directly. We have avoided its use by basing our study on the form p/a and proving an existence theorem for the form $\overrightarrow{pa}$ (Th. 5, Cor. 1).

COROLLARY 5.   Suppose $a \neq b$. Then $x \subset \overrightarrow{ab}$ if and only if (axb) or $x = b$ or (abx).

Proof.   This is essentially a restatement of Corollary 4 in which $\overline{ab}$ and b/a are replaced by their definitions.

COROLLARY 6. (abc) implies $\overrightarrow{ab} = \overrightarrow{ac}$ and $a/b = a/c$.

Proof.   By Corollary 5 (abc) implies $b \subset \overrightarrow{ac}$. Then $\overrightarrow{ab} = \overrightarrow{ac}$ (Th. 5, Cor. 2), and $a/b = a/c$ follows (Th. 5, Cor. 7).

12. Separation of a Segment by One of its Points

Our treatment leans heavily on the Line Separation Theorem and its corollaries.

THEOREM 8. (SEGMENT SEPARATION)   Suppose (apb). Then p separates $\overline{ab}$ into $\overline{ap}$ and $\overline{pb}$.

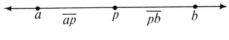

FIGURE 10.12

Proof.   Note $\overline{ap}, \overline{pb} \neq \varnothing$. We must show
   (i) $\overline{ab} = \overline{ap} \cup p \cup \overline{pb}$;
   (ii) p is between each point of $\overline{ap}$ and each point of $\overline{pb}$;
   (iii) p is not between two points of $\overline{ap}$, or two points of $\overline{pb}$;
   (iv) $\overline{ap}, p, \overline{pb}$ are disjoint.
   To establish (i) we derive two decompositions of line ab, equate them and cancel superfluous terms.
   Corollaries 1, 4, and 6 of Theorem 7 imply

$$(1) \qquad ab = \overrightarrow{pa} \cup p \cup \overrightarrow{pb},$$

$$(2) \qquad \overrightarrow{pa} = \overline{pa} \cup a \cup a/p,$$

$$(3) \qquad \overrightarrow{pb} = \overline{pb} \cup b \cup b/p,$$

$$(4) \qquad a/p = a/b,\ b/p = b/a,$$

where the terms in each right-hand side of (1), (2), and (3) are disjoint. If we substitute from (4) into (2) and (3), and then substitute the results in (1) we obtain

$$ab = \overline{pa} \cup a \cup a/b \cup p \cup \overline{pb} \cup b \cup b/a.$$

This implies

$$(5) \qquad ab = (\overline{ap} \cup p \cup \overline{pb}) \cup (a/b \cup a \cup b \cup b/a),$$

where the terms on the right are disjoint. Theorem 4 yields another expression for ab as a union of disjoint terms, namely

$$(6) \qquad ab = \overline{ab} \cup (a/b \cup a \cup b \cup b/a).$$

By equating the right members of (5) and (6) and applying the Cancellation Principle (Section 6) we get

$$\overline{ab} = \overline{ap} \cup p \cup \overline{pb},$$

and (i) is verified. Conclusions (ii), (iii), and (iv) follow from the corresponding properties of linear separation (Theorem 7, Corollary 1) since $\overline{ap} \subset \overrightarrow{pa}$ and $\overline{bp} \subset \overrightarrow{pb}$ by Corollary 4 to Theorem 7.

COROLLARY 1.   If $p \subset \overline{ab}$ then $\overline{ap},\ \overline{pb} \subset \overline{ab}$.

COROLLARY 2.   Any segment is an infinite set.

Proof.   Suppose $\overline{ab}$ is finite and contains exactly n points. Let p_1 be one of them ($\overline{ab} \neq \varnothing$ by Theorem 2). By Corollary 1 $\overline{p_1 b} \subset \overline{ab}$. Moreover $p_1 \subset \overline{ab}$ but $p_1 \not\subset \overline{p_1 b}$ by Theorem 2. Therefore $\overline{p_1 b}$ contains $n - 1$ points at most.

Similarly $\overline{p_1 b}$ has a subset $\overline{p_2 b}$ containing $n - 2$ points at most. We thus have a "decreasing" sequence of segments

$$\overline{ab},\ \overline{p_1 b},\ \overline{p_2 b}, \ldots, \overline{p_n b}$$

where each term has at least one point less than the preceding, and the first, $\overline{ab}$, contains n points. Thus $\overline{p_n b}$ contains no points, contradicting Theorem 2.

COROLLARY 3.   Any line and any ray are infinite sets.

Proof.   Any line or ray contains a segment as a subset.

It follows that a *finite* incidence geometry such as M4 cannot be "ordered." This is a statement of mathematical impossibility. It does not mean that we have not yet succeeded in finding a definition of betweenness for points in M4 which satisfies B1–B5, but rather that it is a logical impossibility that one should exist.

13. Convex Sets

Convex sets form one of the most interesting and familiar types of figure. The idea of convex set is an important unifying concept in classical geometry; it has applications in other parts of mathematics such as game theory and the theory of linear programming and it is related to the notion of convex function and convex arc in the calculus.

DEFINITION.   A set of points S is said to be *convex*, if $x, y \subset S$ and $x \neq y$ always imply $\overline{xy} \subset S$.

Observe that any line is convex since $\overline{ab} \subset ab$. It follows that any plane is convex. The empty set $\varnothing$, and a set consisting of a single point satisfy the definition vacuously (Ch. 8, Sec. 10). It is trivial that the set of all points is convex.

The basic linear or one-dimensional convex sets are rays and segments.

THEOREM 9.   Any ray is convex.

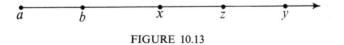

FIGURE 10.13

Proof. Consider p/a. Let $x \neq y$ and $x, y \subset p/a$. We show $\overline{xy} \subset p/a$. Let $z \subset \overline{xy}$. Then (xzy); also (xpa), (ypa) so that (apx), (apy).

Our proof depends on a double application of B4. First we apply B4 to p, a, x, y. We have p collines with a, x, y and $p \neq a, x, y$. Thus (apx) implies (apy) or (xpy) but not both. Hence (xpy) is false.

Now we apply B4 to p, a, x, z. We have p collines with a, x, z and $p \neq a, x, z$. (Why is $p \neq z$?) Thus (apx) implies (apz) or (xpz) but not both. Suppose (xpz). Then, using Corollary 1 of Theorem 8

$$p \subset \overline{xz} \subset \overline{xy}.$$

This implies (xpy) which is false. Thus (xpz) is false and (apz) must hold. Hence $z \subset p/a$ and we conclude $\overline{xy} \subset p/a$.

THEOREM 10.   Any segment is convex.

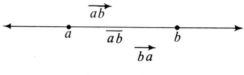

FIGURE 10.14

Proof. Consider $\overline{ab}$. We show first that $\overline{ab}$ is the intersection of the rays $\overrightarrow{ab}, \overrightarrow{ba}$ and then apply Theorem 9. We have (Th. 7, Cor. 4)

(1) $$\overrightarrow{ab} = \overline{ab} \cup b \cup b/a, \qquad \overrightarrow{ba} = \overline{ba} \cup a \cup a/b.$$

Since $\overline{ab} = \overline{ba}$ certainly $\overline{ab} \subset \overrightarrow{ab}, \overrightarrow{ba}$. Further any point common to $\overrightarrow{ab}$ and $\overrightarrow{ba}$ is in $\overline{ab}$, since the terms $b, b/a, a, a/b$ in (1) are disjoint by Theorem 4. Thus $\overline{ab}$ is the intersection of $\overrightarrow{ab}$ and $\overrightarrow{ba}$.

Now let $x, y \subset \overline{ab}$, $x \neq y$. Then $x, y \subset \overrightarrow{ab}$ so that $\overline{xy} \subset \overrightarrow{ab}$ since $\overrightarrow{ab}$ is convex. Similarly $\overline{xy} \subset \overrightarrow{ba}$. Thus $\overline{xy} \subset \overline{ab}$ the intersection of $\overrightarrow{ab}$ and $\overrightarrow{ba}$, which proves $\overline{ab}$ is convex.

14. Uniqueness of Opposite Ray

We have studiously avoided referring to *the* opposite of a ray, since we have not proved that a ray has a unique opposite. We have postponed consideration of the question not because it is unusually difficult, but because it depends on a subtle point: It is necessary to prove that a ray has a unique *endpoint*. Suppose that ray R had distinct endpoints p, p'. Then R could be expressed $R = \overrightarrow{pa} = \overrightarrow{p'a}$. Thus p/a and p'/a would both be opposite R, but it would be doubtful that $p/a = p'/a$. Consequently, we must prove that this situation can not occur. Since it is almost impossible to imagine that a ray could have two distinct endpoints, it may seem absurd to prove that a ray has only one. But consider the situation for a moment. We are not doubting that a ray, as we conceive it in elementary geometry, has a unique endpoint. Quite the contrary—we are insisting that this is an essential property of rays and are showing our good faith by verifying that our postulates do imply this property. The issue is simply whether we have chosen suitable postulates—if not we shall have to change them. Thus we assert and prove

THEOREM 11.   Any ray has a unique endpoint.

Proof.   We make use of the following principles:

(i) $\overrightarrow{ab} \supset \overline{ab}$, *provided $a \neq b$.*
(ii) *Any ray is contained in a unique line.*

Principle (i) follows directly from the Ray Decomposition Principle (Th. 7, Cor. 4). To justify Principle (ii) note $\overrightarrow{ab} \subset ab$ (Th. 5, Cor. 5). It follows that ab is the only line that can contain $\overrightarrow{ab}$, since $\overrightarrow{ab}$ is an infinite set (Th. 8, Cor. 3).

Now let ray R have endpoints p, p'. We show $p = p'$. Suppose $p \neq p'$. Let $a \subset R$. Then (Th. 5, Cor. 2)

(1) $$R = \overrightarrow{pa} = \overrightarrow{p'a}.$$

We have (Th. 5, Cor. 5)

(2) $$\overrightarrow{pa} \subset pa, \qquad \overrightarrow{p'a} \subset p'a.$$

(1) and (2) imply $pa = p'a$ by Principle (ii). By using the Line Decomposition Principle (Th. 7, Cor. 3) we have

(3) $$p \subset pa = p'a = \overrightarrow{p'a} \cup p' \cup p'/a.$$

Suppose $p \subset \overrightarrow{p'a}$. Then (1) implies $p \subset \overrightarrow{pa}$, contrary to Theorem 3(ii). Hence (3) implies $p \subset p'/a$. Then $(pp'a)$, Principle (i) and (1) imply

$$p' \subset \overline{pa} \subset \overrightarrow{pa} = \overrightarrow{p'a},$$

also contrary to Theorem 3(ii). Thus our supposition is false and the theorem holds.

Now it is not hard to prove the main result.

THEOREM 12.   Any ray has a unique opposite ray.

Proof. Consider ray $\overrightarrow{pa}$. We know p/a is opposite $\overrightarrow{pa}$ (Th. 6, Cor. 1). Suppose R opposite $\overrightarrow{pa}$. We show $R = p/a$. By definition R and $\overrightarrow{pa}$ have a common endpoint. By Theorem 11, p is the only endpoint of $\overrightarrow{pa}$. Thus p is a common endpoint of R and $\overrightarrow{pa}$; and the *only* one. Let $b \subset R$. Since $a \subset \overrightarrow{pa}$, we have (bpa) by definition of opposite rays. Thus $b \subset p/a$ so that $R = p/a$ (Th. 5, Cor. 1) and the proof is complete.

Note the role of Theorem 11 in the proof. Since R and $\overrightarrow{pa}$ are opposite, we know by definition that there *exists* a common endpoint of R and $\overrightarrow{pa}$, which is between each point of R and each point of $\overrightarrow{pa}$. Without Theorem 11, we could not assert that such a common endpoint must be p. We could merely get, if p' is such a common endpoint, $(bp'a)$ and conclude R is p'/a, not p/a.

15. Extension of the Concept of Order

It is familiar that a precedence relation can be extended from a two-term order relation to one involving three or more terms. This is done when we order three real numbers in terms of the relation "less than," for example $-5 < 0 < 7$. Similarly our notion of order can be extended from the three-term relation of betweenness to an order relation involving four or more terms. We define a four-term order relation as follows.

DEFINITION. $(abcd)$ means (abc), (abd), (acd), and (bcd). We read $(abcd)$ as "points a, b, c, d are in the *order abcd*."

As you might expect, the basic properties of betweenness can be extended to four-term order. The most important of these for our present purposes is the following version of B3.2: *Suppose $a \neq b$ and that x collines with and is distinct from a, b. Then (xab), (axb) or (abx).* We extend this to four points in

THEOREM 13.   Suppose (abc) and that x collines with and is distinct from a, b, c. Then $(xabc)$, $(axbc)$, $(abxc)$, or $(abcx)$.

Proof. By Theorem 4

(1) $$ac = a/c \cup a \cup \overline{ac} \cup c \cup c/a.$$

By Theorem 8 (abc) implies $\overline{ac} = \overline{ab} \cup b \cup \overline{bc}$. Substituting in (1) we get

(2) $$ac = a/c \cup a \cup \overline{ab} \cup b \cup \overline{bc} \cup c \cup c/a.$$

Since $x \subset ac$ and $x \neq a, b, c$ we have

$$x \subset a/c \cup \overline{ab} \cup \overline{bc} \cup c/a.$$

FIGURE 10.15

Suppose $x \subset a/c$; then (xac). (abc) implies $b \subset \overrightarrow{ac}$ (Th. 7, Cor. 5). Since a/c, $\overrightarrow{ac}$ are opposite rays with endpoint a, a is between x and b or (xab). (xab) implies (bax), so that $x \subset \overrightarrow{ba}$ (Th. 7, Cor. 5). (abc) implies (cba), so that $c \subset b/a$. Since $\overrightarrow{ba}$, b/a are opposite rays with endpoint b, b is between x and c or (xbc). We thus have (xab), (xac), (xbc), and (abc), which is the definition of $(xabc)$. Thus $x \subset a/c$ implies $(xabc)$.

The case $x \subset c/a$ is symmetrical, since the hypothesis is symmetrical in a and c, and in this case we get $(xcba)$, which implies $(abcx)$.

Suppose $x \subset \overline{ab}$; then (axb) and (bxa). Hence $x \subset \overrightarrow{ba}$ (Th. 7, Cor. 5). Thus $x \subset \overrightarrow{ba}$, $c \subset b/a$ imply (xbc), as above. Further (axb) implies $a \subset x/b$ and (xbc) implies $c \subset \overrightarrow{xb}$. Again as above (axc). We have (axb), (axc), (abc), and (xbc) which is the definition of $(axbc)$.

The case $x \subset \overline{bc}$ is symmetrical and yields $(abxc)$.

Theorem 13 is powerful and yields important results on three-term order, as the following corollaries indicate.

COROLLARY 1. If (abc) and (bcd) then $(abcd)$.

Proof. Show that d collines with and is distinct from a, b, c. Then the theorem implies

$$(dabc), (adbc), (abdc), \text{ or } (abcd).$$

The first three relations contradict (*bcd*) and so the fourth holds.

By applying to Corollary 1 the definition of (*abcd*) we obtain

COROLLARY 2.   If (*abc*) and (*bcd*) then (*abd*) and (*acd*).

By similar methods we justify the following corollaries.

COROLLARY 3.   If (*abc*) and (*acd*) then (*abcd*).

COROLLARY 4.   If (*abc*) and (*acd*) then (*abd*) and (*bcd*).

Other properties of betweenness generalize to four terms in a natural way. For example given four points *a*, *b*, *c*, *d* we can state twenty-four different order relations (*abcd*), (*abdc*), (*acbd*), . . . , corresponding to the twenty-four permutations of *a*, *b*, *c*, *d*. Suppose that *a*, *b*, *c*, *d* are distinct and collinear. Then we can assert that two and only two of the twenty-four relations hold, for example (*abcd*) and (*dcba*) or (*badc*) and (*cdab*). This generalizes B3.2 and Theorem 1.

Finally we note that the theory of order can be extended to *n* terms for *n* > 3.

16. Models of the Theory

In developing the theory in this chapter we have introduced a new basic term *between* in addition to the terms *point, line, plane,* of the Theory of Incidence. Thus a model of the theory will be an incidence geometry augmented by the specification of a betweenness relation among its points which satisfies Postulates B1–B5.

E X E R C I S E S   I

1. Interpret "point" to mean real number and "line" to mean the set of all real numbers. Interpret the betweenness relation (*abc*) to mean $a < b < c$ or $c < b < a$, where $<$ denotes "less than." Show that B1–B4 are verified. Is B5 verified? What is the interpretation of "segment"? Of "ray"?

2. The same as Exercise 1, replacing the set of real numbers by any set S in which a precedence relation $<$ is defined and "less than" by the given precedence relation.

3. Interpret "point" to mean real number and "line" to mean the set of real numbers. Interpret (abc) to mean that $a \neq c$, and there exist positive real numbers s, t such that

$$b = sa + tc, \quad s + t = 1.$$

Show that B1–B5 are verified.

4. Assume B1–B5 hold in Euclidean geometry. Consider model M9 of Chapter 8. Interpret (abc) to mean that point b is between points a and c in the sense of Euclidean geometry. Do B1–B5 hold? Why?

5. The same as Exercise 4 for M10–M14 and M23.

6. Consider the following system:

   Point: Each point inside or on a given Euclidean circle.

   Line: Each "closed" chord of the circle; that is, a chord together with its endpoints.

   Plane: The set consisting of all points which are inside or on the circle.

   (i) Show that the system is a model of the Theory of Incidence.

   (ii) Which of the postulates of betweenness does the system satisfy?

DEFINITION.   In the real number system we say b is *between* a and c if $a < b < c$ or $c < b < a$.

7. Interpret "point" to mean dyad, and "line" to mean the set of all dyads. Interpret (abc) where $a = (a_1, a_2)$, $b = (b_1, b_2)$, $c = (c_1, c_2)$ to mean b_1 is between a_1 and c_1, and b_2 is between a_2 and c_2. Which of B1–B5 are verified? What is the interpretation of "segment"? Of "ray"? Does a ray have a unique endpoint? Does a segment have a unique pair of endpoints? Does the Line Separation Theorem hold?

DEFINITION.   In M19, the "dyad linear system" model of Chapter 8, adopt the notation a for the dyad (a_1, a_2), b for (b_1, b_2), et cetera. Interpret (abc) to mean that a and c are distinct dyads and that there exist positive real numbers s, t such that

$$b_1 = sa_1 + tc_1, \, b_2 = sa_2 + tc_2, \, s + t = 1.$$

(This interpretation and notation applies to Exercises 8–18 below.)

8. (B1) Prove: (abc) implies (cba).

9. (B2) Prove: (abc) implies the falsity of (bca).

10. (B3.1) Prove: If (abc) then a, b, c are distinct dyads. (This together with Exercise 13 verifies B3.1.)

11. Prove: If (abc) and $a_1 = c_1$ then $a_1 = b_1 = c_1$; similarly if $a_2 = c_2$ then $a_2 = b_2 = c_2$.

12. Prove: If (abc) then $a_1 = b_1 = c_1$ or b_1 is between a_1 and c_1; similarly $a_2 = b_2 = c_2$ or b_2 is between a_2 and c_2.

13. (B3.1) Prove: If (abc) then a, b, c are contained in a linear system. [*Hint.* a, c are contained in a linear system.]

14. Prove: If a, b, c are distinct and contained in a linear system, then

$$c_1 = sa_1 + tb_1, \; c_2 = sa_2 + tb_2, \; s + t = 1,$$

where s, t are nonzero numbers, not necessarily positive.

15. Prove: If $c_1 = sa_1 + tb_1$, $c_2 = sa_2 + tb_2$, $s + t = 1$, $s < 0$, $t > 0$, and $a \neq b$ then (abc).

16. (B3.2) Prove: If a, b, c are distinct and are contained in a linear system, then (abc), (bca), or (cab).

†17. (B4) Let $p = (p_1, p_2)$ be distinct from and contained in the same linear system as a, b and c. Prove that (apb) implies (bpc) or (apc), but not both.

18. (B5) Prove: If $a \neq b$ there exists $x = (x_1, x_2)$, $y = (y_1, y_2)$, $z = (z_1, z_2)$ such that (xab), (ayb), (abz).

19. Prove: If a, b, c are real numbers then b is between a and c in the sense that $a < b < c$ or $c < b < a$ if and only if (1) $a \neq c$, and (2) there exist positive numbers s, t such that $b = sa + tc$, where $s + t = 1$.

E X E R C I S E S   I I

1. Prove directly from the postulates that the intersection of a/b and b/a is $\varnothing$.

2. Prove: Any segment is contained in a unique line.

3. Prove: If two distinct rays with the same endpoint colline they are opposite.

4. Prove: Two collinear rays can not intersect in a single point.

5. Prove: The intersection of two collinear rays is $\varnothing$, a segment or a ray.

6. Prove: If two collinear rays have a common point, their union is a ray or a line.

7. Prove: If p, q are in a ray with endpoint a then (apq), (aqp), or $p = q$.

8. Prove : $\overrightarrow{ab} \supset \overrightarrow{bc}$ if and only if (abc).

9. Prove: $a/b \subset b/c$ if and only if (abc).

10. Prove: (abc) and (abd) imply $\overrightarrow{ac} = \overrightarrow{ad}$ and $\overrightarrow{bc} = \overrightarrow{bd}$.

11. Prove: (abd) and (acd) imply $\overrightarrow{ab} = \overrightarrow{ac}$ and $a/b = a/c$.

12. Prove that $\overrightarrow{oa}$ and $\overrightarrow{ob}$ are distinct and not opposite if and only if
    (i) $o \notin ab$; or
    (ii) o, a, b are distinct and noncollinear; or

    (iii) $\overrightarrow{oa}$ and $\overrightarrow{ob}$ are noncollinear.

13. Write out in detail the treatment of the cases $x \subset c/a$, $x \subset \overline{bc}$ in the proof of Theorem 13.

14. If $\overline{ab}$ meets $\overline{ac}$, prove one of the segments contains the other.

15. What are the possible intersections of two segments? Prove your answer.

16. What are the possible intersections of a ray and a segment? Prove your answer.

17. Prove that $\overline{ab} \cup a$, $\overline{ab} \cup b$, $\overline{ab} \cup a \cup b$ are convex sets.

18. Is the union of two segments a convex set? The intersection of two segments?

19. Prove: If (abc) and (abd) then $(abcd)$ or $(abdc)$ or $c = d$.

20. Prove: If (abc) and (abd) then (acd) or (adc) or $c = d$.

21. Prove: If (abc) and (abd) then (bcd) or (bdc) or $c = d$.

22. Prove: If (abd) and (acd) then $(abcd)$ or $(acbd)$ or $b = c$.

23. Prove: If (abd) and (acd), $b \neq c$, then either (abc) and (bcd), or else (acb) and (cbd).

24. Prove: (abc) implies
$$ac = a/c \cup a \cup \overline{ab} \cup b \cup \overline{bc} \cup c \cup c/a,$$
where the addends are disjoint.

25. Prove: $(abcd)$ implies
$$ad = a/d \cup a \cup \overline{ab} \cup b \cup \overline{bc} \cup c \cup \overline{cd} \cup d \cup d/a,$$
where the addends are disjoint.

26. State carefully and prove a theorem on the "separation" of a segment by two of its points.

†27. Prove that a segment has a unique pair of endpoints; that is, if $\overline{ab} = \overline{cd}$ then $a = c$ and $b = d$ or $a = d$ and $b = c$. (Compare Exercises I, 7.)

28. (Project) Extend the theory of linear order to 5-term order; n-term order.

E X E R C I S E S   I I I

1. Given ray $\overrightarrow{ox}$, define $a < b$ for points of $\overrightarrow{ox}$ to mean (oab). Prove $<$ is a precedence relation in the set of points $\overrightarrow{ox}$.

2. Given ray $\overrightarrow{ox}$, define $a > b$ for points of $\overrightarrow{ox}$ to mean (oba). Prove $>$ is a precedence relation in the set of points $\overrightarrow{ox}$.

DEFINITION. A precedence relation $<$ defined on $\overrightarrow{ox}$ is *coherent with betweenness* if (abc) is equivalent to the statement, $a < b < c$ or $c < b < a$.

†3. Prove that the only precedence relations on $\overrightarrow{ox}$ that are coherent with betweenness are those defined in Exercises 1 and 2. These are thus singled out from all conceivable precedences on $\overrightarrow{ox}$ and are called the *natural* precedence relations on $\overrightarrow{ox}$.

†4. Define a precedence relation on segment $\overline{ab}$ which is coherent with betweenness. Find a second one. Prove that there are only two such relations.

†5. Show how to construct a precedence relation on a line L which is coherent with betweenness. Find a second one. Prove that there are only two such relations.

DEFINITION. Let R, R' be collinear rays. If one of R, R' contains the other we say R and R' have the *same sense*. In the contrary case they have *opposite sense*.

†6. Prove that if two rays have opposite sense each has the same sense as the ray opposite to the other.

†7. Prove that the relation of having the same sense is an equivalence relation in the family of all rays which are contained in a given line.

DEFINITION. If rays $\overrightarrow{ab}$ and $\overrightarrow{cd}$ have the same sense, we say that the ordered pairs of points (a, b) and (c, d) have the *same sense* (or the *same direction*) or (a, b) has the *same sense* as (c, d).

8. Derive several properties of the relation, (a, b) has the same sense as (c, d).

9. Define the relation (a, b) and (c, d) have *opposite sense*, and derive several of its properties.

DEFINITION. a is *in the same direction from o as* b if $\overline{oa}$ meets $\overline{ob}$.

10. Prove: a is in the same direction from o as b if and only if there is a point p such that (poa) and (pob).
11. Prove that a is in the same direction from o as b if and only if o, a, b colline; $o \neq a, b$; and (aob) is false.
12. Let aRb mean that a is in the same direction from o as b. Prove that relation R is an equivalence relation in the set of all points excluding o.

E X E R C I S E S   I V

DEFINITION.   Let A, B be disjoint sets. The *join* of A and B, written $j(A, B)$ is the set of points each of which is in a segment $\overline{pq}$ where $p \subset A$ and $q \subset B$.

1. If $a \neq b$, prove $j(a, \overline{ab}) = \overline{ab}$.
2. If $a \subset b/c$, prove $j(a, \overline{bc}) = \overline{ac}$.
3. If $(abcd)$, prove $j(ab, \overline{cd}) = \overline{ad}$.
4. If $a \neq b$, prove $j(a, \overrightarrow{ab}) = \overrightarrow{ab}$.
5. If $a \neq b$, prove $j(a/b, \overrightarrow{ab}) = ab$.
6. If K is a convex set and $A, B \subset K$ and A, B are disjoint, prove $j(A, B) \subset K$.

CHAPTER 11

Planar and Spatial
Order Properties

1. Introduction

We have established in Chapter 10 a theory of order which seems to resolve the difficulties exposed in Chapter 1. Certainly it places on a sound basis the order properties of a *line* which are assumed tacitly in Euclid. However we need planar and spatial analogues of some of these properties (such as planar and spatial separation theorems)—also "nonlinear" order properties (for example in the study of angles and triangles) which are not strictly analogues of linear ones. Will B1–B5 be sufficient to yield these?

The answer turns out to be no. To get the desired nonlinear properties of order, an additional postulate must be introduced. The German mathematician M. Pasch (1843–1933) discovered how to do this about 1880. He made a deep analysis of the concept of order and showed that the theory of separation in the plane could be based on the following additional postulate:

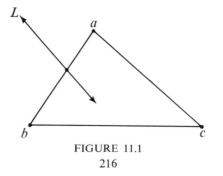

FIGURE 11.1
216

THE POSTULATE OF PASCH.   *Suppose line L is coplanar with $\triangle abc$ and does not contain a vertex of the triangle. Then if L meets one side of the triangle it meets a second, but not all three* (Figure 11.1).

This was a signal intellectual achievement. The greatest minds since the time of Euclid had not realized that there was a problem in this area. To challenge anything so obvious, so intuitively certain, as separation properties —indeed to realize that there was something to challenge—required a critical quality of the highest order.

Pasch's Postulate does not really require the notion of triangle—it is easily generalized to the following form:

B6.   *Let L coplane with and not contain a, b, c. Then if L meets $\overline{ab}$, it meets $\overline{bc}$ or $\overline{ac}$ but not both.*

B6 is a direct generalization of B4 which may be stated:

Let p colline with and be distinct from a, b, c. Then if p is in $\overline{ab}$, it is in $\overline{bc}$ or $\overline{ac}$ but not both.

Observe that B6, in contrast with the original formulation, includes the special or degenerate cases: it holds if a, b, c are distinct and colline, or if $c = a$ or $c = b$.

2. Ordered Incidence Geometries

Postulate B6, when adjoined to B1–B5 gives an adequate basis for the theory of order in the plane and in three-dimensional incidence geometries— it enables us to justify the order properties that are implicit in Euclidean plane and solid geometry. Postulates B1–B6 thus constitute a natural foundation for a geometric study of order in the context of incidence theory. It is appropriate to assign a name to a system which satisfies these postulates, and so we introduce the following

DEFINITION.   An incidence geometry in which a betweenness relation has been specified is called an *ordered incidence geometry* if it satisfies B1–B6.

Important examples of this concept are Euclidean and Lobachevskian geometries. There is an interesting theory of *ordered affine geometries* which shares with Euclidean geometry the incidence, order, and parallel postulates but makes no assumption concerning congruence.

We shall be concerned with ordered incidence geometries in the remainder of the book. This chapter and the two following treat them without restriction; the final chapters study the theory of congruence in terms of ordered incidence geometries which satisfy suitable congruence postulates.

Note that the results of Chapters 7 and 10 on incidence and linear order hold for any ordered incidence geometry.

3. Half-Planes

We have consciously formulated our treatment so as to bring to the surface and exploit an inherent parallelism between order properties of lines, of planes and of three-dimensional incidence geometries. But in developing the theory of order in the plane and in three-dimensional geometry we must not adhere too slavishly to our treatment of linear order. A plane for example is a much richer and more complex figure than a line, and we should not insist that every linear notion have a simple or unique planar analogue. For example, betweenness is essentially a linear or one-dimensional notion, and it has no obvious planar analogue. Similarly for the notion of segment. If the planar analogue of point is taken to be line, then the analogue of a segment would be the region bounded by two lines, but this is puzzling if the lines intersect. From a deeper viewpoint the planar analogue is the interior of a triangle, but this does not lend itself in an obvious way to the study of order in the plane.

The situation is very different for rays. Described intuitively a ray is the set of points of a line that are on a given side of a given point of the line. This generalizes immediately to an intuitive description of *half-plane*: the set of points of a plane that are on a given side of a given line of the plane. Thus the notion of half-plane will play a central role in our planar theory of order. In developing the theory of half-planes we formulate definitions, notation, theorems, and proofs so as to emphasize the analogy with the theory of rays.

We base the formal treatment on the following

DEFINITION.   If $a \notin L$ the set of all points x such that $\overline{xa}$ meets L is called a *half-plane* and is denoted L/a (read "L over a"). L is called the *edge* or *boundary* of the half-plane.

To point up the analogy with ray, consider p/a. This is the set of points x such that (xpa), or equivalently such that $\overline{xa}$ "meets" p, that is contains p (Figure 11.2(a)). The endpoint p of p/a corresponds to the edge L of L/a

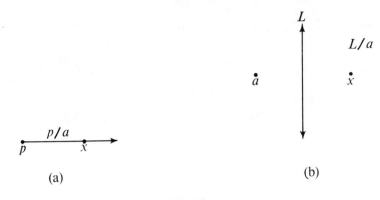

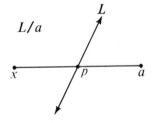

(a)   (b)

FIGURE 11.2

(Figure 11.2(b)). So to speak we step up the dimension by 1 in replacing p by L, but maintain the same defining property.

We begin with an analogue of Theorem 3 of Chapter 10.

THEOREM 1. If $a \not\subset L$ then
 (i) $L/a \subset La$, the plane of L and a;
 (ii) L, L/a, a are disjoint;
 (iii) $L/a \neq \varnothing$.

FIGURE 11.3

Proof. (i) Let $x \subset L/a$. By definition $\overline{xa}$ meets L, say in p. Thus (xpa) and

$$x \subset pa \subset La.$$

Hence $L/a \subset La$.

 (ii) Suppose $x \subset L/a$, L. Then as above $\overline{xa}$ meets L in p and (xpa). Hence

$$a \subset xp = L$$

contrary to hypothesis. Thus L/a and L are disjoint.

$a \not\subset L/a$, for the contrary implies $\overline{aa}$ meets L which is absurd. $a \not\subset L$ by hypothesis.

(iii)   Let $p \subset L$ and $x \subset p/a$. Then $\overline{xa} \supset p$ and $\overline{xa}$ meets L so that $x \subset L/a$. Thus $L/a \neq \varnothing$.

Note that by (ii) a half-plane, like a ray, is "open"—it is disjoint to its edge.

4. Determination of Half-Planes

Our treatment parallels that for determination of rays (Ch. 10, Sec. 8).

THEOREM 2.   If L/a meets L/b then $L/a = L/b$. (Compare Ch. 10, Th. 5.)

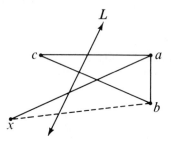

FIGURE 11.4

Proof.   The method is essentially the same as that employed for Theorem 5 of Chapter 10 with B6 replacing B4. Let $c \subset L/a$, L/b. Then $\overline{ca}$ meets L and $\overline{cb}$ meets L. We infer $a \subset L/c$, $b \subset L/c$ by definition of L/c. (A small point: in order to refer to L/c we must know $c \not\subset L$. This holds by Theorem 1, since $c \subset L/a$.) Thus, using Theorem 1

$$a, b, c, L \subset Lc.$$

Moreover $a, b, c \not\subset L$. Consequently, B6 applies to a, b, c, L and we infer $\overline{ab}$ does not meet L, since $\overline{ca}$ and $\overline{cb}$ both do meet L.

Let $x \subset L/a$; then $\overline{xa}$ meets L. By using Theorem 1

$$x \subset L/a \subset La = Lc.$$

We have

$$x, a, b, L \subset Lc; \quad x, a, b \not\subset L,$$

and can apply B6 to x, a, b, L. Since $\overline{xa}$ meets L and $\overline{ab}$ does not, we infer $\overline{xb}$ meets L. Thus $x \subset L/b$ and we have justified $L/a \subset L/b$. Reasoning symmetrically we have $L/b \subset L/a$ and we conclude $L/a = L/b$.

Corollaries corresponding to those of Theorem 5 of Chapter 10 follow; essentially the same methods of proof apply.

COROLLARY 1.   If $a \not\subset L$ there is one and only one half-plane with edge L which contains a.

Proof.   By Theorem 1 there is a point $b \subset L/a$, and $b \not\subset L$. Thus $\overline{ab}$ meets L, so that $a \subset L/b$, proving existence. Suppose $a \subset L/c$ any half-plane with edge L. Then L/b meets L/c, and by the theorem $L/b = L/c$ proving uniqueness.

DEFINITION.   If $a \not\subset L$ we denote by $\overrightarrow{La}$ (read "half-plane La" or "La arrow") the unique half-plane with edge L which contains a.

COROLLARY 2.   Let H be a half-plane with edge L. Then $a \subset H$ implies $H = \overrightarrow{La}$.

Proof.   By Theorem 1, $a \not\subset L$. H is a half-plane with edge L which contains a. By Corollary 1, it is the only such half-plane. Thus by definition H is $\overrightarrow{La}$.

By using the form L/x for H, we have:

$$\textit{If } a \subset L/x \textit{ then } L/x = \overrightarrow{La}.$$

COROLLARY 3.   Any half-plane L/x with edge L is expressible in the form $\overrightarrow{La}$.

Proof.   By Theorem 1 $L/x \neq \varnothing$ and contains a point a. Thus $L/x = \overrightarrow{La}$.

COROLLARY 4.   If $\overline{ab}$ meets L and a, $b \not\subset L$ then

$$\overrightarrow{La} = L/b, \quad \overrightarrow{Lb} = L/a.$$

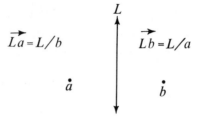

FIGURE 11.5

Proof. $\overline{ab}$ meets L and $b \not\subset L$ imply $a \subset L/b$. By Corollary 2, $L/b = \overrightarrow{La}$. By symmetry $L/a = \overrightarrow{Lb}$.

Remark. The form of the hypothesis in Corollary 4 was chosen because it is symmetrical and easily remembered. But there are useful equivalent forms, namely:

(1) $\overline{ab}$ meets L and $a \not\subset L$;
(2) $b \subset L/a$;
(3) $a \subset L/b$.

Certainly the given form implies (1). (1) implies (2) by definition of L/a. (2) implies the given form. For suppose (2). Then $a \not\subset L$, and by Theorem 1, $b \not\subset L$. By definition of L/a, $\overline{ab}$ meets L and the given form holds. Thus the given form, (1), and (2) are equivalent. By symmetry (3) is equivalent to the given form and so to the others.

We shall use any of the equivalent forms (1), (2), (3) as the hypothesis in Corollary 4 without explicit reference.

COROLLARY 5. $\overrightarrow{La} \subset La$, provided $a \not\subset L$.

Proof. Let $c \subset L/a$. By Corollary 4 and Theorem 1

$$\overrightarrow{La} = L/c \subset Lc = La.$$

COROLLARY 6.   If $\overrightarrow{La} = L/b$ then $\overrightarrow{Lb} = L/a$.

Proof. $a \subset \overrightarrow{La} = L/b$. Hence Corollary 4 implies $\overrightarrow{Lb} = L/a$.

COROLLARY 7. $\overrightarrow{La} = \overrightarrow{Lb}$ if and only if $L/a = L/b$.

Proof. Let $\overrightarrow{La} = \overrightarrow{Lb} = L/x$. By Corollary 6, $\overrightarrow{Lx} = L/a$ and $\overrightarrow{Lx} = L/b$. Thus $L/a = L/b$.
   Conversely, let $L/a = L/b = \overrightarrow{Ly}$. By Corollary 6, $\overrightarrow{La} = L/y$ and $\overrightarrow{Lb} = L/y$ so that $\overrightarrow{La} = \overrightarrow{Lb}$.

5. Opposite Half-Planes

The treatment is almost exactly analogous to that for opposite rays (Ch. 10, Sec. 9).

DEFINITION.   Half-planes H, H' are *opposite* if they have a common edge L, and each point of H is joined to each point of H' by a segment that meets L. (That is $x \subset H$, $y \subset H'$ implies $\overline{xy}$ meets L.)

THEOREM 3.   Let half-planes H, H' have a common edge L. Let there be a point a in H, and a point b in H' such that $\overline{ab}$ meets L. Then each point of H is joined to each point of H' by a segment that meets L, so that H and H' are opposite. (Compare Ch. 10, Th. 6.)

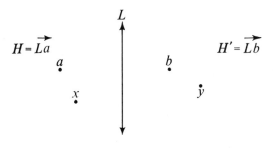

FIGURE 11.6

Proof. Let $x \subset H$, $y \subset H'$; we show $\overline{xy}$ meets L. Since $a \subset H$, $b \subset H'$ we have

(1) $H = \overrightarrow{Lx} = \overrightarrow{La}$,

(2) $H' = \overrightarrow{Ly} = \overrightarrow{Lb}$.

We "eliminate" a, b in (1) and (2) to obtain a relation involving x, y, L. Since $\overline{ab}$ meets L and a, $b \not\subset L$ we have (Th. 2, Cor. 4)

$$(3) \qquad\qquad \overrightarrow{La} = L/b.$$

Equations (1) and (3) yield

$$(4) \qquad\qquad \overrightarrow{Lx} = L/b,$$

and a is eliminated. Equation (4) implies (Th. 2, Cor. 6)

$$(5) \qquad\qquad \overrightarrow{Lb} = L/x.$$

Equations (2) and (5) imply $\overrightarrow{Ly} = L/x$ and b is eliminated. Thus $y \subset L/x$ so that $\overline{xy}$ meets L. Then H, H' are opposite by definition.

COROLLARY 1.   Half-planes $\overrightarrow{La}$ and L/a are opposite and two points of the respective half-planes are always joined by a segment that meets L.

Proof.   Let $x \subset L/a$. Then $\overline{xa}$ meets L. Since $x \subset L/a$ and $a \subset \overrightarrow{La}$ the result is immediate by the theorem.

COROLLARY 2.   Suppose $\overline{ab}$ meets L and a, $b \not\subset L$. Then L/a and L/b are opposite and two points of the respective half-planes are always joined by a segment that meets L.

Proof. $\overline{ab}$ meets L and $b \not\subset L$ imply $a \subset L/b$. By symmetry $b \subset L/a$. Hence the theorem applies.

COROLLARY 3.   Any pair of opposite half-planes are disjoint.

Proof.   Suppose H, H' are opposite half-planes and point x is common to H, H'. Then by definition of opposite half-planes $\overline{xx}$ meets L, for common edge L, which is absurd.

COROLLARY 4.   No half-plane is opposite to itself.

Proof.   By Corollary 3; since no half-plane is an empty set.

6. The Separation Concept Again

We want to formulate and prove planar and spatial analogues of the Line Separation Theorem (Ch. 10, Th. 7). Our formulation of the idea of separation of a set into two sets by a *point* (Ch. 10, Sec. 10) requires a minor adjustment for application to the case of separation by a *line* or a *plane*:

DEFINITION. Let A be a line or a plane. Then we say A *separates* a set of points B *into* the nonempty sets S and S' if the following conditions are satisfied:
(i) $B = S \cup S' \cup A$;
(ii) each segment that joins a point of S to a point of S' meets A;
(iii) each segment that joins two points of S or of S' does not meet A;
(iv) S, S', A are disjoint.

7. Separation of a Plane by One of Its Lines

Our treatment of the Plane Separation Theorem closely parallels that of the Line Separation Theorem (Ch. 10, Th. 7); as you would expect B6 replaces B4 in the proof.

THEOREM 4. (PLANE SEPARATION) Suppose $a, b, L \subset P$; $a, b \notin L$; and $\overline{ab}$ meets L. Then L separates P into L/a and L/b. (Compare Ch. 10, Th. 7.)

Proof. Note L/a, L/b are not empty. We show
(i) $P = L/a \cup L \cup L/b$;
(ii) if $x \subset L/a$ and $y \subset L/b$, then $\overline{xy}$ meets L;
(iii) if $x, y \subset L/a$ or $x, y \subset L/b$, then $\overline{xy}$ does not meet L;
(iv) L/a, L/b, L are disjoint.

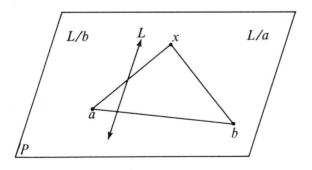

FIGURE 11.7

Proof of (i).   Let $S = L/a \cup L \cup L/b$. Since $a, b, L \subset P$ and $a, b \notin L$,
Theorem 1 implies

$$L/a \subset La = P, \quad L/b \subset Lb = P.$$

Thus

$$S = L/a \cup L \cup L/b \subset P.$$

Conversely, suppose $x \subset P$. If $x \subset L$ clearly $x \subset S$, so take $x \notin L$ (Figure
11.7). Then $a, b, x, L \subset P$ and $a, b, x \notin L$. Since $\overline{ab}$ meets L, B6 implies
$\overline{ax}$ meets L or $\overline{bx}$ meets L. It follows that $x \subset L/a$ or $x \subset L/b$. In either case
$x \subset S$. Thus $P \subset S$, and so $P = S$.

Proof of (ii).   By Corollary 2 to Theorem 3.

Proof of (iii).   Suppose L meets $\overline{xy}$ where $x, y \subset L/a$. By Theorem 3,
L/a is opposite L/a, contradicting Corollary 4 of Theorem 3. Similarly for L/b.

Proof of (iv). L is disjoint to L/a and to L/b by Theorem 1. By (ii) L/a,
L/b are opposite, hence disjoint (Th. 3, Cor. 3).

COROLLARY 1. (PLANE SEPARATION)   The theorem holds if we replace L/a,
L/b by $\overrightarrow{La}$, $\overrightarrow{Lb}$.

Proof. $\overline{ab}$ meets L and $a, b \notin L$ imply $L/a = \overrightarrow{Lb}$, $L/b = \overrightarrow{La}$ (Th. 2,
Cor. 4).

COROLLARY 2. (UNIQUENESS OF SEPARATION)   Let $L \subset P$. Then L separates
P uniquely into two sets; and these sets are half-planes with common edge L.
(Compare Ch. 10, Th. 7, Cor. 2.)

Proof.   Suppose L separates P into S, S'. By definition

$$P = S \cup S' \cup L,$$

where the right-hand terms are disjoint and $S, S' \neq \varnothing$. Let $a \subset S, b \subset S'$.
Then $\overline{ab}$ meets L, and L separates P into $\overrightarrow{La}$, $\overrightarrow{Lb}$ by Corollary 1. Let $x \subset S$.
Then $b \subset S'$ implies $\overline{xb}$ meets L. Hence $x \notin \overrightarrow{Lb}$ and we infer $x \subset \overrightarrow{La}$. Thus
$S \subset \overrightarrow{La}$. Symmetrically $\overrightarrow{La} \subset S$, and we infer $S = \overrightarrow{La}$. By interchanging

a, b in the argument we get $S' = \overrightarrow{Lb}$. Thus S, S' are uniquely determined and are half-planes with edge L. Finally to show that L does separate P into two sets, note that there exist points $c, d \subset P$ such that $c, d \notin L$ and $\overline{cd}$ meets L, and apply the theorem or Corollary 1.

COROLLARY 3. (PLANE DECOMPOSITION)   If $a \notin L$ then

$$La = L/a \cup L \cup \overrightarrow{La}$$

and $L/a, L, \overrightarrow{La}$ are disjoint.

Proof. Let $b \subset L/a$. Then $\overline{ab}$ meets L and $a, b \notin L$. Further $a, b, L \subset La$ and the theorem applies with $P = La$, yielding

$$La = L/a \cup L \cup L/b$$

where $L/a, L, L/b$ are disjoint. The result follows since $L/b = \overrightarrow{La}$ by Corollary 4 to Theorem 2.

The remaining corollaries to the Line Separation Theorem have no simple planar analogues since they involve the ideas of segment or linear order.

8. Convexity of Half-Planes

THEOREM 5.   Any half-plane is convex. (Compare Ch. 10, Th. 9.)

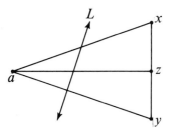

FIGURE 11.8

Proof. Our proof parallels that for the convexity of a ray (Ch. 10, Th. 9). Consider L/a. Let $x \neq y$ and $x, y \subset L/a$. We show $\overline{xy} \subset L/a$. Let $z \subset \overline{xy}$.
First we apply B6 to L, a, x, y. We have $L, a, x, y \subset La$ and $L \not\ni a, x, y$. Moreover L meets $\overline{ax}$ and $\overline{ay}$. Thus by B6, L cannot meet $\overline{xy}$.

Now we apply B6 to L, a, x, z. We have L, a, x, $z \subset La$ and $L \not\supset a$, x, z. (Why?) Thus L meets $\overline{ax}$ implies L meets $\overline{az}$ or $\overline{xz}$ but not both. Suppose L meets $\overline{xz}$. But $\overline{xz} \subset \overline{xy}$. Thus L meets $\overline{xy}$ which is impossible, consequently, L meets $\overline{az}$. Thus $z \subset L/a$ and we conclude $\overline{xy} \subset L/a$.

9. Uniqueness of Opposite Half-Plane

Our treatment closely parallels that for uniqueness of the opposite of a ray (Ch. 10, Sec. 14). First we prove that any half-plane has a unique edge. This rests on two lemmas which correspond to Principles (i), (ii) in the proof that any ray has a unique endpoint (Ch. 10, Th. 11).

LEMMA 1.   If $p \subset L$ then $\overline{pa} \subset \overrightarrow{La}$, provided $a \not\subset L$.

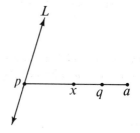

FIGURE 11.9

Proof. Let $x \subset \overline{pa}$. Then, using the Plane Decomposition Principle (Th. 4, Cor. 3)

(1) $$x \subset La = L/a \cup L \cup \overrightarrow{La}.$$

Suppose $x \subset L/a$. Then $\overline{xa}$ meets L, say in q. Thus (Ch. 10, Th. 8, Cor. 1)

$$q \subset \overline{xa} \subset \overline{pa},$$

so that $q \neq p$. Then L meets pa in p, q so that $L = pa \supset a$, contradicting the hypothesis. Suppose $x \subset L$. Then L meets pa in p, x where $p \neq x$ and we get the same contradiction. Thus (1) implies $x \subset \overrightarrow{La}$ and the proof is complete.

LEMMA 2.   Any half-plane is contained in a unique plane.

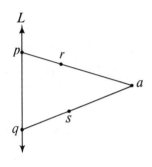

FIGURE 11.10

Proof. Consider half-plane $\overrightarrow{La}$. We have $\overrightarrow{La} \subset La$ (Th. 2, Cor. 5). To prove that La is the only plane which contains $\overrightarrow{La}$, we need merely show that $\overrightarrow{La}$ contains three distinct noncollinear points. Let p, $q \subset L$; $p \neq q$. By Lemma 1

$$\overrightarrow{La} \supset \overline{pa}, \overline{qa}.$$

Let $r \subset \overline{pa}$, $s \subset \overline{qa}$. Then $\overrightarrow{La} \supset a, r, s$. To complete the proof show a, r, s distinct and noncollinear.

THEOREM 6.   Any half-plane has a unique edge. (Compare Ch. 10, Th. 11.)

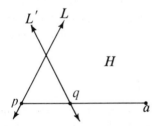

FIGURE 11.11

Proof. Let half-plane H have edges L, L'. We show $L = L'$. Suppose $L \neq L'$. Let $a \subset H$. Then

(1) $$H = \overrightarrow{La} = \overrightarrow{L'a}.$$

We have (Th. 2, Cor. 5)

(2) $$\overrightarrow{La} \subset La, \quad \overrightarrow{L'a} \subset L'a.$$

By Lemma 2, Equations (1) and (2) imply $La = L'a$. Since $L \neq L'$ there exists p such that $p \subset L$, $p \nsubseteq L'$. Then using the Plane Decomposition Principle (Th. 4, Cor. 3)

(3) $$p \subset La = L'a = \overrightarrow{L'a} \cup L' \cup L'/a.$$

Suppose $p \subset \overrightarrow{L'a}$. Then (1) implies $p \subset \overrightarrow{La}$, contrary to the disjointness of L and $\overrightarrow{La}$ (Th. 1 (ii)). Hence (3) implies $p \subset L'/a$ and $\overline{pa}$ meets L', say in q. By Lemma 1 and (1)

$$q \subset \overline{pa} \subset \overrightarrow{La} = \overrightarrow{L'a}$$

contrary to the disjointness of L' and $\overrightarrow{L'a}$. Thus our supposition is false and the theorem holds.

THEOREM 7.   Any half-plane has a unique opposite half-plane. (Compare Ch. 10, Th. 12.)

Proof.   Consider half-plane $\overrightarrow{La}$. We know L/a is opposite $\overrightarrow{La}$ (Th. 3, Cor. 1). Suppose H opposite $\overrightarrow{La}$. We show $H = L/a$. By definition H and $\overrightarrow{La}$ have a common edge. By Theorem 6, L is the only edge of $\overrightarrow{La}$. Thus L is a common edge of H and $\overrightarrow{La}$; and the *only* one. Let $b \subset H$. Since $a \subset \overrightarrow{La}$, we see $\overline{ba}$ meets L by definition of opposite half-planes. Thus $b \subset L/a$ so that $H = L/a$ (Th. 2, Cor. 1) and the proof is complete.

10. Spatial Theory of Order

In order to generalize the theories of linear and planar order to three-dimensional geometry we need an analogue of B4 and B6 which played central roles in those theories. This is provided by

THEOREM 8. (SPATIAL EXTENSION OF PASCH'S POSTULATE)   Let a, b, $c \nsubseteq P$. Then if P meets $\overline{ab}$ it meets $\overline{bc}$ or $\overline{ac}$ but not both.

Proof.   Naturally we try to apply B6. To do this we shall "section" the figure by a plane Q which contains a, b, c. Thus we must prove the existence

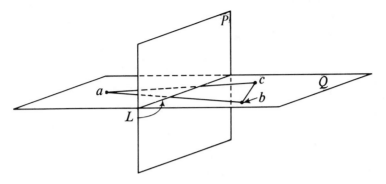

FIGURE 11.12

of such a plane. Note $a \neq b$ by hypothesis. If $c \not\subset ab$ let $Q = abc$. If $c \subset ab$ let $Q = abd$, where d is a point of P not in ab. In either case $Q \supset a, b, c$ and $Q \neq P$. Since $Q \supset \overline{ab}$ it meets P and their intersection is a line L.

We have $a, b, c, L \subset Q$, and $a, b, c \not\subset L$ since $L \subset P$. Moreover L meets $\overline{ab}$. By B6, L meets $\overline{bc}$ or $\overline{ac}$ but not both. The same property holds for P and the theorem is proved.

NOTE: We have used the adjective "spatial" informally in referring to "three-dimensional" generalizations of linear and planar properties, that is, to situations in which there exist at least two planes. We do not introduce the noun "space" in the corresponding sense, since one natural meaning for it is the set of all points—but this might be a plane, a line, or even the null set (Ch. 8, Sec. 10). Our use of the term "spatial" may have led you to suppose that we have assumed the basic ordered incidence geometry studied is three-dimensional. This is not so. We make no existence assumption. We merely apply the term spatial to describe a theorem in which the hypothesis *implies* that the geometry is three-dimensional.

Now our theory will bear an analogy (almost perfect) to the planar theory; this will be borne out in the formulation of definitions, theorems, proofs, and notation.

11. Half-Spaces

We introduce the spatial analogue of rays and half-planes. We take the definition of half-plane and "step up dimension" by 1.

DEFINITION. If plane P does not contain point a, P/a denotes the set of points x such that $\overline{xa}$ meets P. P/a (read "P over a") is called a *half-space*, and P its *face* or *boundary*.

We present a sequence of theorems corresponding to Theorems 1–5. The proofs are not given—sometimes the proof is exactly that of the corresponding planar theorem, sometimes it is a bit simpler insofar as Theorem 8 is slightly simpler than B6.

THEOREM 9. If $a \not\subset P$ then a, P, P/a are disjoint and $P/a \neq \varnothing$. (Compare Th. 1.)

Proof. Use method of Theorem 1 (ii), (iii).

12. Determination of Half-Spaces

THEOREM 10. If P/a meets P/b then $P/a = P/b$. (Compare Th. 2.)

Proof. Use method of Theorem 2.

COROLLARY 1. If $a \not\subset P$ there is one and only one half-space with face P which contains a.

DEFINITION. If $a \not\subset P$ we denote by $\overrightarrow{Pa}$ (read "half-space Pa" or "Pa arrow") the unique half-space with face P which contains a.

COROLLARY 2. Let D be a half-space with face P. Then $a \subset D$ implies $D = \overrightarrow{Pa}$.

NOTE: If $a \subset P/x$ then $P/x = \overrightarrow{Pa}$.

COROLLARY 3. Any half-space P/x with face P is expressible in the form $\overrightarrow{Pa}$.

COROLLARY 4. If $\overline{ab}$ meets P and a, $b \not\subset P$ then $\overrightarrow{Pa} = P/b$, $\overrightarrow{Pb} = P/a$.

Remark. The following statements are equivalent to the hypothesis in Corollary 4:

(1) ab meets P and $a \not\subset P$;

(2) $b \subset P/a$;

(3) $a \subset P/b$.

COROLLARY 5.   If $\overrightarrow{Pa} = P/b$ then $\overrightarrow{Pb} = P/a$.

COROLLARY 6. $\overrightarrow{Pa} = \overrightarrow{Pb}$ if and only if $P/a = P/b$.

13. Opposite Half-Spaces

DEFINITION.   Half-spaces D, D' are *opposite* if they have a common face P, and each point of D is joined to each point of D' by a segment that meets P. (That is $x \subset D$, $y \subset D'$ implies $\overline{xy}$ meets P.)

THEOREM 11.   Let half-spaces D, D' have a common face P. Let there be a point a in D, and a point b in D' such that $\overline{ab}$ meets P. Then each point of D is joined to each point of D' by a segment that meets P, so that D and D' are opposite. (Compare Th. 3.)

Proof.   Use method of Theorem 3.

COROLLARY 1.   Half-spaces $\overrightarrow{Pa}$ and P/a are opposite and two points of the respective half-spaces are always joined by a segment that meets P.

COROLLARY 2.   Suppose $\overline{ab}$ meets P and a, $b \not\subset P$. Then P/a and P/b are opposite and two points of the respective half-spaces are always joined by a segment that meets P.

COROLLARY 3.   Any pair of opposite half-spaces are disjoint.

COROLLARY 4.   No half-space is opposite to itself.

14. Spatial Separation by a Plane

THEOREM 12. (SPATIAL SEPARATION)   Let $\overline{ab}$ meet P and $a, b \notin P$. Then P separates U, the set of all points, into P/a and P/b. (Compare Th. 4.)

Proof. Use method of Theorem 4.

COROLLARY 1. (SPATIAL SEPARATION)   The theorem holds if we replace P/a, P/b by $\overrightarrow{Pa}$, $\overrightarrow{Pb}$.

COROLLARY 2. (UNIQUENESS OF SEPARATION)   Let P be a plane and let U be the set of all points. Suppose $U \neq P$. Then P separates U uniquely into two sets; and these sets are half-spaces with common face P.

COROLLARY 3. (SPATIAL DECOMPOSITION)   Let $a \notin P$. Then U, the set of all points, is given by
$$U = P/a \cup P \cup \overrightarrow{Pa}$$
and P/a, P, $\overrightarrow{Pa}$ are disjoint.

15. Convexity of Half-Spaces

THEOREM 13.   Any half-space is convex. (Compare Th. 5.)

Proof. Use method of Theorem 5.

16. Uniqueness of Opposite Half-Space

LEMMA.   If $p \subset P$ then $\overline{pa} \subset \overrightarrow{Pa}$, provided $a \notin P$. (Compare Sec. 9, Lemma 1.)

Proof. Use method of Section 9, Lemma 1.

THEOREM 14.   Any half-space has a unique face. (Compare Th. 6.)

Proof. Use the method of Theorem 6, but note this is a simpler situation.

THEOREM 15.   Any half-space has a unique opposite half-space. (Compare Th. 7.)

Proof. Use method of Theorem 7.

E X E R C I S E S   I

In the following exercises interpret (abc) to mean that b is between a and c in the sense of Euclidean geometry.

1. Assume that B6 holds in Euclidean geometry. Observe that M9–M14 satisfy B6. Infer that they are ordered incidence geometries. (See Ch. 10, Exercises I, 4, 5.)
2. Show that M23 does not satisfy B6 and so is not an ordered incidence geometry. Show also that the Plane Separation Theorem fails.
3. In M9, what is the interpretation of segment, ray, and half-plane?
4. Find the interpretations of segment, ray, and half-plane in M11 and in M13.
5. Find the interpretations of ray, half-plane, and half-space in M10 and in M12.
6. Find the interpretations of segment, ray, line, and half-plane in M23.

E X E R C I S E S   I I

1. Prove: Every half-plane is infinite.
2. Prove: Two opposite half-planes are coplanar.
3. Prove: If two distinct half-planes with the same edge are coplanar, they are opposite.
4. Prove: The intersection of two coplanar half-planes can not be a point, a ray, or a line. Describe the possible intersections of two coplanar half-planes.
5. If $\overline{ab}$ is contained in a half-plane H, prove that at least one of $\overrightarrow{ab}, \overrightarrow{ba}$ is contained in H.
6. Prove: The intersection of two noncoplanar half-planes is a line, a ray, or the empty set.
7. If $a \notin bc = L$, prove that each of the following is a convex set:

   (i) $L/a \cup b$;   (ii) $L/a \cup \overrightarrow{bc}$;   (iii) $L/a \cup \overline{bc} \cup c$.

8. If K is a convex subset of L (of P), and $a \notin L$, $(a \notin P)$, prove that $L/a \cup K\,(P/a \cup K)$ is a convex set.
9. Prove that Postulate B6 is independent of Postulates I1–I6, B1–B5.
10. Prove: Postulate B4 is deducible from Postulates I1–I6, B1, B2, B3, B5, B6 provided a plane exists.
11. Suppose a, b, c are distinct and noncollinear, (bcd), (cea). Prove that de meets $\overline{ab}$.
12. Suppose a, b, c are distinct and noncollinear, (bcd), (aeb). Prove that de meets $\overline{ac}$.
13. If $a \notin L$, prove that $j(L/a, \overrightarrow{La}) = La$.
   (For the definition of $j(A, B)$, the join of sets A and B, see Ch. 10, Exercises IV.)
14. If $a \notin P$, prove that $j(P/a, \overrightarrow{Pa}) = U$, where U is the set of all points.
15. Is the following definition of half-plane equivalent to the one we have adopted? A half-plane with edge L is a set consisting of a point $p \notin L$ and all points x such that $x \subset Lp$ and $\overline{xp}$ does not meet L. Justify your answer.
16. Prove Theorem 13: Any half-space is convex.
17. Prove Theorem 14: Any half-space has a unique face.
18. Prove Theorem 15: Any half-space has a unique opposite half-space.
†19. Prove: Postulate B6 is equivalent to the following weaker principle: Let L coplane with and not contain a, b, c. Then if L meets $\overline{ab}$ it meets at least one of $\overline{bc}$ or $\overline{ac}$.

E X E R C I S E S   I I I

†1. Formulate precisely and prove that two parallel lines of a plane separate it into three convex sets.
†2. Formulate precisely and prove that two distinct intersecting lines of a plane separate it into four convex sets.
 3. Try to formulate a general concept of separation which will cover the Line, Plane and Spatial Separation Theorems (Ch. 10, Th. 7; Ch. 11, Th. 4, 12) and the separation of a Euclidean plane by a circle, a parabola, or a sine curve.
 4. It is intuitively evident that a point does not separate a plane. Try to formulate this as a precise statement and prove it.
†5. Formulate precisely and prove: If ray R is contained in half-plane H, and the endpoint of R is in the edge of H, then R separates H.
 6. Let L, M be distinct intersecting lines contained in plane P. Prove that each point of P is in some line that joins a point of L and a point of M.

7. Find a "spatial" generalization of Exercise 6 and prove it correct.
8. In Exercise 6 suppose $L \parallel M$. Can you still prove the result? Justify your answer.
9. In Exercise 6, can you prove that each point of P, other than the intersection of L and M, is in some segment that joins a point of L and a point of M? Justify your answer.
10. In Exercise 6, prove that each point of P is in a ray $\overrightarrow{ab}$ such that $a \subset L$, $b \subset M$.
11. Find a "spatial" generalization of Exercise 10 and prove it correct. (Compare Exercise 7.)

E X E R C I S E S   I V

In the following exercises involving M19, betweenness is interpreted in this way: Let $a = (a_1, a_2)$, $b = (b_1, b_2)$, $c = (c_1, c_2)$. Then (abc) means $a \neq c$ and there exist positive real numbers s, t such that

$$b_1 = sa_1 + tc_1, \quad b_2 = sa_2 + tc_2, \quad s + t = 1.$$

1. In M19 let $a = (0, 0)$, $b = (1, 0)$. Show a/b is the set of dyads of the form $(x_1, 0)$ where $x_1 < 0$.

2. In Exercise 1 determine the significance of b/a, $\overrightarrow{ab}$, $\overrightarrow{ba}$.

3. In M19 let $a = (0, 0)$, $b = (1, 1)$. Determine the significance of $\overline{ab}$, a/b, b/a, $\overrightarrow{ab}$.

4. In M19 prove that $x \subset a/b$, $[x = (x_1, x_2), a = (a_1, a_2), b = (b_1, b_2)]$ if and only if $a \neq b$ and there exist s, t such that $s > 0$, $t < 0$ and

$$x_1 = sa_1 + tb_1, \quad x_2 = sa_2 + tb_2, \quad s + t = 1.$$

(Compare Ch. 10, Exercises I, 15.)

5. Assume that M19 satisfies B3 (see Ch. 10, Exercises I, 10, 13, 16). Use Exercise 4 to prove that dyad x is in line ab if and only if $a \neq b$ and there exist real numbers s, t such that

$$x_1 = sa_1 + tb_1, \quad x_2 = sa_2 + tb_2, \quad s + t = 1.$$

6. In M19 let L be the linear system composed of all dyads of the form $(x_1, 0)$. Let $a = (0, 1)$. Show that L/a is the set of dyads (x_1, x_2) such that $x_2 < 0$; and $\overrightarrow{La}$ is the set of dyads (x_1, x_2) such that $x_2 > 0$.

†7. Let $L(x_1, x_2)$ denote the linear expression $dx_1 + ex_2 + f$, $(x_1, x_2, d, e, f$ real; d, e not both zero). Let L' be the linear system determined by $L(x_1, x_2) = 0$. Let $a = (a_1, a_2)$, $b = (b_1, b_2)$. Prove:

(i) If $L(a_1, a_2) > 0$, $L(b_1, b_2) < 0$ then $\overline{ab}$ meets L'. [*Hint.* If $p = L(a_1, a_2)$, $q = L(b_1, b_2)$ then there exist s, t such that

$$sp + tq = 0, \quad s + t = 1, \quad 0 < s, \quad 0 < t.]$$

(ii) If $\overline{ab}$ meets L' and a, $b \notin L'$ then $L(a_1, a_2)$ and $L(b_1, b_2)$ have opposite signs.

8. Use Exercise 7 to prove that M19 satisfies B6.
9. In the notation of Exercise 7 let $L(a_1, a_2) > 0$. Show that L'/a is the set of dyads (x_1, x_2) such that $L(x_1, x_2) < 0$. What is the interpretation of $\overrightarrow{L'a}$? Justify your answer. Interpret your results in the cartesian plane.

EXERCISES V

In the following exercises assume that the basic ordered incidence geometry satisfies the Euclidean parallel postulate (Ch. 9, Sec. 1, Postulate E) and so is an affine geometry in the sense of Chapter 9. Such an incidence geometry is called an *ordered affine geometry*.

1. Suppose $\overrightarrow{op}$, $\overrightarrow{oq}$ are distinct and not opposite. Let line L meet op and oq in distinct points but not meet $\overrightarrow{op}$ or $\overrightarrow{oq}$. Let x be any point of $\overrightarrow{op}$. Prove that there exists a unique point x' such that $xx' \parallel L$ and $x' \subset \overrightarrow{oq}$. Prove that $x \to x'$ is a one-to-one correspondence between $\overrightarrow{op}$ and $\overrightarrow{oq}$. (We call a correspondence (or mapping) of this type a *parallel projection*.)
2. Prove that a parallel projection (Exercise 1) preserves betweenness in the following sense: Suppose

$$a \to a', \quad b \to b', \quad c \to c';$$

then (abc) if and only if $(a'b'c')$.
3. Prove that a parallel projection "maps a segment onto a segment," that is, the set of points corresponding to the points of a segment, themselves form a segment.
4. Given $\overline{ab}$, $\overline{cd}$ noncollinear segments. Show that by means of two parallel projections, applied successively, $\overline{ab}$ can be "projected" onto $\overline{cd}$. Is this correct if $\overline{ab}$ and $\overline{cd}$ colline? [*Hint.* If $c \notin ab$, choose the first projection so that a is projected into c.]

DEFINITION.   The set of all lines of a plane parallel to a given line L of the plane, together with L, is called a *parallel pencil* of lines.

5. Develop a theory of betweenness for lines of a parallel pencil. Formulate a precise definition and see whether B1–B5 (properly interpreted) hold.

DEFINITION.   The set of all lines of a three-dimensional geometry parallel to a given line L, together with L, is called a *parallel bundle* of lines.

†6. Consider the following system:
   Point: Each line of a parallel bundle B.
   Line: Each parallel pencil of lines contained in B.
   Plane: The bundle B.
   Show that this system, with a suitable interpretation of betweenness (see Exercise 5), is an ordered affine geometry.

CHAPTER 12

Angles and Order of Rays

This chapter is devoted to the study of angles and of betweenness of rays. Although the topics are interrelated, the theory of order of rays may be considered more fundamental and we shall spend most of our time on it. The concept of betweenness of rays arises out of betweenness of points and leads directly into the theory of angles. As in Chapter 11, we study ordered incidence geometries and assume I1–I6, and B1–B6. In our proofs we make frequent use of the notion of "sides of a line" which is formalized in Section 2.

1. Introduction

The notion of angle arose historically in connection with attempts to measure differences of direction in space or to determine orientation of objects in space, for example, the angle of elevation of a star. It involves a complex of related ideas: (1) a figure associated with two intersecting lines; (2) a region of the plane bounded by two intersecting lines; (3) a numerical measure of the difference in direction of two intersecting lines. The first roughly describes angle in its simplest sense as a linear or one-dimensional figure. The second, the portion of the plane bounded by an angle in the first sense, that is, the interior of the angle. The third is concerned with measurement of angles. There are even more complicated associated ideas, as when we speak of angles as rotations of unrestricted magnitude, positive or negative.

We shall be concerned in this chapter with angle in its simplest sense as a figure formed by two noncollinear rays with a common endpoint. The interior

240

of an angle will be studied in Chapter 13 and the properties of angle measurement will be introduced in Chapter 14.

The study of nonmetrical properties of angles is related to the study of betweenness of rays, historically a much later and more abstract idea. The two concepts are related in this way: A ray is "between" two rays if the three rays have a common vertex and the first is inside the angle determined by the other two. Roughly speaking, betweenness of rays is related to angles as betweenness of points to segments. Thus we shall make a detailed study of the betweenness relation for rays before formally introducing the concept of angle. This has two advantages: (1) we can base the theory of order of rays on the theory of order of points—the one is a natural outgrowth of the other; (2) configurational properties of angles tend to be assimilated to the formal properties of ray betweenness and are subsumed under a more general and abstract frame of reference.

2. The Sides of a Line

We shall constantly be concerned with a complex of properties associated with planar separation; in practice this can be dealt with simply and graphically, not by appealing to the Plane Separation Theorem (Ch. 11, Th. 4), but by introducing the concept and terminology of "sides of a line."

DEFINITION.   A half-plane with edge L is called a *side* of L. Two opposite half-planes with edge L are called *opposite sides* of L. Two points or two sets of points are *in* or *on* the *same side* of L, if they are contained in the same half-plane with edge L; they are *in* or *on opposite sides* of L if they are contained in opposite half-planes with edge L.

Since each point not in L is in a unique half-plane with edge L, and each half-plane with edge L has a unique opposite half-plane with edge L, the following principles are immediate:

(i) *If a, b are on the same side of L, and b, c are on the same side of L, then a, c are on the same side of L.*

(ii) *If a, b are on the same side of L, and b, c are on opposite sides of L, then a, c are on opposite sides of L.*

(iii) *If a, b are on opposite sides of L, and b, c are on opposite sides of L, then a, c are on the same side of L.*

We begin with two theorems which give elementary criteria that two points be on the same side or on opposite sides of a line.

THEOREM 1.   Two distinct points are on the same side of L if and only if they coplane with L, are not in L, and the segment joining them does not meet L.

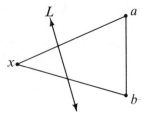

FIGURE 12.1

Proof.   Let a, b be distinct points on the same side of L. By definition there is a half-plane L/x such that a, $b \subset L/x$. Thus $\overline{xa}$ meets L and $\overline{xb}$ meets L. By Theorem 1 of Chapter 11, $L/x \subset Lx$. Thus a, b, x, $L \subset Lx$. Moreover a, b, $x \not\subset L$. Therefore B6 applies to x, a, b, L and yields $\overline{ab}$ does not meet L.

Conversely, suppose a, b, L coplane; a, $b \not\subset L$, and $\overline{ab}$ does not meet L. Let $x \subset L/a$, so that $\overline{xa}$ meets L. By using $L/a \subset La$ we have x, a, b, $L \subset La$; moreover a, b, $x \not\subset L$. Hence B6 applies and yields $\overline{xb}$ meets L. Thus $b \subset L/x$ and $a \subset L/x$; and a, b are on the same side of L by definition.

THEOREM 2.   Two points are on opposite sides of L if and only if they are not in L, and the segment joining them meets L.

Proof.   Let a, b be on opposite sides of L. Then $a \subset L/x$, $b \subset L/y$ where L/x, L/y are opposite half-planes with edge L. By Theorem 1 of Chapter 11, a, $b \not\subset L$. By definition of opposite half-plane $\overline{ab}$ meets L, since a half-plane has a unique edge.

Conversely, suppose a, $b \not\subset L$ and $\overline{ab}$ meets L. Note $a \subset \overrightarrow{La}$, $b \subset \overrightarrow{Lb}$. Hence $\overrightarrow{La}$ and $\overrightarrow{Lb}$ are opposite half-planes with edge L (Ch. 11, Th. 3). Thus a and b are on opposite sides of L, by definition.

We now derive several results which relate rays to half-planes and so to the sides of a line.

THEOREM 3.   If $p \subset L$ and $a \not\subset L$, then $p/a \subset L/a$ and $\overrightarrow{pa} \subset \overrightarrow{La}$.

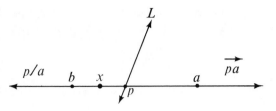

FIGURE 12.2

Proof. Let $x \subset p/a$. Then (xpa), and $p \subset \overline{xa}$, so that $\overline{xa}$ meets L in p. By definition $x \subset L/a$, and we conclude $p/a \subset L/a$.

To complete the proof note there exists b such that (bpa). This implies $\overrightarrow{pa} = p/b$ (Ch. 10, Th. 5, Cor. 4). Note $b \subset L/a$. This implies $\overrightarrow{La} = L/b$ (Ch. 11, Th. 2, Cor. 4). By the first paragraph of the proof, $p/b \subset L/b$. Thus

$$\overrightarrow{pa} = p/b \subset L/b = \overrightarrow{La},$$

or $\overrightarrow{pa} \subset \overrightarrow{La}$.

COROLLARY 1. Suppose $p \subset L$ and $a \not\subset L$. Then

  (i) all points of $\overrightarrow{pa}$ are on the same side of L as a;
  (ii) all points of p/a are on the opposite side of L from a.

Proof. (i) $\overrightarrow{pa} \subset \overrightarrow{La}$ by the theorem.

  (ii) $p/a \subset L/a$ and L/a, $\overrightarrow{La}$ are opposite (Ch. 11, Th. 3, Cor. 1).

COROLLARY 2. If $p \subset L$ and $a \not\subset L$, then all points of $\overline{pa}$ are on the same side of L as a. (Compare Ch. 11, Sec. 9, Lemma 1.)

Proof. By Corollary 1, since $\overline{pa} \subset \overrightarrow{pa}$ (Ch. 10, Th. 7, Cor. 4).

COROLLARY 3. If $p \subset L$ and $a \not\subset L$, then $\overrightarrow{pa}$ (p/a) comprises all points of pa on the same side (opposite side) of L as (from) a.

Proof. We give the argument for $\overrightarrow{pa}$. By Corollary 1 each point of $\overrightarrow{pa}$ is on the same side of L as a, and is in pa since $\overrightarrow{pa} \subset pa$. Conversely, suppose

$x \subset pa$ and is on the same side of L as a. By the Line Decomposition Principle (Ch. 10, Th. 7, Cor. 3)

$$pa = p/a \cup p \cup \overrightarrow{pa}.$$

Suppose $x \subset p/a$. By Corollary 1, x would be on the opposite side of L from a. Since $x \neq p$, we infer $x \subset \overrightarrow{pa}$.

The results of this section though interesting and important in themselves will be found especially useful in the sequel.

3. Betweenness of Rays

The concept of betweenness of points induces in a natural way a similar concept for rays.

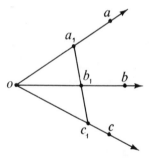

FIGURE 12.3

DEFINITION.   Let rays $\overrightarrow{oa}$, $\overrightarrow{ob}$, $\overrightarrow{oc}$ with common endpoint o be such that $\overrightarrow{oa}$ and $\overrightarrow{oc}$ are distinct and not opposite (Fig. 12.3). Let there exist points a_1, b_1, c_1, such that $a_1 \subset \overrightarrow{oa}$, $b_1 \subset \overrightarrow{ob}$, $c_1 \subset \overrightarrow{oc}$ and $(a_1 b_1 c_1)$. Then we say $\overrightarrow{ob}$ is *between* $\overrightarrow{oa}$ and $\overrightarrow{oc}$, and write $(\overrightarrow{oa}\ \overrightarrow{ob}\ \overrightarrow{oc})$.

NOTE: The condition that $\overrightarrow{oa}$ and $\overrightarrow{oc}$ are distinct and not opposite is essential to insure that the rays in a betweenness relation are distinct. This condition may appear in several equivalent forms: (1) o, a, c are distinct and noncollinear; (2) $o \not\subset ac$; (3) $\overrightarrow{oa}$ and $\overrightarrow{oc}$ are noncollinear (Ch. 10, Exercises II, 12).

We proceed to derive some elementary properties of betweenness of rays.

THEOREM 4. $(\overrightarrow{oa}\ \overrightarrow{ob}\ \overrightarrow{oc})$ implies $(\overrightarrow{oc}\ \overrightarrow{ob}\ \overrightarrow{oa})$.

Proof.   Immediate from definition since $(a_1 b_1 c_1)$ implies $(c_1 b_1 a_1)$.

THEOREM 5. $(\overrightarrow{oa}\ \overrightarrow{ob}\ \overrightarrow{oc})$ implies that each pair of $\overrightarrow{oa},\ \overrightarrow{ob},\ \overrightarrow{oc}$ are distinct and not opposite.

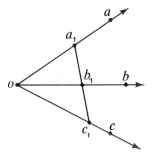

FIGURE 12.4

Proof.   By definition there exist points $a_1 \subset \overrightarrow{oa}$, $b_1 \subset \overrightarrow{ob}$, $c_1 \subset \overrightarrow{oc}$ such that $(a_1 b_1 c_1)$. Thus $\overrightarrow{oa_1} = \overrightarrow{oa}$, $\overrightarrow{ob_1} = \overrightarrow{ob}$, $\overrightarrow{oc_1} = \overrightarrow{oc}$. Also, $\overrightarrow{oa}$ and $\overrightarrow{oc}$ are distinct and not opposite; so, therefore, are $\overrightarrow{oa_1}$ and $\overrightarrow{oc_1}$. Thus by the note above $o \not\subset a_1 c_1$. $(a_1 b_1 c_1)$ implies $a_1 b_1 = a_1 c_1$. Thus $o \not\subset a_1 b_1$. By the note above $\overrightarrow{oa_1},\ \overrightarrow{ob_1}$ are distinct and not opposite; and so therefore are $\overrightarrow{oa}$ and $\overrightarrow{ob}$. Similarly $\overrightarrow{ob},\ \overrightarrow{oc}$ are distinct and not opposite.

THEOREM 6. $(\overrightarrow{oa}\ \overrightarrow{ob}\ \overrightarrow{oc})$ implies
 (i) a, b are on the same side of oc;
 (ii) b, c are on the same side of oa;
 (iii) a, c are on opposite sides of ob.

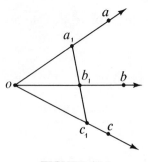

FIGURE 12.5

Proof. $(\overrightarrow{oa}\ \overrightarrow{ob}\ \overrightarrow{oc})$ implies there exist $a_1 \subset \overrightarrow{oa}$, $b_1 \subset \overrightarrow{ob}$, $c_1 \subset \overrightarrow{oc}$ such that $(a_1b_1c_1)$. By Theorem 5 $\overrightarrow{oa}$, $\overrightarrow{oc}$ are distinct and not opposite. Hence o, a, c are noncollinear and $a \not\subset oc$. Thus $a_1 \subset \overrightarrow{oa}$ implies a_1, a are on the same side of oc by Corollary 1 to Theorem 3. Similarly b_1, b are on the same side of oc. But $b_1 \subset \overline{a_1c_1}$. Thus a_1, b_1 are on the same side of oc by Corollary 2 to Theorem 3. Therefore a, b are on the same side of oc. By symmetry, b, c are on the same side of oa, and (i), (ii) are proved.

Continuing we argue as above to show that a_1, a are on the same side of ob; and c_1, c are on the same side of ob. But $\overline{a_1c_1}$ meets ob in b_1. Thus a_1, c_1 are on opposite sides of ob by Theorem 2. We infer that a, c are on opposite sides of ob, and (iii) is proved.

Note that betweenness is an intrinsic relation of three rays—it does not depend on the notation used to specify the rays. Thus we may assert the

COROLLARY.   Suppose $(\overrightarrow{oa}\ \overrightarrow{ob}\ \overrightarrow{oc})$. Then

(i) each point of $\overrightarrow{oa}$ and each point of $\overrightarrow{ob}$ are on the same side of oc;

(ii) each point of $\overrightarrow{ob}$ and each point of $\overrightarrow{oc}$ are on the same side of oa;

(iii) each point of $\overrightarrow{oa}$ and each point of $\overrightarrow{oc}$ are on opposite sides of ob.

THEOREM 7.   Suppose $(\overrightarrow{oa}\ \overrightarrow{ob}\ \overrightarrow{oc})$, $a_1 \subset \overrightarrow{oa}$, $c_1 \subset \overrightarrow{oc}$. Then $\overrightarrow{ob}$ meets $\overline{a_1c_1}$.

Proof.   By the last corollary a_1 and c_1 are on opposite sides of ob. Thus $\overline{a_1c_1}$ meets ob, say in b_1. Points a_1, b_1 are on the same side of oc (Th. 3,

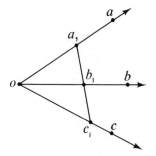

FIGURE 12.6

Cor. 2). And a_1, b are on the same side of oc by the last corollary. Thus b_1, b are on the same side of oc. Since $b_1 \subset ob$, we have $b_1 \subset \overrightarrow{ob}$ (Th. 3, Cor. 3). Thus $\overrightarrow{ob}$ meets $\overline{a_1 c_1}$.

Note that this result makes universal the defining property of $(\overrightarrow{oa} \ \overrightarrow{ob} \ \overrightarrow{oc})$: Now we know that *every* segment joining a point of $\overrightarrow{oa}$ to a point of $\overrightarrow{oc}$ meets $\overrightarrow{ob}$.

4. Relation Between Order of Points and Order of Rays

Theorem 7 enables us to extend certain betweenness properties of points of a line to betweenness properties of rays. Specifically it enables us to relate the order properties of rays that are *between two rays* to the order properties of the points between two points. To do this, choose $\overrightarrow{oa}$, $\overrightarrow{ob}$ to be distinct and not opposite (Fig. 12.7). Then associate with each point x in $\overline{ab}$ the ray $\overrightarrow{ox}$.

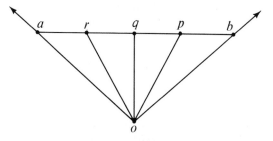

FIGURE 12.7

This establishes a one-to-one correspondence between the set of points in $\overline{ab}$ (the points between a and b) and the set of rays between $\overrightarrow{oa}$ and $\overrightarrow{ob}$. Moreover an order relation (pqr) subsists for three points of $\overline{ab}$ if and only if a corresponding order relation $(\overrightarrow{op}\,\overrightarrow{oq}\,\overrightarrow{or})$ subsists for the associated rays. To indicate this we may say that the correspondence or mapping $x \to \overrightarrow{ox}$ is an *isomorphic mapping* between $\overline{ab}$ considered as an ordered system of points, and the set of rays between $\overrightarrow{oa}$ and $\overrightarrow{ob}$ considered as an ordered system of rays. Thus it should not be surprising that we can carry over a sizable portion of the theory of betweenness for points, to betweenness for rays.

As an example consider the following theorem.

THEOREM 8. $(\overrightarrow{oa}\,\overrightarrow{ob}\,\overrightarrow{oc})$ implies that $(\overrightarrow{ob}\,\overrightarrow{oc}\,\overrightarrow{oa})$ is false. (Compare Postulate B2.)

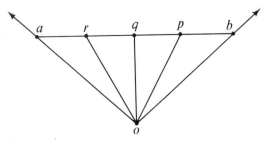

FIGURE 12.8

Proof. By definition, there exist $a_1 \subset \overrightarrow{oa}$, $b_1 \subset \overrightarrow{ob}$, $c_1 \subset \overrightarrow{oc}$ such that $(a_1b_1c_1)$. Suppose $(\overrightarrow{ob}\,\overrightarrow{oc}\,\overrightarrow{oa})$; then by Theorem 7, $\overline{b_1a_1}$ meets $\overrightarrow{oc}$ in a point, which must be c_1. Thus $c_1 \subset \overline{b_1a_1}$, so that $(b_1c_1a_1)$. But $(a_1b_1c_1)$ implies $(b_1c_1a_1)$ is false. Thus our supposition is false.

Note how the result is "induced" by the corresponding betweenness property for points.

COROLLARY.   Suppose (ABC), where A, B, C are rays. Then (CBA) holds, and (BCA), (BAC), (ACB), (CAB), are false. (Compare Ch. 10, Th. 1.)

Proof. (ABC) implies (CBA) by Theorem 4. (ABC), (CBA) imply respectively that (BCA) and (BAC) are false, by Theorem 8. Suppose (ACB); then by Theorem 4, (BCA), which is false. Thus (ACB) is false. Similarly we show (CAB) is false.

Another application of Theorem 7 yields an important order property of four rays.

THEOREM 9. $(\overrightarrow{oa}\ \overrightarrow{ob}\ \overrightarrow{oc})$, $(\overrightarrow{oa}\ \overrightarrow{oc}\ \overrightarrow{od})$ imply $(\overrightarrow{oa}\ \overrightarrow{ob}\ \overrightarrow{od})$, $(\overrightarrow{ob}\ \overrightarrow{oc}\ \overrightarrow{od})$. (Compare Ch. 10, Th. 13, Cor. 4.)

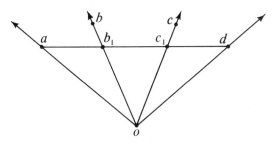

FIGURE 12.9

Proof.   By Theorem 7, $(\overrightarrow{oa}\ \overrightarrow{oc}\ \overrightarrow{od})$ implies $\overrightarrow{oc}$ meets $\overline{ad}$, say in c_1; thus (ac_1d). Similarly $(\overrightarrow{oa}\ \overrightarrow{ob}\ \overrightarrow{oc})$ and $c_1 \subset \overrightarrow{oc}$ implies $\overrightarrow{ob}$ meets $\overline{ac_1}$, say in b_1; thus (ab_1c_1). (ab_1c_1), (ac_1d) imply (ab_1d), (b_1c_1d) (Ch. 10, Th. 13, Cor. 4).

Note $o \not\subset ad$. Thus (ab_1d) yields $(\overrightarrow{oa}\ \overrightarrow{ob}\ \overrightarrow{od})$ by definition. Similarly (b_1c_1d) yields $(\overrightarrow{ob}\ \overrightarrow{oc}\ \overrightarrow{od})$ and the proof is complete.

One should not assume that all order properties of points are valid for rays. Consider the property, (abc) and (bcd) imply (abd) (Ch. 10, Th. 13, Cor. 2). The corresponding property for rays, (ABC) and (BCD) imply (ABD), is false.

5. Order Properties Involving Opposite Rays

Although the theory of order for rays has its genesis in order theory for points and they share a common core of properties, the two theories are not identical. As we indicated at the end of Section 4, there are order properties of points that do not hold for rays. There are also order properties of rays that have no valid analogues for points. We consider now several important examples of such properties which involve opposite rays.

THEOREM 10. $(\overrightarrow{oa}\ \overrightarrow{ob}\ \overrightarrow{oc})$ implies $(\overrightarrow{ob}\ \overrightarrow{oc}\ \overrightarrow{oa'})$, where $\overrightarrow{oa'}$ is the ray opposite $\overrightarrow{oa}$.

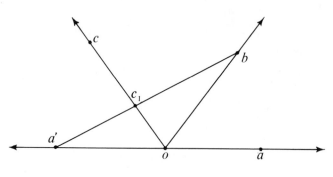

FIGURE 12.10

Proof. $\overrightarrow{oa'}$ is opposite $\overrightarrow{oa}$ implies (aoa'). Thus $a' \subset o/a$ and a' is on the opposite side of oc from a (Th. 3, Cor. 1). But a, b are on the same side of oc (Th. 6). Hence b, a' are on opposite sides of oc, and $\overline{ba'}$ meets oc, say in c_1.

To locate c_1 on oc, note that b, c are on the same side of oa (Th. 6); and b, c_1 are on the same side of oa (Th. 3, Cor. 2). Thus c_1, c are on the same side of oa, and $c_1 \subset \overrightarrow{oc}$ (Th. 3, Cor. 3).

Since $\overrightarrow{oa}, \overrightarrow{ob}$ are distinct and not opposite (Th. 5) it follows that $\overrightarrow{oa'}, \overrightarrow{ob}$ are distinct and not opposite. Thus (bc_1a') yields $(\overrightarrow{ob}\ \overrightarrow{oc}\ \overrightarrow{oa'})$ by definition.

In the following corollaries A, B, C are rays and A', B', C' their respective opposite rays. Using this notation Theorem 10 asserts: (ABC) implies (BCA').

COROLLARY 1.   (CONVERSE OF THEOREM 10) (BCA') implies (ABC).

Proof. By Theorem 4, (BCA') implies $(A'CB)$. By Theorem 10, $(A'CB)$ implies (CBA) since A is opposite A'. By Theorem 4 (ABC).

COROLLARY 2. (ABC) implies $(A'B'C')$, and conversely.

Proof.   Applying Theorem 10 three times: (ABC) implies (BCA'), which implies $(CA'B')$, which in turn implies $(A'B'C')$. By the principle just proved, $(A'B'C')$ implies (ABC), since A, B, C are opposite A', B', C' respectively.

COROLLARY 3. $(AB'C)$ is equivalent to $(BC'A)$; and to $(CA'B)$.

Proof.   It is sufficient to show that $(AB'C)$ implies $(BC'A)$, which implies $(CA'B)$, which in turn implies $(AB'C)$. We have $(AB'C)$ implies $(B'CA')$ by Theorem 10; and $(B'CA')$ implies $(BC'A)$ by Corollary 2. Similarly $(BC'A)$ implies $(CA'B)$ which implies $(AB'C)$.

COROLLARY 4. (ABC) implies the falsity of $(A'BC)$, $(AB'C)$ and (ABC').

Proof.   Suppose (ABC) and $(A'BC)$. By Theorem 10, $(A'BC)$ implies (BCA), which implies (ABC) is false, by the corollary to Theorem 8.

Suppose (ABC) and $(AB'C)$. (ABC) implies (BCA'), by Theorem 10. $(AB'C)$ implies $(CA'B)$ by Corollary 3. But (BCA') and $(CA'B)$ contradict the corollary to Theorem 8.

Suppose (ABC) and (ABC'). By Corollary 1, (ABC') implies (CAB); which, with (ABC) contradicts the corollary to Theorem 8.

COROLLARY 5. $(AB'C)$ implies that A, B, C are not connected by an order relation.

Proof.   By Corollary 3, $(AB'C)$ implies $(BC'A)$ and $(CA'B)$. By Corollary 4, $(AB'C)$ implies (ABC) is false; $(BC'A)$ implies (BCA) is false; and $(CA'B)$ implies (CAB) is false.

The other three possible order relations, (CBA), (ACB), (BAC) are equivalent respectively to (ABC), (BCA), (CAB) by Theorem 4.

THEOREM 11.   Let b and c be on opposite sides of oa, and let $\overrightarrow{oa'}$ be opposite $\overrightarrow{oa}$. Then $(\overrightarrow{ob}\ \overrightarrow{oa}\ \overrightarrow{oc})$, $(\overrightarrow{ob}\ \overrightarrow{oa'}\ \overrightarrow{oc})$ or $\overrightarrow{ob}$ and $\overrightarrow{oc}$ are opposite.

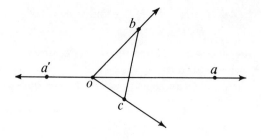

FIGURE 12.11

Proof. The theorem holds if $\overrightarrow{ob}$, $\overrightarrow{oc}$ are opposite, so we may suppose the contrary. Clearly, $\overrightarrow{ob} \neq \overrightarrow{oc}$. Since $\overrightarrow{oa}$ and $\overrightarrow{oa'}$ are opposite, (aoa'). By the Line Separation Principle (Ch. 10, Th. 7, Cor. 1)

(1) $$oa = aa' = \overrightarrow{oa} \cup \overrightarrow{oa'} \cup o.$$

Since b, c are on opposite sides of oa, $\overline{bc}$ meets oa and Equation (1) implies (a) $\overline{bc}$ meets $\overrightarrow{oa}$ or (b) $\overline{bc}$ meets $\overrightarrow{oa'}$ or (c) $\overline{bc} \supset o$. If (a), $(\overrightarrow{ob}\ \overrightarrow{oa}\ \overrightarrow{oc})$ by definition. Similarly (b) yields $(\overrightarrow{ob}\ \overrightarrow{oa'}\ \overrightarrow{oc})$. If (c) then (boc) so that $\overrightarrow{ob}$, $\overrightarrow{oc}$ are opposite contrary to supposition.

COROLLARY 1.   Let b and c be on the same side of oa, and $\overrightarrow{oa'}$ be opposite $\overrightarrow{oa}$. Then $(\overrightarrow{oa}\ \overrightarrow{oc}\ \overrightarrow{ob})$, $(\overrightarrow{oa'}\ \overrightarrow{oc}\ \overrightarrow{ob})$ or $\overrightarrow{ob} = \overrightarrow{oc}$.

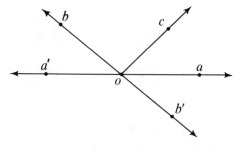

FIGURE 12.12

Proof. Let $\overrightarrow{ob'}$ be opposite $\overrightarrow{ob}$. Then $(b'ob)$ and $b' \subset o/b$. Thus b' is on the opposite side of oa from b (Th. 3, Cor. 1). Hence b' and c are on opposite sides of oa. By the theorem $(\overrightarrow{ob'}\ \overrightarrow{oa}\ \overrightarrow{oc})$, $(\overrightarrow{ob'}\ \overrightarrow{oa'}\ \overrightarrow{oc})$ or $\overrightarrow{ob'}$ and $\overrightarrow{oc}$ are opposite. By using Theorem 10 we have $(\overrightarrow{oa}\ \overrightarrow{oc}\ \overrightarrow{ob})$, $(\overrightarrow{oa'}\ \overrightarrow{oc}\ \overrightarrow{ob})$ or $\overrightarrow{ob} = \overrightarrow{oc}$.

COROLLARY 2.   Let b, c be on the same side of oa. Then $(\overrightarrow{oa}\ \overrightarrow{ob}\ \overrightarrow{oc})$, $(\overrightarrow{oa}\ \overrightarrow{oc}\ \overrightarrow{ob})$ or $\overrightarrow{ob} = \overrightarrow{oc}$.

Proof. Let $\overrightarrow{oa'}$ be opposite $\overrightarrow{oa}$. By the preceding corollary, $(\overrightarrow{oa}\ \overrightarrow{oc}\ \overrightarrow{ob})$, $(\overrightarrow{oa'}\ \overrightarrow{oc}\ \overrightarrow{ob})$ or $\overrightarrow{ob} = \overrightarrow{oc}$. By Theorem 10 $(\overrightarrow{oa'}\ \overrightarrow{oc}\ \overrightarrow{ob})$ implies $(\overrightarrow{oc}\ \overrightarrow{ob}\ \overrightarrow{oa})$; and by Theorem 4, $(\overrightarrow{oc}\ \overrightarrow{ob}\ \overrightarrow{oa})$ implies $(\overrightarrow{oa}\ \overrightarrow{ob}\ \overrightarrow{oc})$. Thus the corollary follows.

Observe that Corollary 2 has a simple analogue for points of a line: Suppose b, c are in a ray with endpoint a (or as we might say b, c are on the *same side* of point a). Then (abc), (acb) or $b = c$. (See Ch. 10, Th. 7, Cor. 5.)

NOTE: In Theorem 11 and in its first two corollaries, only one of the three conditions in each conclusion holds. If either of the order relations for rays holds, then in the theorem $\overrightarrow{ob}$ is not opposite $\overrightarrow{oc}$, and in the corollaries $\overrightarrow{ob} \neq \overrightarrow{oc}$, in each case by Theorem 5. Furthermore, in each case, the two order relations for rays cannot hold simultaneously; in the theorem and in Corollary 1 by Corollary 4 to Theorem 10, and in Corollary 2 by the corollary to Theorem 8.

COROLLARY 3.   (A converse of Theorem 6)   If a, b are on the same side of oc, and b, c are on the same side of oa, then $(\overrightarrow{oa}\ \overrightarrow{ob}\ \overrightarrow{oc})$.

Proof.   Corollary 2 implies

(1) $\qquad\qquad\qquad (\overrightarrow{oc}\ \overrightarrow{oa}\ \overrightarrow{ob})$, $(\overrightarrow{oc}\ \overrightarrow{ob}\ \overrightarrow{oa})$ or $\overrightarrow{oa} = \overrightarrow{ob}$;

(2) $\qquad\qquad\qquad (\overrightarrow{oa}\ \overrightarrow{ob}\ \overrightarrow{oc})$, $(\overrightarrow{oa}\ \overrightarrow{oc}\ \overrightarrow{ob})$ or $\overrightarrow{ob} = \overrightarrow{oc}$.

$\overrightarrow{oa} \neq \overrightarrow{ob}$ and $\overrightarrow{ob} \neq \overrightarrow{oc}$ since $b \not\subseteq oa, oc$. Now $(\overrightarrow{oa} \; \overrightarrow{ob} \; \overrightarrow{oc})$ must hold since the alternative is $(\overrightarrow{oc} \; \overrightarrow{oa} \; \overrightarrow{ob})$ and $(\overrightarrow{oa} \; \overrightarrow{oc} \; \overrightarrow{ob})$, contrary to the corollary of Theorem 8.

NOTE: This result is interesting and important since it relates order of rays to planar separation properties which seem more basic. Observe that the criterion does not even refer to the concept of ray.

Theorem 11 and its corollaries are important and useful results in the foundations of geometry despite the relative ease with which they have been obtained. As an application we prove that the diagonals of a parallelogram intersect (see Ch. 1, Sec. 5). Since we have not formally defined parallelogram we state the principle in the following equivalent form:

THEOREM 12.   If $ab \parallel dc$ and $ad \parallel bc$ then $\overline{ac}$ meets $\overline{bd}$.

Proof. $ab \parallel dc$ implies a, b, c, d coplane, and ab does not meet dc. Thus c, d are on the same side of ab. Similarly, b, c are on the same side of ad. By Corollary 3 of Theorem 11, $(\overrightarrow{ab} \; \overrightarrow{ac} \; \overrightarrow{ad})$. Thus by Theorem 7, $\overrightarrow{ac}$ meets $\overline{bd}$. By the same reasoning $\overrightarrow{bd}$ meets $\overline{ac}$. Since $ac \neq bd$, $\overline{ac}$ meets $\overline{bd}$ follows.

6. The Notion of Angle

Our study of betweenness of rays suggests the study of an "angular segment" or "segment of rays," that is, the set of rays between two given rays. Such objects correspond to angles and have related properties. We shall however define angle in the conventional manner, but study its properties as far as convenient in terms of betweenness of rays.

DEFINITION.   Suppose $\overrightarrow{oa}, \overrightarrow{ob}$ are distinct and not opposite (or equivalently o, a, b are distinct and noncollinear). Then the point set $\overrightarrow{oa} \cup \overrightarrow{ob} \cup o$ is called an *angle* and is denoted $\angle aob$. Rays $\overrightarrow{oa}, \overrightarrow{ob}$ are the *sides* of the angle, and o is its *vertex*.

From the definition it is immediate that $\angle aob = \angle boa$. Also $\angle aob \subset aob$. Thus an angle is a plane figure (that is, a subset of a plane) and is contained in a unique plane.

NOTE.   If $a' \subset \overrightarrow{oa}$ and $b' \subset \overrightarrow{ob}$ (and $\overrightarrow{oa}$, $\overrightarrow{ob}$ are distinct and not opposite), then $\angle aob = \angle a'ob'$, since $\overrightarrow{oa'} = \overrightarrow{oa}$, $\overrightarrow{ob'} = \overrightarrow{ob}$.

Intuitively we feel that if a ray is between two rays it is "inside" the angle which they form. This suggests the following

DEFINITION.   The *interior* of $\angle aob$, denoted $I(\angle aob)$, is the set of points x such that $\overrightarrow{ox}$ is between $\overrightarrow{oa}$ and $\overrightarrow{ob}$ (Fig. 12.13). The *exterior* of $\angle aob$, denoted $E(\angle aob)$, is the set of points of its plane which are neither in the angle nor in its interior. A point or set of points is *inside* (*outside*) an angle if it is contained in its interior (exterior).

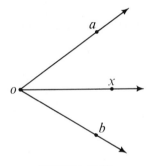

FIGURE 12.13

NOTE: If $x \subset I(\angle aob)$, then $\overrightarrow{ox} \subset I(\angle aob)$. Thus the interior of an angle is "completely covered" by rays emanating from its vertex. More precisely, $I(\angle aob)$ is the union of the family of rays which are between $\overrightarrow{oa}$ and $\overrightarrow{ob}$. Similarly, the exterior of an angle is the union of a certain family of rays. Observe that a half-plane and a "punctured" plane (a plane with one point deleted) also have the property of being completely covered by a family of rays which emanate from a common point.

Our definition of angle excludes the case of a straight angle. There are some advantages in introducing the concept of straight angle but we feel the disadvantages outweigh the advantages. If we wished to include it we would omit the condition that $\overrightarrow{oa}$, $\overrightarrow{ob}$ are not opposite in the definition of $\angle aob$. Then a straight angle would be a set of points of the form $\overrightarrow{oa} \cup \overrightarrow{ob} \cup o$ where

$\overrightarrow{oa}$, $\overrightarrow{ob}$ are opposite. Thus a straight angle would be a line L, and could be restructured as a straight angle with any point of L as vertex. Consequently a straight angle would not have a unique vertex or a unique pair of sides. Moreover it would not be contained in a unique plane, nor have an interior or an exterior. A straight angle would be so exceptional in these ways that it is more convenient to dispense with the idea. In excluding the concept of straight angle we do not give up its core—the idea of a pair of opposite rays—this is indispensible in the theory of order.

We conclude the chapter with several ideas which are used in the theory of congruence (Ch. 14).

DEFINITION. Two angles (with a common vertex) form a pair of *vertical angles* if the sides of one are opposite to the sides of the other. Two angles form a *linear pair* if they have a side in common and the other two sides are opposite rays. We extend this idea to three angles as follows: The angles $\angle aob$, $\angle boc$, $\angle cod$ form a *linear triple* provided the rays $\overrightarrow{oa}$, $\overrightarrow{ob}$, $\overrightarrow{oc}$, $\overrightarrow{od}$ satisfy (1) $(\overrightarrow{oa}\ \overrightarrow{ob}\ \overrightarrow{oc})$, (2) $\overrightarrow{oa}$, $\overrightarrow{od}$ are opposite (Fig. 12.14).

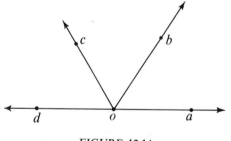

FIGURE 12.14

Note (1), (2) imply $(\overrightarrow{ob}\ \overrightarrow{oc}\ \overrightarrow{od})$ by Theorem 10 so that the lack of symmetry in the definition of linear triple is only apparent.

DEFINITION. An angle is *formed* by two lines L, M if its vertex is the intersection of L and M and its sides are contained in L, M respectively.

THEOREM 13.   Two distinct intersecting lines form exactly four angles.

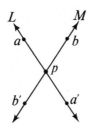

FIGURE 12.15

Proof. Let L, M intersect in p; $L \neq M$. Choose a, $a'. \subset L$ such that (apa'); b, $b' \subset M$ such that (bpb'). Then a, p, b are not collinear so that $\overrightarrow{pa}$, $\overrightarrow{pb}$ are distinct and not opposite. Thus we may significantly refer to $\angle apb$, and we see that it is formed by L, M. Similarly for $\angle apb'$, $\angle a'pb$, $\angle a'pb'$. These four angles are distinct. To show this suppose for example, $\angle apb = \angle apb'$. Then

(1) $$b \subset \angle apb' = \overrightarrow{pa} \cup p \cup \overrightarrow{pb'}.$$

$b \not\subset \overrightarrow{pa}$ since a, p, b are not collinear. $b \neq p$ since (bpb'). Finally $b \not\subset \overrightarrow{pb'}$ since $\overrightarrow{pb}$, $\overrightarrow{pb'}$ are opposite and so disjoint. Thus (1) is false and we conclude $\angle apb \neq \angle apb'$. The other cases are similar and we conclude that L, M form at least four angles.

To complete the proof, suppose $\angle cpd$ is formed by L, M. It is not restrictive to suppose $\overrightarrow{pc} \subset L$, $\overrightarrow{pd} \subset M$. By the Line Separation Principle (Ch. 10, Th. 7, Cor. 1)

$$c \subset L = \overrightarrow{pa} \cup p \cup \overrightarrow{pa'}.$$

Thus $c \subset \overrightarrow{pa}$ or $c \subset \overrightarrow{pa'}$ so that $\overrightarrow{pc} = \overrightarrow{pa}$ or $\overrightarrow{pc} = \overrightarrow{pa'}$. Similarly $\overrightarrow{pd} = \overrightarrow{pb}$ or $\overrightarrow{pd} = \overrightarrow{pb'}$. It follows that $\angle cpd$ is one of the angles $\angle apb$, $\angle apb'$, $\angle a'pb$, $\angle a'pb'$ and these are the *only* angles formed by L, M.

E X E R C I S E S   I

DEFINITION.   A *side* of plane P is a half-space with face P. *Opposite sides* of P are opposite half-spaces with face P.

Prove each of the following:

1. Two distinct points are in the same side of P if and only if they are not in P, and the segment joining them does not meet P.
2. Two points are in opposite sides of P if and only if they are not in P, and the segment joining them meets P.
3. If $p, L \subset P$, and $a \notin P$, then
     (i) $p/a \subset P/a$;
     (ii) $L/a \subset P/a$;
     (iii) $\overrightarrow{pa} \subset \overrightarrow{Pa}$;
     (iv) $\overrightarrow{La} \subset \overrightarrow{Pa}$.
4. If $p, L \subset P$, and $a \notin P$, then
     (i) all points of $\overrightarrow{pa}$ and all points of $\overrightarrow{La}$ are in the same side of P as a;
     (ii) all points of p/a and all points of L/a are in the opposite side of P from a.
5. If $p, L \subset P$, and $a \notin P$, then
     (i) $\overrightarrow{pa}$ comprises all points of pa in the same side of P as a;
     (ii) $\overrightarrow{La}$ comprises all points of La in the same side of P as a;
     (iii) p/a comprises all points of pa in the opposite side of P from a;
     (iv) L/a comprises all points of La in the opposite side of P from a.

EXERCISES II

1. Suppose a, b, c are distinct and noncollinear, (bdc) and (cea). Prove $\overline{ad}$ meets $\overline{be}$.
2. Suppose a, b, c are distinct and noncollinear, (aed) and (bdc). Prove there is a point f such that (bef) and (cfa).
3. Suppose a, b, c are distinct and noncollinear, (axb), (ayc), and (bzc). Prove $\overline{xy}$ meets $\overline{az}$.
4. Suppose a, b, c are distinct and noncollinear, (axb), (ayc), and (xzy). Prove there is a point w such that (azw) and (bwc).
5. Suppose o, a, b are distinct and noncollinear, (aoa'), (bob'), and (axb). Prove there is a point x' such that (xox') and $(a'x'b')$.

EXERCISES III

1. If $\overrightarrow{oa}, \overrightarrow{ob}$ are distinct and not opposite, prove that there exist rays $\overrightarrow{ox}, \overrightarrow{oy}, \overrightarrow{oz}$ such that $(\overrightarrow{ox}\ \overrightarrow{oa}\ \overrightarrow{ob})$, $(\overrightarrow{oa}\ \overrightarrow{oy}\ \overrightarrow{ob})$, and $(\overrightarrow{oa}\ \overrightarrow{ob}\ \overrightarrow{oz})$.

2. Suppose $\overrightarrow{oa}$, $\overrightarrow{ob}$, $\overrightarrow{oc}$ are distinct, coplanar, and not connected by an order relation. Prove that $\overrightarrow{oa}$, $\overrightarrow{ob}$, $\overrightarrow{oc}$ have no common transversal, that is no line meets all three.

3. Suppose $\overrightarrow{oa}$, $\overrightarrow{ob}$, $\overrightarrow{oc}$ distinct, coplanar, no pair opposite, and not connected by an order relation. Prove that $\overrightarrow{ob}$ and $\overrightarrow{oc}$ are on opposite sides of oa.

4. Suppose $\overrightarrow{oa}$, $\overrightarrow{ob}$, $\overrightarrow{oc}$ distinct, coplanar, no pair opposite, and not connected by an order relation. Prove $(\overrightarrow{oa}\ \overrightarrow{ob'}\ \overrightarrow{oc})$, where $\overrightarrow{ob'}$ is opposite $\overrightarrow{ob}$.

5. Let a, b, c be distinct and noncollinear. Prove a/b, b/c, a/c have a common transversal.

6. Let a, b, c be distinct and noncollinear. Prove a/b, b/c, c/a have no common transversal.

DEFINITION.   Let half-planes $\overrightarrow{La}$, $\overrightarrow{Lb}$, $\overrightarrow{Lc}$ with common edge L be such that $\overrightarrow{La}$, $\overrightarrow{Lc}$ are distinct and not opposite; and let there be points $a_1 \subset \overrightarrow{La}$, $b_1 \subset \overrightarrow{Lb}$ $c_1 \subset \overrightarrow{Lc}$ such that $(a_1 b_1 c_1)$. Then we say $\overrightarrow{Lb}$ is *between* $\overrightarrow{La}$ and $\overrightarrow{Lc}$, and write $(\overrightarrow{La}\ \overrightarrow{Lb}\ \overrightarrow{Lc})$.

7. An analogue of Theorem 4 in terms of half-planes is: $(\overrightarrow{La}\ \overrightarrow{Lb}\ \overrightarrow{Lc})$ implies $(\overrightarrow{Lc}\ \overrightarrow{Lb}\ \overrightarrow{La})$. Prove this statement correct.

8. Write analogues of Theorems 5–11 and of their corollaries, in terms of half-planes. Which of these analogues hold, and which do not? Prove your answers.

9. Can the notion of betweenness be extended to half-spaces with a common face? Why?

E X E R C I S E S   I V

DEFINITION. $(\overrightarrow{oa}\ \overrightarrow{ob}\ \overrightarrow{oc}\ \overrightarrow{od})$ means $(\overrightarrow{oa}\ \overrightarrow{ob}\ \overrightarrow{oc})$, $(\overrightarrow{oa}\ \overrightarrow{ob}\ \overrightarrow{od})$, $(\overrightarrow{oa}\ \overrightarrow{oc}\ \overrightarrow{od})$ and $(\overrightarrow{ob}\ \overrightarrow{oc}\ \overrightarrow{od})$.

1. Suppose $(\overrightarrow{oa}\ \overrightarrow{ob}\ \overrightarrow{oc}\ \overrightarrow{od})$. Prove each two of $\overrightarrow{oa}$, $\overrightarrow{ob}$, $\overrightarrow{oc}$, $\overrightarrow{od}$ are distinct and not opposite.

2. An analogue of Theorem 13 of Chapter 10 in terms of rays is:

Suppose $(\overrightarrow{oa}\ \overrightarrow{ob}\ \overrightarrow{oc})$, and $\overrightarrow{ox}$ coplanes with and is distinct from $\overrightarrow{oa}$, $\overrightarrow{ob}$, $\overrightarrow{oc}$. Then $(\overrightarrow{ox}\ \overrightarrow{oa}\ \overrightarrow{ob}\ \overrightarrow{oc})$, $(\overrightarrow{oa}\ \overrightarrow{ox}\ \overrightarrow{ob}\ \overrightarrow{oc})$, $(\overrightarrow{oa}\ \overrightarrow{ob}\ \overrightarrow{ox}\ \overrightarrow{oc})$ or $(\overrightarrow{oa}\ \overrightarrow{ob}\ \overrightarrow{oc}\ \overrightarrow{ox})$. Prove that this statement is false.

3. An analogue of Corollary 1 of Theorem 13 of Chapter 10 in terms of rays is:

If $(\overrightarrow{oa}\ \overrightarrow{ob}\ \overrightarrow{oc})$ and $(\overrightarrow{ob}\ \overrightarrow{oc}\ \overrightarrow{od})$, then $(\overrightarrow{oa}\ \overrightarrow{ob}\ \overrightarrow{oc}\ \overrightarrow{od})$. Prove that this statement is false.

4. An analogue of Corollary 3 of Theorem 13 of Chapter 10 in terms of rays is:

If $(\overrightarrow{oa}\ \overrightarrow{ob}\ \overrightarrow{oc})$ and $(\overrightarrow{oa}\ \overrightarrow{oc}\ \overrightarrow{od})$, then $(\overrightarrow{oa}\ \overrightarrow{ob}\ \overrightarrow{oc}\ \overrightarrow{od})$. Prove that this statement is true.

5. The following are analogues of Exercises II, 19–23 of Chapter 10:

(i) If $(\overrightarrow{oa}\ \overrightarrow{ob}\ \overrightarrow{oc})$ and $(\overrightarrow{oa}\ \overrightarrow{ob}\ \overrightarrow{od})$, then $(\overrightarrow{oa}\ \overrightarrow{ob}\ \overrightarrow{oc}\ \overrightarrow{od})$, $(\overrightarrow{oa}\ \overrightarrow{ob}\ \overrightarrow{od}\ \overrightarrow{oc})$ or $\overrightarrow{oc} = \overrightarrow{od}$.

(ii) If $(\overrightarrow{oa}\ \overrightarrow{ob}\ \overrightarrow{oc})$ and $(\overrightarrow{oa}\ \overrightarrow{ob}\ \overrightarrow{od})$, then $(\overrightarrow{oa}\ \overrightarrow{oc}\ \overrightarrow{od})$, $(\overrightarrow{oa}\ \overrightarrow{od}\ \overrightarrow{oc})$ or $\overrightarrow{oc} = \overrightarrow{od}$.

(iii) If $(\overrightarrow{oa}\ \overrightarrow{ob}\ \overrightarrow{oc})$ and $(\overrightarrow{oa}\ \overrightarrow{ob}\ \overrightarrow{od})$, then $(\overrightarrow{ob}\ \overrightarrow{oc}\ \overrightarrow{od})$, $(\overrightarrow{ob}\ \overrightarrow{od}\ \overrightarrow{oc})$ or $\overrightarrow{oc} = \overrightarrow{od}$.

(iv) If $(\overrightarrow{oa}\ \overrightarrow{ob}\ \overrightarrow{od})$ and $(\overrightarrow{oa}\ \overrightarrow{oc}\ \overrightarrow{od})$, then $(\overrightarrow{oa}\ \overrightarrow{ob}\ \overrightarrow{oc}\ \overrightarrow{od})$, $(\overrightarrow{oa}\ \overrightarrow{oc}\ \overrightarrow{ob}\ \overrightarrow{od})$ or $\overrightarrow{ob} = \overrightarrow{oc}$.

(v) If $(\overrightarrow{oa}\ \overrightarrow{ob}\ \overrightarrow{od})$, $(\overrightarrow{oa}\ \overrightarrow{oc}\ \overrightarrow{od})$ and $\overrightarrow{ob} \neq \overrightarrow{oc}$, then either $(\overrightarrow{oa}\ \overrightarrow{ob}\ \overrightarrow{oc})$ and $(\overrightarrow{ob}\ \overrightarrow{oc}\ \overrightarrow{od})$, or else $(\overrightarrow{oa}\ \overrightarrow{oc}\ \overrightarrow{ob})$ and $(\overrightarrow{oc}\ \overrightarrow{ob}\ \overrightarrow{od})$.

Which of these statements are true, and which are false? Prove your answers.

6. If $(\overrightarrow{oa}\ \overrightarrow{ob}\ \overrightarrow{oc}\ \overrightarrow{od})$, prove $\overrightarrow{oa}$, $\overrightarrow{ob}$, $\overrightarrow{oc}$, $\overrightarrow{od}$ have a common transversal.

DEFINITION.   In a plane P, the set of all rays with a common endpoint o is called a *pencil of rays*. If $o \subset L \subset P$, the set of all rays in P with endpoint o which are in one side of L is called a *half-pencil of rays*. o is called the *vertex* of the half-pencil; L is called its *edge* or *boundary*.

7. Let $\overrightarrow{oa}$, $\overrightarrow{ob}$, $\overrightarrow{oc}$ be distinct rays in a half-pencil of rays with vertex o. Prove that one and only one of the following holds:

$$(\overrightarrow{oa}\ \overrightarrow{ob}\ \overrightarrow{oc}),\ (\overrightarrow{ob}\ \overrightarrow{oc}\ \overrightarrow{oa}),\ (\overrightarrow{oc}\ \overrightarrow{oa}\ \overrightarrow{ob}).$$

8. Let $\overrightarrow{oa}$, $\overrightarrow{ob}$, $\overrightarrow{oc}$ be distinct rays in a half-pencil of rays with edge op and vertex o, and $(\overrightarrow{oa}\ \overrightarrow{ob}\ \overrightarrow{oc})$. Prove that either $(\overrightarrow{op}\ \overrightarrow{oa}\ \overrightarrow{ob}\ \overrightarrow{oc})$ or $(\overrightarrow{op}\ \overrightarrow{oc}\ \overrightarrow{ob}\ \overrightarrow{oa})$.

9. Let $\overrightarrow{oa}$, $\overrightarrow{ob}$, $\overrightarrow{oc}$ be distinct rays in a half pencil of rays with vertex o and edge op. Prove that there is a ray $\overrightarrow{px}$, with endpoint p, which intersects $\overrightarrow{oa}$, $\overrightarrow{ob}$, and $\overrightarrow{oc}$.

10. Let $(\overrightarrow{oa}\ \overrightarrow{ob}\ \overrightarrow{oc}\ \overrightarrow{od})$. Prove $\overrightarrow{oa}$, $\overrightarrow{ob}$, $\overrightarrow{oc}$, $\overrightarrow{od}$ are in a half-pencil of rays with vertex o.

11. Let $\overrightarrow{op}$ be distinct from, and in the same half-pencil of rays with vertex o as $\overrightarrow{oa}$, $\overrightarrow{ob}$, $\overrightarrow{oc}$, and let $(\overrightarrow{oa}\ \overrightarrow{ob}\ \overrightarrow{oc})$. Prove $(\overrightarrow{op}\ \overrightarrow{oa}\ \overrightarrow{ob}\ \overrightarrow{oc})$, $(\overrightarrow{oa}\ \overrightarrow{op}\ \overrightarrow{ob}\ \overrightarrow{oc})$, $(\overrightarrow{oa}\ \overrightarrow{ob}\ \overrightarrow{op}\ \overrightarrow{oc})$ or $(\overrightarrow{oa}\ \overrightarrow{ob}\ \overrightarrow{oc}\ \overrightarrow{op})$.

E X E R C I S E S   V

DEFINITION.   If a, b, c, d are distinct coplanar points, no three of which colline, then the point set

$$\overline{ab} \cup \overline{bc} \cup \overline{cd} \cup \overline{da} \cup a \cup b \cup c \cup d$$

is called a *plane quadrilateral*, and is denoted $Q(abcd)$. a, b, c, d are called its *vertices*; $\overline{ab}$, $\overline{bc}$, $\overline{cd}$, $\overline{da}$ its *sides*; ab, bc, cd, da its *sidelines*; $\overline{ac}$ and $\overline{bd}$ its *diagonals*. If no two sides meet, $Q(abcd)$ is said to be *simple*. If $Q(abcd)$ is such that vertices not in a sideline are on the same side of that sideline, then $Q(abcd)$ is called a *convex quadrilateral*.

1. Prove: $Q(abcd) = Q(bcda)$, but $Q(abcd) \neq Q(abdc)$.
2. If $Q(abcd)$ is a convex quadrilateral, prove it is simple.
3. If $Q(abcd)$ is a convex quadrilateral, prove $(\overrightarrow{ab}\ \overrightarrow{ac}\ \overrightarrow{ad})$.
4. Prove: A simple plane quadrilateral is a convex quadrilateral if and only if its diagonals intersect.
5. If $\overline{pq}$ meets $\overline{rs}$ and $pq \neq rs$, prove that $Q(prqs)$ is a convex quadrilateral.
6. Let $Q(abcd)$ be a convex quadrilateral, $p \subset \overline{ab}$, $q \subset \overline{bc}$, $r \subset \overline{cd}$, $s \subset \overline{da}$. Prove that each of the following is a convex quadrilateral:
   (i) $Q(abcs)$; (ii) $Q(abqs)$; (iii) $Q(abrs)$; (iv) $Q(apqr)$; (v) $Q(pqrs)$.
7. Let $Q(abcd)$ be a convex quadrilateral, $p \subset \overline{ab}$, $q \subset \overline{bc}$, $x \subset \overline{cd} \cup d \cup \overline{da}$. Prove that $\overline{pq}$ intersects $\overline{bx}$.

8. Prove: A simple plane quadrilateral is a convex quadrilateral if and only if its intersection with each sideline is a closed segment, that is, a segment united with its endpoints. Does this condition hold for a nonsimple plane quadrilateral?

9. Prove: In a plane, if a line meets one side of a quadrilateral, and does not contain a vertex, then it meets a second side.

10. Prove: In a plane, if a line meets three sides of a quadrilateral, then it meets the fourth.

CHAPTER 13

Separation Properties of Angles and Triangles*

In Chapter 1 we observed that the notions of interior and exterior of an angle and interior and exterior of a triangle are involved, at least in an implicit sense, in school geometry. This chapter, which is the culmination of our study of nonmetrical geometry, makes a detailed study of these notions, relating them to associated planes and half-planes, and to each other.

1. The Interior of an Angle

In this section we derive a few elementary properties of angle interiors. The key result, Theorem 1, expresses an angle interior, in a natural way, as an intersection of half-planes.

We shall frequently have to refer to a half-plane whose edge is given in the notation "line pq" rather than "line L." Thus we introduce the following variation of the "arrow" notation for half-planes (Ch. 11, Sec. 4).

NOTATION. The half-plane with edge pq which contains point a is denoted $\overrightarrow{(pq)}a$.

Our definition of $I(\angle aob)$, the interior of $\angle aob$ (Ch. 12, Sec. 6), as the set of points x such that $(\overrightarrow{oa}\ \overrightarrow{ox}\ \overrightarrow{ob})$, is a natural and simple way to

---

* The succeeding chapters are independent of the present one and may be read before it.

introduce the idea. We derive now a characterization which shifts the emphasis from order relations of rays to a configurational property involving the intersection of half-planes. So to speak we get the interior of an angle—not by a simple pointwise construction in terms of its sides—but by taking figures which are obviously too big and cutting off the excess.

NOTATION. $S \cap T$ denotes the intersection of sets S, T.

THEOREM 1. $I(\angle aob) = \overrightarrow{(oa)b} \cap \overrightarrow{(ob)a}.$*

Proof. We show

(1) $x \subset I(\angle aob)$

is equivalent to

(2) $x \subset \overrightarrow{(oa)b} \cap \overrightarrow{(ob)a}.$

By definition (1) is equivalent to $(\overrightarrow{oa}\, \overrightarrow{ox}\, \overrightarrow{ob})$. This holds if and only if x, a are on the same side of ob and x, b are on the same side of oa (Ch. 12, Th. 6; Th. 11, Cor. 3). The latter is equivalent to

$$x \subset \overrightarrow{(ob)a}, \quad x \subset \overrightarrow{(oa)b}$$

and so to (2).

COROLLARY 1. $I(\angle aob)$ is the set of points which are simultaneously on the same side of oa as b, and on the same side of ob as a.

COROLLARY 2. $\angle aob$ and $I(\angle aob)$ are disjoint.

Proof. Let $x \subset \angle aob = \overrightarrow{oa} \cup \overrightarrow{ob} \cup o$. If $x = o$ or $x \subset \overrightarrow{oa}$, then $x \subset oa$ and so $x \not\subset \overrightarrow{(oa)b}$ (Ch. 11, Th. 1). Thus $x \not\subset I(\angle aob) = \overrightarrow{(oa)b} \cap \overrightarrow{(ob)a}$. Similarly, $x \subset \overrightarrow{ob}$ implies $x \not\subset I(\angle aob)$. Thus $\angle aob$ and $I(\angle aob)$ are disjoint.

COROLLARY 3.   The interior of an angle is a convex set. (Compare Ch. 10, Th. 10.)

---

* We have omitted the implicit hypothesis that o, a, b are distinct and noncollinear, which is needed to insure that the terms in the equation are significant. This practice is followed throughout the chapter.

Proof. Let $x, y \subset I(\angle aob)$, $x \neq y$. Then $x, y \subset \overrightarrow{(oa)b}$ and $x, y \subset \overrightarrow{(ob)a}$. Since $\overrightarrow{(oa)b}$ and $\overrightarrow{(ob)a}$ are convex (Ch. 11, Th. 5) it follows that $\overline{xy} \subset \overrightarrow{(oa)b}$ and $\overline{xy} \subset \overrightarrow{(ob)a}$. Thus

$$\overline{xy} \subset \overrightarrow{(oa)b} \cap \overrightarrow{(ob)a} = I(\angle aob),$$

proving $I(\angle aob)$ is convex.

NOTE: The proof of Corollary 3 reminds one of the proof of the convexity of a segment (Ch. 10, Th. 10). The latter was based on the principle

(1) $$\overline{ab} = \overrightarrow{ab} \cap \overrightarrow{ba},$$

and used the convexity of a ray. Thus if we consider a half-plane as the planar analogue of a ray, Theorem 1 is the analogue of Equation (1) and the convexity proofs for segment and angle interior are exactly analogous.

2. Formula for a Half-Plane

The study of angle interiors yields insight into the structure of other plane figures. For half-planes we have

THEOREM 2. $\overrightarrow{(oa)b} = I(\angle aob) \cup I(\angle a'ob) \cup \overrightarrow{ob}$, where $\overrightarrow{oa'}$ is opposite $\overrightarrow{oa}$.

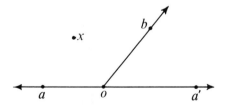

FIGURE 13.1

Proof. Let $A = I(\angle aob) \cup I(\angle a'ob) \cup \overrightarrow{ob}$, assuming $\overrightarrow{oa'}$ is opposite $\overrightarrow{oa}$. Let $x \subset \overrightarrow{(oa)b}$. Then x, b are on the same side of oa. By Corollary 1 to Theorem 11 of Chapter 12 $(\overrightarrow{oa}\ \overrightarrow{ox}\ \overrightarrow{ob})$, $(\overrightarrow{oa'}\ \overrightarrow{ox}\ \overrightarrow{ob})$ or $\overrightarrow{ob} = \overrightarrow{ox}$; which imply $x \subset I(\angle aob)$, $x \subset I(\angle a'ob)$ or $x \subset \overrightarrow{ob}$. Thus $x \subset A$, proving $\overrightarrow{(oa)b} \subset A$.

Since $\overrightarrow{oa'}, \overrightarrow{oa}$ are opposite, $(a'oa)$. Thus $oa' = oa$, so that $\overrightarrow{(oa')b} = \overrightarrow{(oa)b}$. By Theorem 1, $I(\angle aob) \subset \overrightarrow{(oa)b}$, and $I(\angle a'ob) \subset \overrightarrow{(oa')b} = \overrightarrow{(oa)b}$. Finally $\overrightarrow{ob} \subset \overrightarrow{(oa)b}$, by Theorem 3 of Chapter 12. Thus $A \subset \overrightarrow{(oa)b}$ and the theorem follows.

Exercise 1.   Prove that the terms in the expression for $\overrightarrow{(oa)b}$ are disjoint.

Exercise 2.   Prove that $\overrightarrow{ob}$ "separates" $I(\angle aob)$ and $I(\angle a'ob)$, that is, any segment joining a point of $I(\angle aob)$ and a point of $I(\angle a'ob)$ meets $\overrightarrow{ob}$.

3. Formula for a Plane

By using the last theorem we derive a similar formula for a plane.

THEOREM 3. $oab = I(\angle aob) \cup I(\angle aob') \cup I(\angle a'ob) \cup I(\angle a'ob') \cup$ $oa \cup ob$, where $\overrightarrow{oa'}, \overrightarrow{ob'}$ are opposite $\overrightarrow{oa}, \overrightarrow{ob}$, respectively.

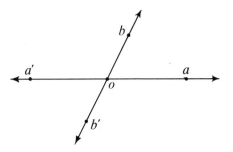

FIGURE 13.2

Proof.   Since $\overrightarrow{ob}, \overrightarrow{ob'}$ are opposite, (bob'). Thus $\overline{bb'}$ meets oa. Further $b, b' \notin oa$. Hence the Plane Separation Principle (Ch. 11, Th. 4, Cor. 1) implies

(1) $oab = \overrightarrow{(oa)b} \cup oa \cup \overrightarrow{(oa)b'}$.

By Theorem 2 we have

(2) $\overrightarrow{(oa)b} = I(\angle aob) \cup I(\angle a'ob) \cup \overrightarrow{ob}$,

(3) $\overrightarrow{(oa)b'} = I(\angle aob') \cup I(\angle a'ob') \cup \overrightarrow{ob'}$.

By substituting (2) and (3) into (1) and rearranging terms, we get

$$oab = I(\angle aob) \cup I(\angle aob') \cup I(\angle a'ob) \cup I(\angle a'ob') \cup oa \cup ob,$$

where we have replaced $oa \cup \overrightarrow{ob} \cup \overrightarrow{ob'}$ by its equal $oa \cup ob$.

Exercise 1.   Prove that each pair of sets in the expression for *oab* is disjoint except, of course, the pair *oa* and *ob*.

Exercise 2.   Prove that the set $oa \cup ob$ "separates" the four angle interiors in the expression for *oab;* that is, any segment joining points in two of these angle interiors intersects $oa \cup ob$.

4. The Exterior of an Angle

Since our formula for plane *oab* involves $I(\angle aob)$, it is not hard to derive from it a formula for $E(\angle aob)$, the exterior of $\angle aob$.

THEOREM 4. $E(\angle aob) = I(\angle aob') \cup I(\angle a'ob) \cup I(\angle a'ob') \cup \overrightarrow{oa'} \cup \overrightarrow{ob'}$,

where $\overrightarrow{oa'}$, $\overrightarrow{ob'}$ are opposite $\overrightarrow{oa}$, $\overrightarrow{ob}$ respectively.

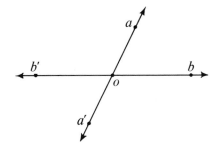

FIGURE 13.3

Proof.   By the preceding theorem

(1) $oab = I(\angle aob) \cup I(\angle aob') \cup I(\angle a'ob) \cup I(\angle a'ob') \cup (oa \cup ob)$.

Note that the five components indicated are disjoint. We have $(a'oa)$, $(b'ob)$. Hence the Line Separation Principle (Ch. 10, Th. 7, Cor. 1) implies

(2) $oa = \overrightarrow{oa} \cup o \cup \overrightarrow{oa'}, \qquad ob = \overrightarrow{ob} \cup o \cup \overrightarrow{ob'}$.

By substituting (2) into (1), rearranging terms, and suppressing one of the terms *o*, we get

(3) $oab = I(\angle aob) \cup (\overrightarrow{oa} \cup \overrightarrow{ob} \cup o) \cup$

$\qquad\qquad [I(\angle aob') \cup I(\angle a'ob) \cup I(\angle a'ob') \cup \overrightarrow{oa'} \cup \overrightarrow{ob'}];$

and the three components are disjoint. For the second component we have

$$\overrightarrow{oa} \cup \overrightarrow{ob} \cup o = \angle aob.$$

Hence by definition of the exterior of an angle

(4) $E(\angle aob) = I(\angle aob') \cup I(\angle a'ob) \cup I(\angle a'ob') \cup \overrightarrow{oa'} \cup \overrightarrow{ob'}.$

The components of the expression just derived for $E(\angle aob)$ are disjoint. We next derive a formula for $E(\angle aob)$ in terms of "overlapping" components which is neater and quite useful. To state it simply we introduce the following

NOTATION.   If $L = pq$ we denote L/a by pq/a.

THEOREM 5. $E(\angle aob) = oa/b \cup ob/a.$

Proof. We rearrange the terms in Formula (4) above for $E(\angle aob)$ and insert the term $I(\angle a'ob')$ a second time (which does not, of course, affect the set union):

(1) $E(\angle aob) = [I(\angle aob') \cup I(\angle a'ob') \cup \overrightarrow{ob'}] \cup$

$\qquad\qquad\qquad\qquad [I(\angle a'ob) \cup I(\angle a'ob') \cup \overrightarrow{oa'}].$

By Theorem 2,

(2) $\overrightarrow{(oa)b'} = I(\angle aob') \cup I(\angle a'ob') \cup \overrightarrow{ob'},$

(3) $\overrightarrow{(ob)a'} = I(\angle boa') \cup I(\angle b'oa') \cup \overrightarrow{oa'}.$

We substitute (2) and (3) in (1), and obtain

(4) $\qquad\qquad E(\angle aob) = \overrightarrow{(oa)b'} \cup \overrightarrow{(ob)a'}.$

Since $\overrightarrow{ob'}, \overrightarrow{ob}$ are opposite it follows that $\overrightarrow{(oa)b'}, \overrightarrow{(oa)b}$ are opposite. Hence $\overrightarrow{(oa)b'} = oa/b.$ Similarly $\overrightarrow{(ob)a'} = ob/a.$ By substituting in (4) we have

$$E(\angle aob) = oa/b \cup ob/a.$$

5. Separation and the Notion of Path

In the study of the separation of a plane by a line or of spatial separation by a plane (Ch. 11, Secs. 7 and 14), the natural "path" to use in "joining" two points is a segment. Certainly a segment is the simplest "path" that joins its endpoints, and in treating the "separation sets"—half-planes and half-spaces—there is no need for a more complex type of path. For any two points of a half-plane or a half-space are joined by a segment which is wholly contained in the given set, that is, these sets are convex. In order to study separation when the separation sets are not convex, we must choose a more complicated type of "path," since the segment joining two points of the set need not be contained in the set. This is our immediate reason for introducing the following definition, which has wide application in geometry.

DEFINITION.   If $a_1, a_2, \ldots, a_n$ ($n \geq 2$) are distinct points, the set union

$$\overline{a_1 a_2} \cup \overline{a_2 a_3} \cup \cdots \cup \overline{a_{n-1} a_n} \cup a_1 \cup a_2 \cup \cdots \cup a_n$$

is called a *path* or a *broken line*, and is said to *join* a_1 and a_n.

Note that such a path is not necessarily "simple"; for example $\overline{a_1 a_2}$ and $\overline{a_3 a_4}$ may intersect. Note also, that if $n = 2$, the path reduces to $\overline{a_1 a_2} \cup a_1 \cup a_2$ which is not a segment, but a closed segment. This may be disconcerting momentarily, since we think of a path as a generalization of a segment. However it causes no difficulty in our work, since the properties we are concerned with hold equally well for a segment or a closed segment. We could have avoided this distinction by deleting the terms a_1 and a_n in the definition. But the definition as stated is neater and more symmetrical, and generally more useful.

Our definition of separation by a line or a plane (Ch. 11, Sec. 6) can now easily be modified to cover the cases we are interested in: The separation of a plane by an angle or a triangle. We state the definition for arbitrary separation of a plane.

DEFINITION.   We say set A separates plane P into the nonempty sets S, S' if the following conditions are satisfied:
   (i) $P = S \cup S' \cup A$;
   (ii) each segment that joins a point of S to a point of S' meets A;
   (iii) there exists in P a path joining any two given points of S, or of S', which does not meet A;
   (iv) S, S', A are disjoint.

6. Separation of a Plane by an Angle

THEOREM 6.   Any angle in a plane separates the plane into its interior and its exterior.

Proof. Let $P \supset \angle abc$. Let I, E denote the interior, exterior of $\angle abc$. We show

(i) $P = I \cup E \cup \angle abc$;

(ii) the segment joining any point of I and any point of E meets $\angle abc$;

(iii) the segment joining any two points of I does not meet $\angle abc$;

(iii') there exists in P a path joining any two given points of E which does not meet $\angle abc$;

(iv) I, E, $\angle abc$ are disjoint.

Proof of (i).   Note $P = abc$. Then using Theorem 1:

$$I = \overrightarrow{(ab)c} \cap \overrightarrow{(bc)a} \subset abc = P.$$

Then (i) follows by definition of E.

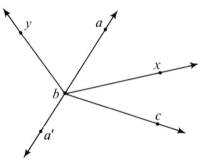

FIGURE 13.4

Proof of (ii).   Let $x \subset I$, $y \subset E$ (Fig. 13.4). By Theorems 1 and 5

$$x \subset \overrightarrow{(ab)c} \cap \overrightarrow{(bc)a}, \qquad y \subset ab/c \cup bc/a.$$

Thus $y \subset ab/c$ or $y \subset bc/a$, so that x and y are on opposite sides of ab, or x and y are on opposite sides of bc. It is not restrictive to assume x, y are on opposite sides of ab. Let $\overrightarrow{ba'}$ be opposite $\overrightarrow{ba}$. Then $(\overrightarrow{bx}\ \overrightarrow{ba}\ \overrightarrow{by})$, $(\overrightarrow{bx}\ \overrightarrow{ba'}\ \overrightarrow{by})$ or $\overrightarrow{bx}$ and $\overrightarrow{by}$ are opposite (Ch. 12, Th. 11).

Suppose $(\overrightarrow{bx}\ \overrightarrow{ba}\ \overrightarrow{by})$. Then $\overline{xy}$ meets $\overrightarrow{ba}$ (Ch. 12, Th. 7). Suppose next $(\overrightarrow{bx}\ \overrightarrow{ba'}\ \overrightarrow{by})$. Since $x \subset I$, $(\overrightarrow{ba}\ \overrightarrow{bx}\ \overrightarrow{bc})$. Hence $(\overrightarrow{bx}\ \overrightarrow{bc}\ \overrightarrow{ba'})$ (Ch. 12, Th. 10). This and $(\overrightarrow{bx}\ \overrightarrow{ba'}\ \overrightarrow{by})$ imply $(\overrightarrow{bx}\ \overrightarrow{bc}\ \overrightarrow{by})$ (Ch. 12, Th. 9). Thus $\overline{xy}$ meets $\overrightarrow{bc}$ (Ch. 12, Th. 7). Suppose finally $\overrightarrow{bx}$ and $\overrightarrow{by}$ are opposite. Then (xby) so that $\overline{xy} \supset b$. In each case $\overline{xy}$ meets $\angle abc$ and (ii) is established.

Proof of (iv). E is disjoint to $\angle abc$ and to I by definition of E; $\angle abc$ and I are disjoint by Corollary 2 to Theorem 1.

Proof of (iii).   This is immediate from (iv) and the convexity of I (Th. 1, Cor. 3).

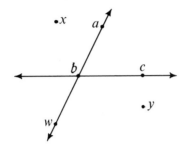

FIGURE 13.5

Proof of (iii′).   Let $w \subset b/a$ (Fig. 13.5). We show first $x \subset E$, $x \neq w$ implies $\overline{xw} \subset E$. We have (Th. 5)

(1) $E = ab/c \cup bc/a.$

Thus $x \subset ab/c$ or $x \subset bc/a$. Suppose $x \subset ab/c$. Then all points of $\overline{xw}$ are on the same side of ab as x (Ch. 12, Th. 3, Cor. 2). Thus by (1)

$$\overline{xw} \subset ab/c \subset E.$$

Now suppose $x \subset bc/a$. By Theorem 3 of Chapter 12 and (1)

$$w \subset b/a \subset bc/a \subset E.$$

By the convexity of bc/a (Ch. 11, Th. 5) and (1)

$$\overline{xw} \subset bc/a \subset E.$$

Finally let $x, y \subset E$; $x \neq y$. Then x, y are joined by the path

$$\overline{xw} \cup \overline{wy} \cup x \cup w \cup y,$$

which by the preceding paragraph is a subset of E. In view of (iv), the path does not meet $\angle abc$ and (iii′) is proved.

Finally note $I \neq \varnothing$ since $I \supset \overline{ac}$ (Why?); and $E \neq \varnothing$ since $E \supset ab/c$.

In the proof of (iii′) above we showed that any two points of E are joined by a path which is a subset of E. This suggests the following definition.

DEFINITION.   Let set A have the property that any two distinct points of A are joined by a path which is wholly contained in A. Then we say A is *pathwise connected*.

Note that pathwise connectedness is a generalization of convexity. In view of the definition, Theorem 6 has the following

COROLLARY.   The exterior of an angle is pathwise connected.

7. Triangles

We are now prepared to study triangles.

DEFINITION.   If a, b, c are distinct and noncollinear, the set union

$$\overline{ab} \cup \overline{bc} \cup \overline{ca} \cup a \cup b \cup c$$

is called *triangle abc* (written $\triangle abc$). Points a, b, c are its *vertices;* segments $\overline{ab}$, $\overline{bc}$, $\overline{ca}$ are its *sides*; lines ab, bc, ca are its *side-lines;* angles $\angle abc$, $\angle bca$ and $\angle cab$ are its *angles*. Vertex a and side $\overline{bc}$ (b and $\overline{ca}$, c and $\overline{ab}$) are said to be *opposite*. A point is *inside* a triangle if it is between a vertex and a point of the opposite side. A point is *outside* a triangle if it is in the plane of the triangle, but is neither inside the triangle nor a point of the triangle. The *interior* (*exterior*) of a triangle is the set of points inside (outside) the triangle. $I(\triangle abc)$, $E(\triangle abc)$ denote the interior, exterior respectively of $\triangle abc$.

8. The Interior of a Triangle

We prove in Theorem 7 that the interior of a triangle is the intersection of three half-planes, and derive several consequences. The result is analogous to Theorem 1, that an angle interior is the intersection of two half-planes.

THEOREM 7. $I(\triangle abc) = \overrightarrow{(ab)c} \cap \overrightarrow{(bc)a} \cap \overrightarrow{(ca)b}$.

Proof.   Let $x \subset I(\triangle abc)$. It is not restrictive to assume that x is between a and a point y of $\overline{bc}$ (Fig. 13.6(a)). Thus $x \subset \overline{ay}$ and x, a are on the same side of bc (Ch. 12, Th. 3, Cor. 2). Similarly, each pair y, b and x, y are on the

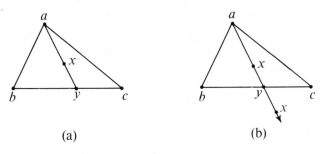

FIGURE 13.6

same side of *ca*. Thus *x*, *b* are on the same side of *ca*. Similarly, we can show *x*, *c* are on the same side of *ab*. Thus we may assert

(1) $$x \subset \overrightarrow{(ab)c} \cap \overrightarrow{(bc)a} \cap \overrightarrow{(ca)b}.$$

Conversely, let *x* satisfy Equation (1). Then, using Theorem 1

$$x \subset \overrightarrow{(ab)c} \cap \overrightarrow{(ca)b} = I(\angle bac).$$

By definition of angle interior $(\overrightarrow{ab}\ \overrightarrow{ax}\ \overrightarrow{ac})$, and $\overrightarrow{ax}$ meets $\overline{bc}$, say in *y* (Fig. 13.6(*b*)), by Theorem 7 of Chapter 12. Thus by the Ray Decomposition Principle (Ch. 10, Th. 7, Cor. 4)

(2) $$y \subset \overrightarrow{ax} = \overline{ax} \cup x \cup x/a.$$

Suppose $y \subset \overline{ax}$. Then $\overline{ax}$ meets *bc* and *x*, *a* are on opposite sides of *bc* (Ch. 12, Th. 2). This contradicts (1). Suppose $y = x$. Then $x \subset bc$ contrary to (1). Hence (2) implies $y \subset x/a$. Thus (*yxa*). Since $y \subset \overline{bc}$, by definition $x \subset I\,(\triangle abc)$. This completes the proof.

COROLLARY 1. $I(\triangle abc)$ is the set of points which are simultaneously on the same side of *ab* as *c*, on the same side of *bc* as *a*, and on the same side of *ca* as *b*.

COROLLARY 2.   The interior of a triangle is a convex set.

Proof.   The intersection of several convex sets is convex.

COROLLARY 3.   Let $x \subset I(\triangle abc)$. Then *x* is between each vertex and a point of its opposite side.

Proof. By the theorem x satisfies Equation (1) above. In the proof of the theorem we showed that (1) implies x is between a and a point of $\overline{bc}$. Since (1) is symmetric in a, b, c it follows that x is between b and a point of $\overline{ca}$, and that x is between c and a point of $\overline{ab}$.

This immediately yields

COROLLARY 4.   Let a, b, c be distinct and noncollinear, (axy) and (byc). Then there exists z such that (bxz) and (cza) (Fig. 13.7).

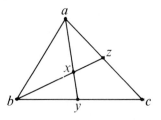

FIGURE 13.7

COROLLARY 5. $\triangle abc$ and $I(\triangle abc)$ are disjoint.

Proof. We have

(1) $$\triangle abc \subset ab \cup bc \cup ca,$$

(2) $$I(\triangle abc) = \overrightarrow{(ab)c} \cap \overrightarrow{(bc)a} \cap \overrightarrow{(ca)b}.$$

Since ab is disjoint to $\overrightarrow{(ab)c}$, bc to $\overrightarrow{(bc)a}$ and ca to $\overrightarrow{(ca)b}$, the result follows from (1) and (2).

9. The Exterior of a Triangle

We derive in Theorem 8 a decomposition of a plane which readily yields a formula for the exterior of a triangle.

We begin with two lemmas which are not uninteresting in themselves.

LEMMA 1. $abc = I(\triangle abc) \cup ab \cup bc \cup ca \cup [ab/c \cup bc/a \cup ca/b]$.

Proof.   Let S be the right member of the equation to be derived. We show $S \subset abc$. Theorem 7 asserts

(1) $I(\triangle abc) = \overrightarrow{(ab)c} \cap \overrightarrow{(bc)a} \cap \overrightarrow{(ca)b}$,

so that

$$I(\triangle abc) \subset \overrightarrow{(ab)c} \subset abc.$$

The other components of S are seen to be subsets of abc, and we conclude $S \subset abc$.

To show conversely $abc \subset S$, suppose $x \subset abc$. If $x \subset I(\triangle abc)$ certainly $x \subset S$. Suppose $x \nsubseteq I(\triangle abc)$. In view of the symmetry in (1) it is not restrictive to suppose

(2) $x \nsubseteq \overrightarrow{(ab)c}$.

We have (Ch. 11, Th. 4, Cor. 3)

(3) $abc = ab/c \cup ab \cup \overrightarrow{(ab)c}$.

Equations (2) and (3) imply

$$x \subset ab/c \cup ab \subset S.$$

Thus $abc \subset S$ and we conclude $abc = S$.

LEMMA 2. $\triangle abc \cup I(\triangle abc)$ is disjoint to ab/c.

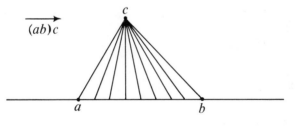

ab/c

FIGURE 13.8

Proof.   Show $\triangle abc \cup I(\triangle abc) \subset \overrightarrow{(ab)c} \cup ab$ which is disjoint to ab/c (Fig. 13.8).

Note the concrete geometrical significance of Lemma 2: The "closed" triangular region with vertices a, b, c lies wholly in the "closed" half-plane with edge ab that contains c, and is disjoint to the opposite half-plane.

THEOREM 8. $abc = I(\triangle abc) \cup \triangle abc \cup [ab/c \cup bc/a \cup ca/b]$, and the indicated components are disjoint.

Proof.   Let $K = ab/c \cup bc/a \cup ca/b$. By Lemma 1

(1) $abc = I(\triangle abc) \cup ab \cup bc \cup ca \cup K$.

Naturally we express the lines ab, bc, ca in terms of the segments $\overline{ab}$, $\overline{bc}$, $\overline{ca}$ and hope to eliminate superfluous terms. We have

(2) $ab = \overline{ab} \cup a \cup b \cup a/b \cup b/a$.

We shall substitute (2) into (1) and "absorb" $a/b \cup b/a$ into K. To justify this we have (Ch. 12, Th. 3)

$$a/b \subset ca/b, \qquad b/a \subset bc/a;$$

so that

$$a/b \cup b/a \subset ca/b \cup bc/a \subset K.$$

Thus (1) and (2) yield

(3) $abc = I(\triangle abc) \cup (\overline{ab} \cup a \cup b) \cup bc \cup ca \cup K$.

Similarly, we substitute in (3) for bc in terms of $\overline{bc}$, and for ca in terms of $\overline{ca}$ and get

$$abc = I(\triangle abc) \cup \triangle abc \cup K.$$

To complete the proof note first $I(\triangle abc)$ and $\triangle abc$ are disjoint (Th. 7, Cor. 5). Further by Lemma 2, $I(\triangle abc) \cup \triangle abc$ is disjoint to ab/c, symmetrically to bc/a and to ca/b, and so to their union K.

COROLLARY. $E(\triangle abc) = ab/c \cup bc/a \cup ca/b$.

Proof.   By definition from Theorem 8.

Exercise.   Prove: $\triangle abc \cup I(\triangle abc) \subset \overrightarrow{(ab)c} \cup ab$.

10. Separation of a Plane by a Triangle

We derive two properties of triangle interiors which are used to prove the principal theorem.

THEOREM 9.   Let $x \subset I(\triangle abc)$, $y \subset \triangle abc$. Then $\overline{xy} \subset I(\triangle abc)$.

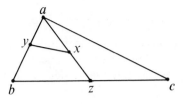

FIGURE 13.9

Proof. It suffices to consider two cases, namely $y \subset \overline{ab}$ and $y = a$. Suppose $y \subset \overline{ab}$. By Theorem 7

(1) $$x \subset I(\triangle abc) = \overrightarrow{(ab)c} \cap \overrightarrow{(bc)a} \cap \overrightarrow{(ca)b}.$$

Thus x is on the side of ab containing c. Hence the same is true of each point of $\overline{xy}$ (Ch. 12, Th. 3, Cor. 2). Thus $\overline{xy} \subset \overrightarrow{(ab)c}$. Further, $y \subset \overline{ab}$ implies that y, a are on the same side of bc. That is $y \subset \overrightarrow{(bc)a}$. Since $x \subset \overrightarrow{(bc)a}$ by (1), we have $\overline{xy} \subset \overrightarrow{(bc)a}$ by the convexity of a half-plane. Similarly, $\overline{xy} \subset \overrightarrow{(ca)b}$. Thus

$$\overline{xy} \subset \overrightarrow{(ab)c} \cap \overrightarrow{(bc)a} \cap \overrightarrow{(ca)b} = I(\triangle abc).$$

Suppose $y = a$. By Corollary 3 to Theorem 7, $x \subset I(\triangle abc)$ implies x is between a and a point z in $\overline{bc}$, so that (axz). Thus $\overline{ax} \subset \overline{az}$ (Ch. 10, Th. 8, Cor. 1). By definition $\overline{az} \subset I(\triangle abc)$; thus $\overline{ax} \subset I(\triangle abc)$ and the proof is complete.

THEOREM 10.   Let $x \subset I(\triangle abc)$, $y \subset abc$, $y \neq x$. Then $\overrightarrow{xy}$ meets $\triangle abc$.

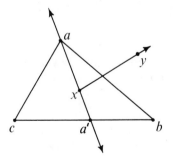

FIGURE 13.10

Proof. By Corollary 3 to Theorem 7 (axa') for some point $a' \subset \overline{bc}$. Hence $(ba'c)$. Thus $\overrightarrow{xa}$ and $\overrightarrow{xa'}$ are opposite rays (Ch. 10, Th. 6). If $\overrightarrow{xy} = \overrightarrow{xa}$

or $\overrightarrow{xy} = \overrightarrow{xa'}$, the conclusion certainly holds. Suppose $\overrightarrow{xy} \neq \overrightarrow{xa}, \overrightarrow{xa'}$. Then $y \notin xa = \overrightarrow{xa} \cup \overrightarrow{xa'} \cup x$. Since $\overline{bc}$ meets xa, it follows that b, c are on opposite sides of xa (Ch. 12, Th. 2). Thus y is on the same side of xa as b or c. It is not restrictive to suppose y, b are on the same side of xa. Then

$$(\overrightarrow{xa}\ \overrightarrow{xy}\ \overrightarrow{xb}), \quad (\overrightarrow{xa'}\ \overrightarrow{xy}\ \overrightarrow{xb}) \quad \text{or} \quad \overrightarrow{xy} = \overrightarrow{xb}$$

(Ch. 12, Th. 11, Cor. 1). $(\overrightarrow{xa}\ \overrightarrow{xy}\ \overrightarrow{xb})$ implies $\overrightarrow{xy}$ meets $\overline{ab}$ (Ch. 12, Th. 7). Similarly $(\overrightarrow{xa'}\ \overrightarrow{xy}\ \overrightarrow{xb})$ implies $\overrightarrow{xy}$ meets $\overline{a'b}$. But $\overline{a'b} \subset \overline{bc}$, so that $\overrightarrow{xy}$ meets $\overline{bc}$. Finally $\overrightarrow{xy} = \overrightarrow{xb}$ implies $\overrightarrow{xy} \supset b$. In each case $\overrightarrow{xy}$ meets $\triangle abc$ and the theorem is proved.

THEOREM 11.   Any triangle in a plane separates the plane into its interior and its exterior.

Proof.   Let $P \supset \triangle abc$. Let I, E denote the interior, exterior of $\triangle abc$. We show
  (i) $P = I \cup E \cup \triangle abc$;
  (ii) the segment joining any point of I and any point of E meets $\triangle abc$;
  (iii) the segment joining any two points of I does not meet $\triangle abc$;
  (iii′) there exists in P a path joining any two given points of E which does not meet $\triangle abc$;
  (iv) I, E, $\triangle abc$ are disjoint.

Proof of (i).   Note $P = abc$. By Theorem 8, $P \supset I$. Then (i) follows by definition of E.

Proof of (iv). E is disjoint to $\triangle abc$ and to I by definition of E; $\triangle abc$ and I are disjoint by Corollary 5 to Theorem 7.

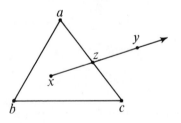

FIGURE 13.11

Proof of (ii).   Let $x \subset I$, $y \subset E$ (Fig. 13.11). By Theorem 10, $\overrightarrow{xy}$ meets $\triangle abc$, say in z. Then

(1) $$z \subset \overrightarrow{xy} = \overline{xy} \cup y \cup y/x.$$

If $z = y$, then $\triangle abc$ and E have a common point, namely z or y, contrary to (iv). Suppose $z \subset y/x$; then (zyx), so that $y \subset \overline{xz}$. But $\overline{xz} \subset I$ by Theorem 9. Thus I and E have a common point y, contrary to (iv). Thus the only possibility permitted by (1) is $z \subset \overline{xy}$, and (ii) is proved.

Proof of (iii).   This follows from the convexity of I and the disjointness of I and $\triangle abc$.

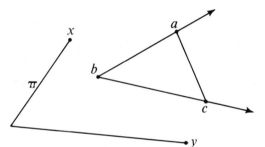

FIGURE 13.12

Proof of (iii′).   Let x, $y \subset E = ab/c \cup bc/a \cup ca/b$; $x \neq y$ (Fig. 13.12). Then of the three indicated half-planes, x and y are either in the same one or in two different ones. In either case, x and y are in the union of two of the half-planes, and it is not restrictive to assume

$$x, y \subset ab/c \cup bc/a.$$

But

$$ab/c \cup bc/a = E(\angle abc),$$

by Theorem 5. Since $E(\angle abc)$ is pathwise connected (Th. 6, Cor.), there is a path π joining x and y such that

$$\pi \subset ab/c \cup bc/a.$$

Thus $\pi \subset E$, so that by (iv), π does not meet $\triangle abc$.

Finally, note $I \neq \varnothing$ since by definition it contains any point between a and a point of $\overline{bc}$; and $E \neq \varnothing$ since $E \supset ab/c$.

COROLLARY.   The exterior of a triangle is pathwise connected (Compare Th. 6, Cor.).

EXERCISES I

1. If $x \subset \angle abc$ and $y \subset I(\angle abc)$, prove $\overline{xy} \subset I(\angle abc)$.

2. If a, b, c are distinct and noncollinear, $p \subset \overrightarrow{ba}$ and $q \subset \overrightarrow{bc}$, prove $\overline{pq} \subset I(\angle abc)$.

3. If $x \subset I(\angle aob)$, prove $I(\angle aob) = I(\angle aox) \cup \overrightarrow{ox} \cup I(\angle box)$.

4. Let x, $y \subset I(\angle aob)$, $x \neq y$. Prove that one and only one of the following holds:

$$(\overrightarrow{oa}\ \overrightarrow{ox}\ \overrightarrow{oy}\ \overrightarrow{ob}), \quad (\overrightarrow{oa}\ \overrightarrow{oy}\ \overrightarrow{ox}\ \overrightarrow{ob}), \quad \overrightarrow{ox} = \overrightarrow{oy}.$$

(For the definition of the four-term order relation of rays, see Ch. 12, Exercises IV.)

5. Let a, b, c be distinct and noncollinear, $p \subset \overline{ab} \cup a$, $q \subset \overline{bc}$. Prove that all points of $\overline{pq}$ are in $I(\triangle abc)$, and no other points of pq are in $I(\triangle abc)$.

6. If line L contains an interior point p of $\triangle abc$ and coplanes with it, prove that L meets $\triangle abc$ in precisely two points x, y and that (xpy).

7. Prove: $I(\triangle abc) = I(\angle abc) \cap I(\angle bca) \cap I(\angle cab)$.

8. Prove: $I(\triangle abc) = I(\angle abc) \cap I(\angle bca)$.

9. If a, b, c are distinct and noncollinear, prove

$$j(a, \overline{bc}) = j(b, \overline{ca}) = j(c, \overline{ab}) = I(\triangle abc).$$

(For the definition of $j(A, B)$, the join of sets A and B, see Ch. 10, Exercises IV.)

10. If a, b, c are distinct and noncollinear, prove

$$j(\overline{ab}, \overline{bc}) = j(\overline{bc}, \overline{ca}) = j(\overline{ca}, \overline{ab}) = I(\triangle abc).$$

11. If a, b, c are distinct and noncollinear, prove that

$$j(\overline{ab}, j(\overline{bc}, \overline{ca})) = I(\triangle abc).$$

12. If a, b, c are distinct and noncollinear, prove that $j(\overrightarrow{ba}, \overrightarrow{bc}) \subset I(\angle abc)$. Is $j(\overrightarrow{ba}, \overrightarrow{bc}) = I(\angle abc)$? Justify your answer.

13. If K is a convex set, and $a \notin K$, prove that $j(a, K)$ is a convex set.

14. Let a, b, c be distinct and noncollinear, and (bcb'). Prove $a/b \subset I(\angle acb')$.

15. Let K be a convex set, and let B be the set of all points $x \notin K$ such that $\overline{xk} \subset K$ for some $k \subset K$. Prove that $B \cup K$ is a convex set.

16. Prove: If $\overrightarrow{ox} \subset I(\angle aob)$, $\overrightarrow{oy} \subset I(\angle boc)$, and $\angle aob$, $\angle boc$ form a linear pair, then $(\overrightarrow{ox}\ \overrightarrow{ob}\ \overrightarrow{oy})$.

17. Prove: If a ray is in the interior of an angle and its endpoint is the vertex of the angle, then its opposite is in the interior of the vertical angle.

18. Prove: An angle has a unique vertex and a unique pair of sides. That is, if $\angle abc = \angle def$ then $b = e$ and either $\overrightarrow{ba} = \overrightarrow{ed}$ and $\overrightarrow{bc} = \overrightarrow{ef}$ or $\overrightarrow{ba} = \overrightarrow{ef}$ and $\overrightarrow{bc} = \overrightarrow{ed}$. (Compare Ch. 10, Exercises II, 27.)

19. Prove: A triangle has a unique set of vertices and a unique set of sides.

20. Prove: The union of an angle and its exterior is pathwise connected.

21. Is the following definition of angle interior equivalent to the one we have adopted? The interior of $\angle abc$ consists of all points which are in segments joining a point of $\overrightarrow{ba}$ and a point of $\overrightarrow{bc}$. Justify your answer.

22. Is the following definition of angle interior equivalent to the one we have adopted? Given $\angle abc$. Let $p \subset \overline{ac}$. The interior of $\angle abc$ consists of p and all points $x \subset abc$ such that $\overline{px}$ does not meet $\angle abc$. Justify your answer.

EXERCISES II

(For the definition of plane quadrilateral $Q(abcd)$, simple quadrilateral and convex quadrilateral, see Ch. 12, Exercises V.)

DEFINITION.   Let $Q(abcd)$ be a convex quadrilateral. The interior of $Q(abcd)$, denoted $IQ(abcd)$ is given by

$$IQ(abcd) = \overrightarrow{(ab)c} \cap \overrightarrow{(bc)d} \cap \overrightarrow{(cd)a} \cap \overrightarrow{(da)b}.$$

1. Prove that the interior of a convex quadrilateral is a nonempty convex set.

2. If $Q(abcd)$ is a convex quadrilateral, prove that

$$Q(abcd) \cap IQ(abcd) = \varnothing.$$

3. If $Q(abcd)$ is a convex quadrilateral, prove

$$IQ(abcd) = I(\angle abc) \cap I(\angle bcd) \cap I(\angle cda) \cap I(\angle dab).$$

4. If $Q(abcd)$ is a convex quadrilateral, prove

$$IQ(abcd) = I(\angle abc) \cap I(\angle cda).$$

5. If $Q(abcd)$ is a convex quadrilateral, and $x \subset IQ(abcd)$, prove that $(\overrightarrow{ab}\ \overrightarrow{ax}\ \overrightarrow{ad})$.

6. Let $Q(abcd)$ be a convex quadrilateral, $p \subset \overline{ab}, q \subset \overline{bc}, r \subset \overline{cd}$.
   Prove: (i) $\overline{ac} \subset IQ(abcd)$;   (ii) $\overline{pq} \subset IQ(abcd)$;
           (iii) $\overline{pr} \subset IQ(abcd)$;   (iv) $\overline{aq} \subset IQ(abcd)$.

7. Let $Q(abcd)$ be a convex quadrilateral, $x \subset Q(abcd)$, $y \subset IQ(abcd)$. Prove $\overline{xy} \subset IQ(abcd)$.

8. If $Q(abcd)$ is a convex quadrilateral, prove that $Q(abcd) \cup IQ(abcd)$ is a convex set.

9. If $Q(abcd)$ is a convex quadrilateral, prove that no one of a, b, c, d is in the interior of the triangle whose vertices are the other three.

10. Let $Q(abcd)$ be a simple nonconvex plane quadrilateral. Prove that one of a, b, c, d is in the interior of the triangle whose vertices are the other three.

11. Let $d \subset I(\triangle abc)$. Prove $a \not\subset I(\triangle bcd)$.

Exercises 9 and 10 yield that a simple plane quadrilateral is a nonconvex quadrilateral if and only if one of its vertices is in the interior of the triangle whose vertices are the other three vertices of the given quadrilateral. Exercise 11 implies that if one of the vertices of a simple plane quadrilateral is in the interior of the triangle determined by the other three vertices, then only one is. This suggests the

DEFINITION.   Let $Q(abcd)$ be a simple quadrilateral which is not a convex quadrilateral, and let $d \subset I(\triangle abc)$. Then the interior of $Q(abcd)$, denoted $IQ(abcd)$ is given by

$$IQ(abcd) = I(\triangle abd) \cup I(\triangle cbd) \cup \overline{bd}.$$

12. If $Q(abcd)$ is a convex quadrilateral, prove that

$$IQ(abcd) = I(\triangle abd) \cup I(\triangle cbd) \cup \overline{bd}.$$

†13. If the interior of a simple plane quadrilateral is a convex set, prove the quadrilateral is a convex quadrilateral.

14. Let $Q(abcd)$ be a convex quadrilateral, $x \subset IQ(abcd)$, $x \not\subset \overline{bd}$. Prove that one of the quadrilaterals $Q(dabx)$, $Q(bcdx)$ is a convex quadrilateral, and that the other is not.

15. Prove: A simple plane quadrilateral is a convex quadrilateral if and only if any segment joining points of two sides is in its interior.

16. If line L contains a point p of the interior of a convex quadrilateral, prove that L meets the quadrilateral in two distinct points x, y such that (xpy), and that the intersection of L with the interior of the quadrilateral is the segment $\overline{xy}$.

17. A simple plane quadrilateral is a convex quadrilateral if and only if any line, not a side line, meets it in at most two points.

†18. Let $Q(abcd)$ be a convex quadrilateral. Prove that

$$j(\overline{ab}, \overline{cd}) = j(\overline{bc}, \overline{da}) = IQ(abcd).$$

(For the definition of $j(A, B)$, the join of sets A and B, see Ch. 10, Exercises IV.)

†19. If A and B are coplanar, disjoint, convex sets, prove that $j(A, B)$ is a convex set.

20. If a, b, c, d are distinct, noncollinear and noncoplanar, prove that

$$j(a, I(\triangle bcd)) = j(b, I(\triangle cda))$$
$$= j(c, I(\triangle dab)) = j(d, I(\triangle abc)).$$

21. If a, b, c, d are distinct, noncollinear and noncoplanar, prove that

$$j(a, I(\triangle bcd)) = j(\overline{ab}, \overline{cd}),$$

and deduce that

$$j(\overline{ab}, \overline{cd}) = j(\overline{ac}, \overline{bd}) = j(\overline{ad}, \overline{bc}),$$

and that these are equal to the joins listed in Exercise 20.

22. If A and B are disjoint convex sets, prove that $j(A, B)$ is a convex set.

23. Prove: The exterior of a simple plane quadrilateral is pathwise connected. (The exterior of a simple plane quadrilateral Q is the set of all points of its plane which are neither in Q nor in its interior.)

E X E R C I S E S   I I I

1. If a, b, c are distinct and noncollinear, and $p \subset \overline{bc}$, prove

$$I(\triangle abc) = I(\triangle abp) \cup I(\triangle acp) \cup \overline{ap},$$

where the addends on the right are disjoint.

2. If $d \subset I(\triangle abc)$, prove

$$I(\triangle abc) = I(\triangle abd) \cup I(\triangle acd) \cup I(\triangle bcd) \cup \overline{ad} \cup \overline{bd} \cup \overline{cd} \cup d,$$

where the addends on the right are disjoint.

3. If $\triangle abc \subset I(\triangle def)$, prove $I(\triangle abc) \subset I(\triangle def)$.

4. Prove: $\triangle abc \cup I(\triangle abc)$ is a convex set.

†5. Prove: $I(\triangle abc) \subset I(\triangle def)$ if and only if

$$\triangle abc \subset \triangle def \cup I(\triangle def).$$

6. If $(\overrightarrow{oa}\, \overrightarrow{ob}\, \overrightarrow{oc})$, prove that

$$I(\angle aoc) = I(\angle aob) \cup I(\angle boc) \cup \overrightarrow{ob}.$$

7. If a, b, c are distinct and noncollinear, prove

$$ba/c \cap cb/a \cap ac/b = \varnothing.$$

8. If a, b, c are distinct and noncollinear, prove

$$ab/c \cap bc/a \subset \overrightarrow{(ca)b}.$$

DEFINITION.   If A and B are disjoint sets, A/B, the *extension* of A from B, is the set of points each of which is in a ray of the form a/b, where $a \subset A$ and $b \subset B$.*

9. If a, b, c are distinct and noncollinear, prove

$$(a/b)/c = a/\overline{bc}.$$

10. If a, b, c are distinct and noncollinear, (bab') and (cac'), prove that $a/\overline{bc} = I(\angle b'ac')$.

11. If A, B are sets and $c \not\subset A \cup B$, prove that

$$(A \cup B)/c = A/c \cup B/c.\dagger$$

12. If A_1, A_2, ..., A_n are sets, and $c \not\subset A_1 \cup A_2 \cup \cdots \cup A_n$, prove that

$$(A_1 \cup A_2 \cup \cdots \cup A_n)/c = A_1/c \cup A_2/c \cup \cdots \cup A_n/c.$$

NOTE: If $c \not\subset ab = a/b \cup a \cup \overline{ab} \cup b \cup b/a$, it follows immediately from Exercise 12 and Exercise 9 that

$$ab/c = (a/b \cup a \cup \overline{ab} \cup b \cup b/a)/c$$
$$= (a/b)/c \cup a/c \cup \overline{ab}/c \cup b/c \cup (b/a)/c$$
$$= a/\overline{bc} \cup a/c \cup \overline{ab}/c \cup b/c \cup b/\overline{ac}.$$

13. Prove that the sets $a/\overline{bc}$, a/c, $\overline{ab}/c$, b/c, $b/\overline{ac}$ in the note are disjoint.
14. If a, b, c are distinct and noncollinear, prove

$$\overline{ac}/b = \overrightarrow{(ab)c} \cap \overrightarrow{(bc)a} \cap ca/b.$$

15. If a, b, c are distinct and noncollinear, prove that

$$b/\overline{ac} = ab/c \cap bc/a.$$

---

* For a systematic study of the operations "join of sets," $j(A, B)$, and "extension of sets," A/B, see W. Prenowitz, *A Contemporary Approach to Classical Geometry*, American Mathematical Monthly, Vol. 68 no. 1, Part II, 1961.

† The right member stands for $(A/c) \cup (B/c)$, we adopt the convention that portions of expressions separated by $\cup$ signs are to be considered enclosed in parentheses.

16. Prove: $I(\angle abc) = I(\triangle abc) \cup \overline{ac} \cup \overline{ac}/b$, where the terms on the right are disjoint.

17. If a, b, c are distinct and noncollinear and (bcb'), prove

$$I(\angle acb') = \overrightarrow{ca}/b = \overline{ac}/b \cup a/b \cup a/\overline{cb},$$

where the three terms on the right are disjoint.

18. If a, b, c are distinct and noncollinear, prove

$$E(\triangle abc) = a/\overline{bc} \cup b/\overline{ca} \cup c/\overline{ab} \cup \overline{ab}/c \cup \overline{bc}/a \cup \overline{ca}/b \cup$$
$$a/b \cup a/c \cup b/a \cup b/c \cup c/a \cup c/b,$$

where the terms on the right are disjoint.

19. Prove: Any point of plane abc is in a line joining a vertex of $\triangle abc$ to a point of a side.

E X E R C I S E S   I V

DEFINITION. Let line L be given. Then a subset A of L is an *open* subset of L or briefly is *open in L* if for each point p of A there is a segment containing p which is a subset of A. Similarly, let plane P be given. Then a subset A of P is an *open* subset of P or is *open in P* if for each point p of A there is a triangle interior containing p which is a subset of A. The empty set $\varnothing$ is open in any line and in any plane.

1. Let $L = ab$. Prove that each of the following subsets of L is open in L:

   (i) ab;   (ii) $\overrightarrow{ab}$;   (iii) $\overline{ab}$;   (iv) $a/b \cup b/a$.

2. Let $L = ab$. Prove that each of the following subsets of L is not open in L:

   (i) The set whose sole element is b; (ii) $\overrightarrow{ab} \cup a$; (iii) $\overline{ab} \cup a$.

3. Let $P = abc$. Prove that each of the following subsets of P is open in P:

   (i) abc;   (ii) $\overrightarrow{(ab)c}$;   (iii) $I(\triangle abc)$;   (iv) $E(\triangle abc)$.

4. Let $P = abc$. Prove that each of the following subsets of P is not open in P:

   (i) Any nonempty subset of ab;   (ii) $ab/c \cup ab$;   (iii) $I(\triangle abc) \cup \overline{ab}$.

5. Prove that the union of any number, even infinitely many, of sets open in L (in P) is a set which is open in L (in P).*

6. Prove that the intersection of two sets open in L is a set which is open in L. Generalize to a finite number.

---

* The union of an arbitrary collection M of sets is the set which contains each element of each set of M and no other element.

†7. Prove that the intersection of two sets open in P is a set which is open in P. Generalize to a finite number.

8. Prove that the word "finite" is essential in Exercises 6 and 7 by finding examples of infinitely many sets open in L (in P) whose intersection is a set which is not open in L (in P).

9. Project: Generalize the notions above to set *open* in a three-dimensional (ordered incidence) geometry. List properties analogous to those above and try to prove or disprove them.

CHAPTER 14

Theory of Congruence —
Introduction to
Euclidean Geometry*

In this chapter a tentative treatment of congruence is given based on a proposal of G. D. Birkhoff (1884–1944) that the real number system should be assumed in the treatment of Euclidean geometry at an elementary level. Birkhoff's development was modified and simplified by the School Mathematics Study Group. Our treatment is an adaptation of theirs and assumes a modification of their Ruler Postulate employed by MacLane (see Postulate C2 below).† We do not consider this a "final" treatment of congruence, which we think should be intrinsic and not require the assumption of the existence of measures of segments and angles as real numbers. However Birkhoff's idea does enable the student to study the familiar theory of congruence in rigorous form relatively early, by by-passing the difficulties involved in the proof of the existence of measures of segments and angles. Indeed he probably will better appreciate the latter after having studied a treatment of the type presented here. In Chapter 15 we give an introduction to an intrinsic theory of congruence.

---

* Although Chapter 15 is logically independent of this chapter, it is generally advisable to read Chapter 14 first, since Chapter 15 presents a deeper treatment and has occasion to refer to the present chapter for details of definition and methods of proof.

† See and compare: G. D. Birkhoff and R. Beatley, *Basic Geometry*, Scott Foresman and Co., 1940, reprinted by Chelsea Publishing Co., New York; School Mathematics Study Group, *Geometry*, Student's Text Part I, Yale University Press, New Haven, 1961; S. MacLane, *Metric Postulates for Plane Geometry*, American Mathematical Monthly, Vol. 66, pp. 543–555, 1959.

We operate in the framework of an ordered incidence geometry and assume Postulates C1–C8 below. The basic theory assumes no parallel postulate and is applicable to Euclidean and to Lobachevskian geometry. In the final section (Sec. 17) we give an introduction to Euclidean geometry, carrying the development through the theorem that the sum of the 'degree' measures of the angles of a triangle is 180.

1. Introduction

The conventional theory of congruence in school geometry is not logically satisfactory. The underlying idea seems simple: Two figures are congruent if one is an exact copy or replica of the other. The usual treatment seems to capture the flavor of this notion in the familiar principles:

(A) Two figures are congruent if they can be made to coincide by a rigid motion.

(B) A rigid motion preserves size and shape.

Although these principles might possibly become the basis of a satisfactory theory of congruence if clarified and supplemented, as they stand and are employed in a conventional proof of congruence by superposition, they seem hardly more than physical notions which have not been completely mathematicized. They seem more appropriate to a description of a physical experiment—for example, the superposition of two cardboard triangles whose corresponding edges fit—than to the final characterization of a mathematical concept. It is not clear how motion is to be conceived—is it comparable to a physical process that takes place in an interval of time? Even in physical terms how can one conceive a rigid motion which would bring two congruent billiard balls into coincidence? The reference to *size* and *shape* indicates the complexity of the principles since these are complicated notions which involve the theories of measurement and similarity.

Principles (A) and (B) taken together suggest the following principle:

(C) Two figures are congruent if they have the same size and the same shape.

Principle (C) indicates the possibility that congruence may be definable in terms of metrical notions such as measures of length, angles, area. Specifically, two segments can be considered congruent (usually called "equal") if their lengths are equal. Similarly two angles are congruent if they have equal measures. Then two triangles will be congruent if their "corresponding" sides and angles are congruent.

Thus we adopt the following approach: We introduce as primitive notions, the distance between two points (or equivalently the length or measure of a segment) and the measure of an angle; we define congruence of segments and of angles in terms of these notions; and derive their properties from suitable postulates on distance and angle measures.

Such a treatment simplifies the theory of congruence since it postulates the existence of lengths of segments and measures of angles and makes algebraic methods available from the start. Of course the simplification is obtained by avoiding the problem of constructing measures in geometry on a purely geometrical basis.

2. Postulates for Distance

We begin by stating three postulates for distance between two points.

C1. (DISTANCE POSTULATE) *To each pair of distinct points* (a, b) *there corresponds a unique positive real number Dab called the distance between a and b, or the length of $\overline{ab}$, or the measure of $\overline{ab}$.*

Since there is no difference between the pair (a, b) and the pair (b, a) we note $Dab = Dba$.

A familiar operation in school geometry is to "lay off" a segment on a line, or to "extend" a segment by a given length. Such operations suggest—and are clarified by—the following postulate:

C2. (POINT LOCATION POSTULATE) *Let $\overrightarrow{ab}$ be any ray and x any positive real number. Then there is exactly one point $p \subset \overrightarrow{ab}$ such that $Dap = x$.*

Finally, betweenness and distance are related in the natural way:

C3. (ADDITION POSTULATE) *If (abc) then $Dab + Dbc = Dac$.*

3. Distance and Order of Points

We have several theorems which relate betweenness of points and order properties of distance.

THEOREM 1.   If (apq) then $Dap < Daq$.

Proof. $Dap + Dpq = Daq$ by C3. The conclusion follows since Dpq is positive by C1.

THEOREM 2. (A CONVERSE OF THEOREM 1)   Let $p, q \subset \overrightarrow{ab}$. Then $Dap < Daq$ implies (apq).

Proof. $p \subset \overrightarrow{ab}$ implies $\overrightarrow{ab} = \overrightarrow{ap}$. Thus $q \subset \overrightarrow{ap}$, which implies (aqp) or $q = p$ or (apq) (Ch. 10, Th. 7, Cor. 5). If (aqp) then $Daq < Dap$ by Theorem 1. If $q = p$ then $Daq = Dap$. Thus (apq) must hold.

Directly from Theorems 1, 2 we get the following

COROLLARY.   Let p, q be in a ray with endpoint a. Then (apq) if and only if $Dap < Daq$.

THEOREM 3.   Let $p, q, r \subset \overrightarrow{ab}$. Then (pqr) if and only if $Dap < Daq < Dar$ or $Dap > Daq > Dar$.

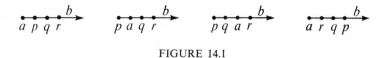

FIGURE 14.1

Proof.   Suppose (pqr). Since p, q, r, a are distinct and collinear, Theorem 13 of Chapter 10 implies

$$(apqr), (paqr), (pqar) \text{ or } (pqra).$$

$(paqr)$ is false since it implies

$$a \subset \overline{pq} \subset \overrightarrow{ab}$$

by the convexity of a ray. Similarly $(pqar)$ is false. Thus $(apqr)$ or $(pqra)$. $(apqr)$ implies (apq), (aqr) so that $Dap < Daq < Dar$, by Theorem 1. $(pqra)$ implies (pqa), (qra) so that (arq), (aqp) which implies $Dar < Daq < Dap$, and therefore $Dap > Daq > Dar$.

Conversely, consider the case

(1) $Dap < Daq < Dar$

By Theorem 2 (apq), (aqr). These imply (pqr) (Ch. 10, Th. 13, Cor. 4).

The second case $Dap > Daq > Dar$ may be written $Dar < Daq < Dap$ and is symmetrical to the first.

This theorem is interesting in that it exhibits a coherence of two between-nesses: geometric betweenness of three points of a ray and algebraic be-

tweenness of their distances from its endpoint. The theorem indicates the existence of an isomorphism between $\overrightarrow{ab}$ considered as a system of points ordered by geometric betweenness, and the set of positive reals as a system of numbers ordered by algebraic betweenness.

Suppose $\overline{ab}$ is shorter than $\overline{cd}$, that is $Dab < Dcd$. Can we give this a degree of geometric significance? The answer lies in

THEOREM 4. Suppose $Dab < Dcd$. Then there exists a point p such that $Dab = Dcp$ and (cpd).

FIGURE 14.2

Proof. By C2 there is a point p in $\overrightarrow{cd}$ such that $Dcp = Dab$. Then (cpd), $p = d$ or (cdp). If $p = d$ then $Dcd = Dab$, contrary to hypothesis. Suppose (cdp). By Theorem 1, $Dcd < Dcp = Dab$, contrary to hypothesis. Thus (cpd).

Note that the converse of Theorem 4 holds: *If $Dab = Dcp$ and (cpd) then $Dab < Dcd$.*

Finally we have a modified converse of Postulate C3:

THEOREM 5. Let a, b, c colline. Then $Dab + Dbc = Dac$ implies (abc).

Proof. Points a, b, c are distinct and collinear. Thus (abc), (bca), or (cab). Suppose (bca); then C3 implies $Dbc + Dca = Dba$ or $Dbc + Dac = Dab$, which contradicts the hypothesis $Dab + Dbc = Dac$. Thus (bca) is false. Similarly (cab) is false, and (abc) must hold.

4. Distance Functions

Since we have had some experience in proving theorems with the distance postulates, it may be good at this point to examine them a bit more closely. Postulates of this type are rather unusual. C1 seems to say that there is only one way to measure distance—this does not jibe with our experience in elementary mathematics, where we have different measures based on different choices of units. In effect C1 says that a number is assigned to each pair of points. This suggests the function concept. Let us define formally a function

$d(p, q)$ which assigns to each *ordered* pair of distinct points (p, q) the distance between p and q. Thus $d(p, q) = Dpq$. What are the essential properties of such a function? Upon restating C1, C2, C3 we have:

(i) $d(p, q) = d(q, p) > 0$.

(ii) *If $\overrightarrow{ab}$ is any ray and x any positive real number there is exactly one point $p \subset \overrightarrow{ab}$ such that $d(a, p) = x$.*

(iii) *If (pqr) then $d(p, q) + d(q, r) = d(p, r)$.*

This suggests the following:

DEFINITION.   Let G be an ordered incidence geometry and let $d(p, q)$ be a real-valued function defined for $p, q \subset G, p \neq q$. Then if $d(p, q)$ satisfies (i), (ii), (iii) above we call it a *distance function* (*on G*).

Thus Postulates C1, C2, C3 are equivalent to the assertion of the existence of a distance function. Clearly, if one distance function exists many others can be formed, since the product of a distance function by a positive constant is again a distance function. This suggests the following question. Our theory of congruence depends on the existence of a specific distance function. If we choose a new distance function, will the theory be affected? The answer is no. The theorems depend on the properties of a distance function, not on the specific choice made.

5. Postulates for Angle Measure

Now we give a treatment of angle measure analogous to that for distance. The analogy is of course not perfect. For all positive real numbers are distances, but each familiar system of angle measure excludes certain values. Degree measure, for example, excludes all numbers not less than 180; radian measure those not less than π. Thus a system of angle measure, as we conceive it, will have a minimum positive excluded value. We take this value, for the sake of familiarity, to be 180—thus our measure will have the familiar properties of degree measure.

We introduce four postulates for angle measure; the first three are analogous to C1, C2, C3.

C4. (ANGLE MEASURE POSTULATE)   *To each angle $\angle abc$ there corresponds a unique real number x, $0 < x < 180$, called the measure of the angle, and denoted $m\angle abc$.*

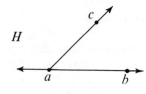

FIGURE 14.3

C5. (RAY LOCATION POSTULATE)   *Given a half-plane H, a ray $\overrightarrow{ab}$ contained in its edge, and a real number x, $0 < x < 180$. Then there exists exactly one ray $\overrightarrow{ac}$ such that $\overrightarrow{ac} \subset H$ and $m\angle bac = x$* (Fig. 14.3).

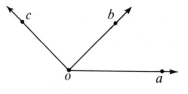

FIGURE 14.4

C6. (ANGLE ADDITION POSTULATE)   *If $(\overrightarrow{oa}\ \overrightarrow{ob}\ \overrightarrow{oc})$ then $m\angle aob + m\angle boc = m\angle aoc$* (Fig. 14.4).

A postulate about opposite rays which has no analogue for points is:

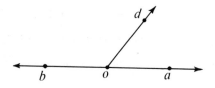

FIGURE 14.5

C7. (SUPPLEMENT POSTULATE)   *If $\overrightarrow{oa}$ is opposite $\overrightarrow{ob}$, and $d \notin oa$ then $m\angle aod + m\angle bod = 180$* (Fig. 14.5).

The remarks of Section 4 concerning the distance postulates C1–C3 apply in essence to the angle measure postulates C4–C7. We can easily define the notion of an *angle measure function*. That is a real-valued function $m(A)$, defined for all angles A, which satisfies properties corresponding to C4–C7 modified so that 180 in C4, C5 and C7 is replaced by a fixed positive number K. Postulates C4–C7 are equivalent to the assertion of the existence of an angle measure function.

6. Angle Measure and Order of Rays

Each result of Section 3 on distance and betweenness of points has a valid analogue for angle measure and betweenness of rays. We present formally analogues of Theorems 1, 2, and 4, and include the others among the exercises at the end of the section.

THEOREM 6.   If $(\overrightarrow{oa}\ \overrightarrow{ob}\ \overrightarrow{oc})$ then $m\angle aob < m\angle aoc$.

Proof.   Use the method of Theorem 1.

THEOREM 7. (A CONVERSE OF THEOREM 6)   Let b, c be on the same side of oa. Then $m\angle aob < m\angle aoc$ implies $(\overrightarrow{oa}\ \overrightarrow{ob}\ \overrightarrow{oc})$.

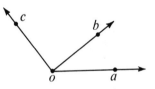

FIGURE 14.6

Proof.   We use the method of Theorem 2. Since b, c are on the same side of oa $(\overrightarrow{oa}\ \overrightarrow{ob}\ \overrightarrow{oc})$, $(\overrightarrow{oa}\ \overrightarrow{oc}\ \overrightarrow{ob})$ or $\overrightarrow{ob} = \overrightarrow{oc}$ (Ch. 12, Th. 11, Cor. 2). Suppose $(\overrightarrow{oa}\ \overrightarrow{oc}\ \overrightarrow{ob})$. Then $m\angle aoc < m\angle aob$ by Theorem 6, contrary to hypothesis. Suppose $\overrightarrow{ob} = \overrightarrow{oc}$. Then $\angle aob = \angle aoc$ and $m\angle aob = m\angle aoc$. Thus $(\overrightarrow{oa}\ \overrightarrow{ob}\ \overrightarrow{oc})$ must hold.

COROLLARY.   Let b, c be on the same side of oa. Then $(\overrightarrow{oa}\ \overrightarrow{ob}\ \overrightarrow{oc})$ if and only if $m\angle aob < m\angle aoc$.

THEOREM 8.   Suppose $m\angle abc < m\angle def$. Then there exists a ray $\overrightarrow{ep}$ such that $m\angle abc = m\angle dep$ and $(\overrightarrow{ed}\ \overrightarrow{ep}\ \overrightarrow{ef})$. (Compare Th. 4.)

Proof.   Use the method of Theorem 4.

EXERCISES

1. Let p, q, r be on the same side of oa. Then $(\overrightarrow{op}\ \overrightarrow{oq}\ \overrightarrow{or})$ if and only if $m\angle aop < m\angle aoq < m\angle aor$ or $m\angle aop > m\angle aoq > m\angle aor$. (Compare Th. 3.)

2. Prove Theorem 8.

3. Let rays $\overrightarrow{oa}$, $\overrightarrow{ob}$, $\overrightarrow{oc}$ with common endpoint be coplanar. Then $m\angle aob + m\angle boc = m\angle aoc$ implies $(\overrightarrow{oa}\ \overrightarrow{ob}\ \overrightarrow{oc})$. (Compare C6, Th. 5.)

7. Congruence of Segments

Since our immediate object is to rigorize the conventional treatment of congruence we do not give a general theory of congruent figures but restrict ourselves to the simplest figures—segments, angles, triangles—and define congruence for them piecemeal.*

DEFINITION.   If $Dab = Dpq$, equivalently if $\overline{ab}$ and $\overline{pq}$ have the same length, we say $\overline{ab}$ is *congruent* to $\overline{pq}$ and write $\overline{ab} \cong \overline{pq}$.

The basic properties of congruence of segments are stated in

THEOREM 9.   Congruence of segments has the following properties:

(i) $\overline{ab} \cong \overline{ab}$.

(ii) If $\overline{ab} \cong \overline{cd}$ then $\overline{cd} \cong \overline{ab}$.

(iii) If $\overline{ab} \cong \overline{cd}$, $\overline{cd} \cong \overline{ef}$ then $\overline{ab} \cong \overline{ef}$.

(iv) If $\overline{ab} \cong \overline{a'b'}$, $\overline{bc} \cong \overline{b'c'}$, (abc) and $(a'b'c')$ then $\overline{ac} = \overline{a'c'}$.

(v) Let $\overrightarrow{ab}$ be any ray and $\overline{cd}$ any segment. Then there is exactly one point $p \subset \overrightarrow{ab}$ such that $\overline{ap} \cong \overline{cd}$.

Proof.   (i), (ii), (iii) are immediate by the corresponding equality properties of real numbers. (iv), (v) are essentially restatements of C3, C2.

---

* For the general concept of congruence see, for example, E. Moise, *Elementary Geometry From An Advanced Standpoint*, Addison-Wesley Publishing Co., Reading, 1963, Chapter 17.

Note that these properties are implicit in school geometry. We may describe them by saying that congruence of segments is an equivalence relation ((i), (ii), (iii)) which satisfies the Segment Additivity Principle (iv) and the Point Location Principle (v).

The notion of midpoint of a segment arises naturally in the present context.

DEFINITION.   Point n is a midpoint of $\overline{ab}$ if $n \subset \overline{ab}$ and $\overline{an} \cong \overline{bn}$.

THEOREM 10.   A segment has one and only one midpoint.

Proof.   Let $\overline{ab}$ be given. By C2 there is a point $n \subset \overrightarrow{ab}$ such that $Dan = \frac{1}{2}Dab$. Since $Dan < Dab$, Theorem 2 implies (anb) and $n \subset \overline{ab}$. By C3 $Dan + Dnb = Dab$ so that

$$Dnb = Dab - Dan = 2\,Dan - Dan = Dan.$$

Thus $\overline{an} \cong \overline{bn}$ and n is a midpoint of $\overline{ab}$ by definition.

Suppose n' is also a midpoint of $\overline{ab}$. Then $n' \subset \overline{ab}$, $Dan' = Dbn'$. Thus $(an'b)$ and $n' \subset \overrightarrow{ab}$. By C3, $Dan' + Dn'b = Dab$. Thus $Dan' = \frac{1}{2}Dab = Dan$. By C2 $n' = n$, proving uniqueness.

Exercise.   Prove: If $n \subset ab$ and $\overline{an} \cong \overline{bn}$ then n is the midpoint of $\overline{ab}$.

8. Congruence of Angles

Our treatment parallels that for congruence of segments.

DEFINITION.   If $m\angle abc = m\angle pqr$ we say $\angle abc$ is *congruent* to $\angle pqr$ and write $\angle abc \cong \angle pqr$.

THEOREM 11.   Congruence of angles has the following properties:
  (i) $\angle abc \cong \angle abc$.
  (ii) If $\angle abc \cong \angle def$ then $\angle def \cong \angle abc$.
  (iii) If $\angle abc \cong \angle def$, $\angle def \cong \angle ghk$ then $\angle abc \cong \angle ghk$.

  (iv) If $\angle aob \cong \angle a'o'b'$, $\angle boc \cong \angle b'o'c'$, $(\overrightarrow{oa}\ \overrightarrow{ob}\ \overrightarrow{oc})$ and $(\overrightarrow{o'a'}\ \overrightarrow{o'b'}\ \overrightarrow{o'c'})$ then $\angle aoc \cong \angle a'o'c'$.

(v) Given a half-plane H, a ray $\overrightarrow{ab}$ contained in its edge, and an angle $\angle pqr$. Then there exists exactly one ray $\overrightarrow{ac}$ such that $\overrightarrow{ac} \subset H$ and $\angle bac \cong \angle pqr$.

DEFINITION.   Ray $\overrightarrow{op}$ is a *bisector* of $\angle aob$ if $(\overrightarrow{oa}\ \overrightarrow{op}\ \overrightarrow{ob})$ and $\angle aop \cong \angle bop$.

THEOREM 12.   An angle has one and only one bisector.

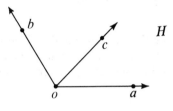

FIGURE 14.7

Proof.   Let $\angle aob$ be given. Let H be the half-plane with edge oa that contains b. By C5 there is a ray $\overrightarrow{oc} \subset H$ such that $m\angle aoc = \frac{1}{2}m\angle aob$. Then c, b are on the same side of oa and $m\angle aoc < m\angle aob$. Hence $(\overrightarrow{oa}\ \overrightarrow{oc}\ \overrightarrow{ob})$ by Theorem 7. By C6, $m\angle aoc + m\angle boc = m\angle aob$. Thus

$$m\angle boc = m\angle aob - m\angle aoc = 2m\angle aoc - m\angle aoc = m\angle aoc.$$

Thus $\angle aoc \cong \angle boc$ and $\overrightarrow{oc}$ is a bisector of $\angle aob$ by definition.

Uniqueness is proved essentially as for midpoint of a segment in Theorem 10.

Now we make our first use of the Supplement Postulate C7.

THEOREM 13.   Vertical angles are congruent.

Proof.   Let $\angle aob$, $\angle a'ob'$ be vertical angles such that $\overrightarrow{oa}, \overrightarrow{oa'}$ are opposite and $\overrightarrow{ob}, \overrightarrow{ob'}$ are opposite. By the Supplement Postulate C7 we have

(1) $m\angle aob + m\angle a'ob = 180,$

(2) $m\angle b'oa' + m\angle boa' = 180.$

By equating the left members of (1) and (2), and observing $\angle a'ob = \angle boa'$, $\angle b'oa' = \angle a'ob'$, we get $m\angle aob = m\angle a'ob'$. Thus $\angle aob \cong \angle a'ob'$.

9. Supplementary Angles

We define supplementary angles in terms of the notion of linear pair and of linear triple (Ch. 12, Sec. 6).

DEFINITION.   Two angles are *supplementary* and each is a *supplement* of the other if they are respectively congruent to the angles of a linear pair. Similarly, we define three angles to be *supplementary* if they are respectively congruent to the angles of a linear triple.

It easily follows, using the Supplement Postulate C7, that two angles are supplementary if and only if the sum of their measures is 180.

Further we have the following versions of two familiar principles.

THEOREM 14.   Supplements of congruent angles are congruent.

Proof. Let $\angle abc \cong \angle def$, and let $\angle a'b'c'$, $\angle d'e'f'$ be supplements of $\angle abc$, $\angle def$ respectively. Then

$$m\angle a'b'c' = 180 - m\angle abc, \ m\angle d'e'f' = 180 - m\angle def.$$

Since $m\angle abc = m\angle def$, we have $m\angle a'b'c' = m\angle d'e'f'$ and $\angle a'b'c' \cong \angle d'e'f'$.

THEOREM 15.   Let $\angle aob$, $\angle aoc$ be supplementary, where b and c are on opposite sides of oa. Then $\overrightarrow{ob}$, $\overrightarrow{oc}$ are opposite, and $\angle aob$, $\angle aoc$ form a linear pair.

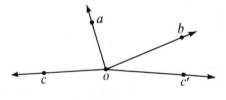

FIGURE 14.8

Proof. Let $\overrightarrow{oc'}$ be opposite $\overrightarrow{oc}$. Then b, c' are on the same side of oa. By hypothesis $\angle aob$ is supplementary to $\angle aoc$; $\angle aoc'$ is supplementary to $\angle aoc$ since they form a linear pair. Thus Theorem 14 implies $\angle aob \cong$

$\angle aoc'$. By the Ray Location Principle (Th. 11(v)) $\overrightarrow{ob} = \overrightarrow{oc'}$, that is, $\overrightarrow{ob}$ is opposite $\overrightarrow{oc}$.

THEOREM 16.   Three angles are supplementary if and only if the sum of their measures is 180.

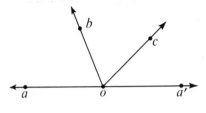

FIGURE 14.9

Proof.   Let A, B, C be three supplementary angles. By definition (Ch. 12, Sec. 6) they are respectively congruent to the angles of a linear triple, which may be assigned the following notation: $\angle aob$, $\angle boc$, $\angle coa'$ where $(\overrightarrow{oa}\ \overrightarrow{ob}\ \overrightarrow{oc})$ and $\overrightarrow{oa}$, $\overrightarrow{oa'}$ are opposite (Fig. 14.9). C6 and C7 imply

$$m\angle aob + m\angle boc = m\angle aoc,$$
$$m\angle aoc + m\angle a'oc = 180.$$

By eliminating $m\angle aoc$ we get

$$m\angle aob + \angle boc + m\angle aoc' = 180,$$

and the sum of the measures of A, B, C is 180.

Conversely, suppose A, B, C are angles whose measures $m(A)$, $m(B)$, $m(C)$ satisfy

$$m(A) + m(B) + m(C) = 180.$$

Let $A = \angle aob$, and let $\overrightarrow{oa'}$ be opposite $\overrightarrow{oa}$ (Fig. 14.9). Then

$$m(B) + m(C) = 180 - m(A) = m\angle boa'.$$

Thus $m(B) < m\angle boa'$. By Theorem 8 there is a ray $\overrightarrow{oc}$ such that $m(B) = m\angle boc$ and $(\overrightarrow{ob}\ \overrightarrow{oc}\ \overrightarrow{oa'})$. Further

$$m\angle coa' = m\angle boa' - m\angle boc = m(B) + m(C) - m(B) = m(C).$$

Thus A, B, C are congruent to $\angle aob$, $\angle boc$, $\angle coa'$. The latter form a linear triple since $(\overrightarrow{ob}\ \overrightarrow{oc}\ \overrightarrow{oa'})$ implies $(\overrightarrow{oa}\ \overrightarrow{ob}\ \overrightarrow{oc})$ (Ch. 12, Th. 10, Cor. 1). Thus by definition A, B, C are supplementary.

10. Right Angles and Perpendicularity

Our definitions are essentially the conventional ones:

DEFINITION.   An angle is a *right angle* if it is congruent to one of its supplements. If $\angle aob$ is a right angle we say the lines oa, ob are *perpendicular* (at o) or oa is *perpendicular* to ob (at o) and we write $oa \perp ob$.

The following principles follow without difficulty:
(i) *An angle is a right angle if and only if its measure is 90.*
(ii) *Any supplement of a right angle is a right angle.*
(iii) *Any two right angles are congruent.*
(iv) *An angle congruent to a right angle is a right angle.*
(v) *A right angle is congruent to each of its supplements.*
(vi) *No line is perpendicular to itself.*
Observe that two lines are perpendicular if and only if one of the angles they form (Ch. 12, Sect. 6) is a right angle. This suggests

THEOREM 17.   If one of the angles formed by two intersecting lines is a right angle, then all of the angles formed by them are right angles.

Proof.   Let oa, ob be such that $\angle aob$ is a right angle. Suppose (aoa'), (bob'). Then $\angle aob$, $\angle a'ob'$ are vertical and so congruent by Theorem 13. Thus $\angle a'ob'$ is a right angle. Further $\angle a'ob$ is a supplement of $\angle aob$ and so is a right angle. Similarly, $\angle aob'$ is a right angle. The result follows since two distinct intersecting lines form exactly four angles (Ch. 12, Th. 13), and the four angles mentioned above are distinct.

COROLLARY.   If two lines are perpendicular all the angles they form are right angles.

THEOREM 18.   In plane P, if $a \subset L$ there is one and only one line M such that $M \supset a$ and $M \perp L$.

Proof.   Let $L = ab$ (Figure 14.10). Let $p \subset P$, $p \not\subset ab$. By the Ray Location Postulate C5 there exists a ray $\overrightarrow{ac}$ such that $m\angle bac = 90$ and $\overrightarrow{ac} \subset \overrightarrow{Lp}$. Thus $ac \perp ab$ and $M = ac$ is a line with the desired properties.

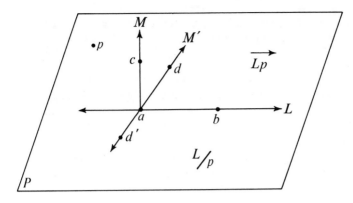

FIGURE 14.10

Suppose also $M' \subset P$, $M' \perp L$, $M' \supset a$. Let $d \subset M'$, $d \neq a$ and let d' satisfy $(d'ad)$. Then d, d' are on opposite sides of L. But

$$d, d' \subset P = \overrightarrow{Lp} \cup L \cup L/p.$$

Thus d or d' is in $\overrightarrow{Lp}$, say d. Note $\overrightarrow{ad} \subset \overrightarrow{Lp}$. Since $ad \perp ab$, all of the angles formed by ad and ab are right angles (Th. 17, Cor.), and $\angle bad$ is a right angle. Thus $\overrightarrow{ad} = \overrightarrow{ac}$ by C5, and $M' = M$.

11. Congruence of Triangles

First we state the definition of a triangle*:

DEFINITION.   If a, b, c are distinct and noncollinear, the set union

$$\overline{ab} \cup \overline{bc} \cup \overline{ca} \cup a \cup b \cup c$$

is called *triangle abc* (written $\triangle abc$). Points a, b, c are its *vertices*; segments $\overline{ab}$, $\overline{bc}$, $\overline{ca}$ are its sides; lines ab, bc, ca are its *side-lines*; angles $\angle abc$, $\angle bca$, $\angle cab$ are its *angles*.

We want to make precise the familiar principle: *Two triangles are congruent if their corresponding parts are "equal"* (that is "congruent" in our termin-

---

* The definition is restated from Chapter 13 of which this chapter is independent.

ology). To do this we must clarify the use of the term "corresponding." Thus we introduce the following

DEFINITION. A one-to-one correspondence between the sets of vertices of two triangles is called a *correspondence between the vertices* of the two triangles. For example $a \rightarrow d$, $b \rightarrow e$, $c \rightarrow f$ is a correspondence between the vertices of $\triangle abc$ and $\triangle def$. We write this correspondence compactly so: $abc \rightarrow def$.

Such a correspondence $abc \rightarrow def$ between the vertices of two triangles induces a correspondence between their sides:

(1) $\overline{ab} \rightarrow \overline{de}$, $\overline{bc} \rightarrow \overline{ef}$, $\overline{ac} \rightarrow \overline{df}$, in which two sides correspond provided their endpoints correspond. The pairs of sides in (1) are said to *correspond* under the correspondence $abc \rightarrow def$ or simply to be *corresponding sides* of the two triangles if it is clear which correspondence is referred to.

Similarly, the correspondence $abc \rightarrow def$ induces a correspondence between the angles of the two triangles:

(2) $\angle abc \rightarrow \angle def$, $\angle bca \rightarrow \angle efd$, $\angle cab \rightarrow \angle fde$; and the pairs of angles in (2) are said to *correspond* under $abc \rightarrow def$, or simply to be *corresponding angles* of the two triangles.

If the correspondence $abc \rightarrow def$ is such that corresponding sides are congruent, and corresponding angles are congruent, the correspondence is called a *congruence* and we write $\triangle abc \cong \triangle def$. Note that the order in which the vertices are written is essential. Finally we say two triangles are *congruent* if there exists a congruence between their vertices.

We introduce some familiar terminology of school geometry and then state our final congruence postulate.

DEFINITION. In $\triangle abc$, we say $\overline{ab}$ and $\angle bca$ are *opposite*, and that either is *opposite* the other. Similarly, for $\overline{bc}$ and $\angle cab$, and for ca and $\angle abc$. We say $\angle abc$ is *included* by $\overline{ab}$ and $\overline{bc}$, and that $\overline{ab}$ is *included* by $\angle cab$ and $\angle abc$. Similarly, $\angle bca$, $\angle cab$, $\overline{bc}$, $\overline{ca}$ are respectively included by $\overline{bc}$ and $\overline{ca}$, $\overline{ca}$ and $\overline{ab}$, $\angle abc$ and $\angle bca$, and $\angle bca$ and $\angle cab$.

C8. (SAS POSTULATE) *If a correspondence between the vertices of two triangles is such that two sides and the included angle of one triangle are respectively congruent to the corresponding two sides and angle of the other, then the correspondence is a congruence.*

This completes our list of postulates. We have assumed for incidence: I1–I6; for order: B1–B6; and for congruence: C1–C8. Let us call an incidence geometry which satisfies these postulates a *Congruence Geometry*. We have in this concept a precise formulation of the notion of Neutral Geometry of Chapter 3. A congruence geometry may be described as an ordered incidence geometry in which there exists a distance function (Sec. 4) and an angle measure function (Sec. 5) that are related by the SAS Postulate. Model M19 with order as interpreted in Chapter 11 Exercises IV, and suitable interpretations of distance and angle is a model of our theory, that is, a congruence geometry.*

12. Some Elementary Consequences of the SAS Postulate

Our theories of congruence of segments and congruence of angles have so far been unrelated. Now with the assumption of the SAS Postulate they begin to affect each other.

THEOREM 19.   If two sides of a triangle are congruent, then the angles opposite these sides are congruent.

Proof. In $\triangle abc$, let $\overline{ab} \cong \overline{ac}$. Consider the correspondence $abc \to acb$. Since $\overline{ab} \cong \overline{ac}$, $\overline{ac} \cong \overline{ab}$ and $\angle cab \cong \angle bac$, it follows by the SAS Postulate C8 that the correspondence is a congruence. Therefore $\angle abc \cong \angle acb$ and the theorem is proved.

DEFINITION.   If two sides of a triangle are congruent, the triangle is said to be *isosceles*. If all three sides of a triangle are congruent, the triangle is *equilateral*; if all three angles of a triangle are congruent, the triangle is *equiangular*.

COROLLARY.   An equilateral triangle is equiangular.

THEOREM 20. (ASA THEOREM)   If a correspondence between the vertices of two triangles is such that two angles and the included side of the first triangle

---

* For a treatment of this, on a somewhat different postulational basis, see E. Moise, *Elementary Geometry From An Advanced Standpoint*, Addison-Wesley Publishing Co., Reading, 1963, Chapter 26.

are congruent respectively to the corresponding two angles and side of the
second triangle, then the correspondence is a congruence.

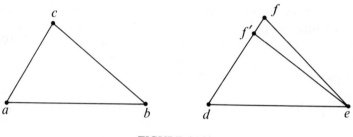

FIGURE 14.11

Proof. Let $abc \rightarrow def$ be such that $\angle cab \cong \angle fde$, $\angle abc \cong \angle def$ and
$\overline{ab} \cong \overline{de}$. By the Point Location Principle (Th. 9(v)) there is a point $f' \subset \overrightarrow{df}$
such that $\overline{df'} \cong \overline{ac}$. Note $\angle edf = \angle edf'$. Consider the correspondence
$abc \rightarrow def'$. Since $\overline{ac} \cong \overline{df'}$, $\overline{ab} \cong \overline{de}$ and $\angle cab \cong \angle f'de$, the correspondence
$abc \rightarrow def'$ is a congruence by the SAS Postulate. Thus $\angle abc \cong \angle def'$. Since
by hypothesis $\angle abc \cong \angle def$, it follows that $\angle def \cong \angle def'$.

Since $f' \subset \overrightarrow{df}$, f and f' are on the same side of de (Ch. 12, Th. 3, Cor. 1.)
Thus $\overrightarrow{ef}$, $\overrightarrow{ef'}$ are on the same side of de (Ch. 12, Th. 3, Cor. 1). Hence $\overrightarrow{ef'} = \overrightarrow{ef}$
by the Ray Location Principle (Th. 11(v)). Thus f' is common to df and ef
so that $f' = f$. Hence the congruence $abc \rightarrow def'$ becomes $abc \rightarrow def$.

COROLLARY.   If two angles of a triangle are congruent, then the sides
opposite these angles are congruent.

Proof.   In $\triangle abc$, let $\angle abc \cong \angle acb$. Since $\angle abc \cong \angle acb$, $\angle acb \cong abc$
and $\overline{bc} \cong \overline{cb}$, the correspondence $abc \rightarrow acb$ is a congruence, by the theorem.
Thus $\overline{ac} \cong \overline{ab}$.

13. Equidistance and Perpendicular Bisectors of Segments

The study of the set (or locus) of points which are equidistant from two
given points is in itself of interest and importance; it sheds light on the theory
of perpendicularity; it serves as a unifying idea in geometry. In this section we
derive the familiar characterization of the set of points of a plane equidistant

from two of its points, and use it to prove the SSS congruence theorem. We begin with two familiar definitions.

DEFINITION.   Point p is said to be *equidistant* from a and b if $\overline{pa} \cong \overline{pb}$.

DEFINITION.   Line L is a *perpendicular bisector* of segment $\overline{ab}$ if it contains a midpoint of $\overline{ab}$ and is perpendicular to line ab.

Suppose plane $P \supset \overline{ab}$. Certainly $\overline{ab}$ has a unique midpoint n (Th. 10). Also there is a unique line L such that $n \subset L \subset P$ and $L \perp ab$ (Th. 18). Thus *in plane P, $\overline{ab}$ has a unique perpendicular bisector.*

THEOREM 21.   In a plane, a point is equidistant from the endpoints of a segment if and only if it is in the perpendicular bisector of the segment.

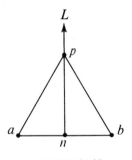

FIGURE 14.12

Proof.   In plane P, let $\overline{ab}$ be given and let L be the perpendicular bisector of $\overline{ab}$. Let n be the midpoint of $\overline{ab}$. Then $n \subset \overline{ab}$ and $\overline{an} \cong \overline{bn}$.

Assume $p \subset L$. If $p = n$, clearly $\overline{pa} \cong \overline{pb}$. Suppose $p \neq n$. Then $\angle pna$ and $\angle pnb$ are right angles and so $\angle pna \cong \angle pnb$. Since $\overline{pn} \cong \overline{pn}$, $\overline{na} \cong \overline{nb}$, the correspondence $pna \to pnb$ is a congruence by the SAS Postulate. Hence $\overline{pa} \cong \overline{pb}$.

Conversely, assume $q \subset P$ and $\overline{qa} \cong \overline{qb}$. We show $q \subset L$. First consider the case $q \subset ab$. We have q, a, b are distinct and collinear, so that (qab), (abq) or (bqa). Suppose (qab). Then a, $b \subset \overrightarrow{qb}$, and $\overline{qa} \cong \overline{qb}$ implies $a = b$ by the uniqueness condition in the Point Location Principle (Th. 9 (v)). But (qab) implies $a \neq b$. Thus (qab) is false. Similarly (abq) is false. Thus (bqa) and $q \subset \overline{ab}$. Hence q is the midpoint of $\overline{ab}$ and $q = n$. Certainly $q \subset L$.

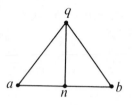

FIGURE 14.13

Now suppose $q \notin ab$ (Fig. 14.13). Then $\overline{qa} \cong \overline{qb}$ implies $\angle qab \cong \angle qba$ (Th. 19). Thus $\angle qan \cong \angle qbn$, since $n \subset \overrightarrow{ab},\ \overrightarrow{ba}$. Therefore the correspondence $qna \to qnb$ is a congruence by the SAS Postulate and $\angle qna \cong \angle qnb$. Since these angles form a linear pair, $\angle qna$ is a right angle and $qn \perp an$ at n. Thus $qn = L$ and $q \subset L$.

As an application we have a relatively simple proof of

THEOREM 22. (SSS THEOREM) If a correspondence between the vertices of two triangles is such that the sides of one are congruent respectively to the corresponding sides of the other, then the correspondence is a congruence.

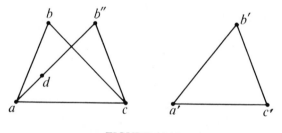

FIGURE 14.14

Proof. Suppose $abc \to a'b'c'$ is such that $\overline{ab} \cong \overline{a'b'}$, $\overline{bc} \cong \overline{b'c'}$ and $\overline{ca} \cong \overline{c'a'}$. Let H be the half-plane with edge ac that contains b. By the Ray Location Principle (Th. 11(v)) there is a ray $\overrightarrow{ad}$ such that $\overrightarrow{ad} \subset H$ and $\angle cad \cong \angle c'a'b'$. By the Point Location Principle (Th. 9(v)) there is a point $b'' \subset \overrightarrow{ad}$ such that $\overline{ab''} \cong \overline{a'b'}$. $\angle cad = \angle cab''$; hence $\angle cab'' \cong \angle c'a'b'$. Since $\overline{ac} \cong \overline{a'c'}$, the correspondence $ab''c \to a'b'c'$ is a congruence by the SAS Postulate. Thus $\overline{b''c} \cong \overline{b'c'}$. Since $\overline{ab} \cong \overline{a'b'}$ and $\overline{bc} \cong \overline{b'c'}$, it follows that $\overline{ab} \cong \overline{ab''}$ and $\overline{bc} \cong \overline{b''c}$; or each of a, c is equidistant from b and b''.

Suppose $b \neq b''$. Then by Theorem 21, each of a, c is in the perpendicular bisector of $\overline{bb''}$ in plane abc, and ac is this perpendicular bisector of $\overline{bb''}$. Thus $\overline{bb''}$ meets ac and b, b'' are on opposite sides of ac. This is impossible since $b'' \subset \overrightarrow{ad} \subset H$. The supposition $b \neq b''$ is therefore false. Hence $b = b''$, and the congruence $ab''c \to a'b'c'$ becomes $abc \to a'b'c'$.

14. The Exterior Angle Theorem

We are now prepared to prove the Exterior Angle Theorem. The proof uses—in addition to the theory of congruence—incidence properties (Ch. 7), linear and planar order properties (Ch. 10, Ch. 11), and the theory of betweenness for rays (Ch. 12).

First we formalize the notion of exterior angle.

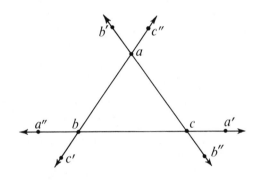

FIGURE 14.15

DEFINITION.   Given $\triangle abc$, let a' satisfy (bca') (Fig. 14.15). Then $\angle aca'$ is called an *exterior angle* of $\triangle abc$. Similarly, if (cab') then $\angle bab'$ is an exterior angle of $\triangle abc$; and if (abc') then $\angle cbc'$ is also one. Finally, if (cba'') then $\angle aba''$ is an exterior angle of $\triangle abc$ and so is $\angle bcb''$ if (acb''), and $\angle cac''$ if (bac''). To distinguish the three angles of $\triangle abc$ as already defined from these exterior angles we call them interior angles of $\triangle abc$.

To each exterior angle of $\triangle abc$ we associate two interior angles of the triangle, called the *remote interior* angles of the exterior angle. They are the two interior angles whose vertices are different from the vertex of the exterior angle. Thus $\angle abc$ and $\angle bca$ are the remote interior angles of exterior angle $\angle bab'$ and of exterior angle $\angle cac''$; $\angle bca$ and $\angle cab$ are the remote interior angles of $\angle aba''$ and of $\angle cbc'$; and $\angle cab$ and $\angle abc$ are the remote interior angles of $\angle bcb''$ and of $\angle aca'$.

Note that the six exterior angles can be grouped into three pairs of vertical angles and so into three pairs of congruent angles.

THEOREM 23.   Any exterior angle of a triangle has greater measure than either of its remote interior angles.

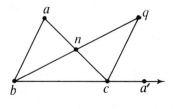

FIGURE 14.16

Proof.   Given $\triangle abc$, suppose (bca'). Then $\angle aca'$ is an exterior angle. We prove $m\angle aca' > m\angle bac$ and $m\angle aca' > m \angle abc$.

Let n be the midpoint of $\overline{ac}$. Then (anc) and $\overline{an} \cong \overline{cn}$. Show n, a, b are distinct and noncollinear, and n, b, c are distinct and noncollinear. By the Point Location Principle (Th. 9 (v)) there is a point $q \subset n/b$ such that $\overline{nq} \cong \overline{bn}$. Then (qnb). Hence $\angle anb$ and $\angle cnq$ are vertical; thus $\angle anb \cong \angle cnq$. From above, $\overline{an} \cong \overline{cn}$ and $\overline{bn} \cong \overline{qn}$. Thus by the SAS Postulate the correspondence $anb \rightarrow cnq$ is a congruence. Hence $\angle ban \cong \angle qcn$. (anc) implies $\overrightarrow{an} = \overrightarrow{ac}$ and $\overrightarrow{cn} = \overrightarrow{ca}$ (Ch. 10, Th. 7, Cor. 6). Thus $\angle ban = \angle bac$, $\angle qcn = \angle qca$, so that $\angle bac \cong \angle qca$.

Since n, b, c are distinct and noncollinear so are q, b, c. Hence $\overrightarrow{cb}$ and $\overrightarrow{cq}$ are distinct and not opposite. (anc) implies $n \subset \overline{ac} \subset \overrightarrow{ca}$. Since $b \subset \overrightarrow{cb}$, $q \subset \overrightarrow{cq}$ and (bnq), by definition $(\overrightarrow{cb}\ \overrightarrow{ca}\ \overrightarrow{cq})$. (bca') implies $\overrightarrow{cb}$ and $\overrightarrow{ca'}$ are opposite. Hence $(\overrightarrow{cb}\ \overrightarrow{ca}\ \overrightarrow{cq})$ implies $(\overrightarrow{ca}\ \overrightarrow{cq}\ \overrightarrow{ca'})$ (Ch. 12, Th. 10). Therefore $m\angle aca' > m\angle acq$ (Th. 6). Since $\angle acq = \angle qca$, and from above $\angle qca \cong \angle bac$, it follows that $m\angle aca' > m\angle bac$.

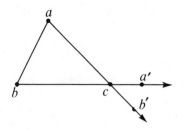

FIGURE 14.17

To prove $m\angle aca' > m\angle abc$ suppose (acb') (Fig. 14.17). Note $\angle aca' \cong$ $\angle bcb'$. By symmetry the argument above yields $m\angle bcb' > m\angle abc$ and the result follows.

15. The Perpendicular to a Line from an External Point

We have shown (Th. 18) the existence and uniqueness, in a plane, of a perpendicular to a line at one of its points. Now we consider the case where the point is not in the line.

THEOREM 24. If $c \not\subset L$ there is one and only one line which contains c and is perpendicular to L.

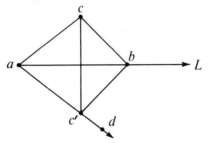

FIGURE 14.18

Proof. Let $L = ab$. Consider half-plane L/c. By the Ray Location Principle (Th. 11(v)) there is a ray $\overrightarrow{ad} \subset L/c$ such that $\angle bad \cong \angle bac$. By the Point Location Principle (Th. 9(v)) there is a point $c' \subset \overrightarrow{ad}$ such that $\overline{ac'} \cong \overline{ac}$. Note $\angle bac \cong \angle bac'$ and $\overline{ab} \cong \overline{ab}$. Hence the correspondence $abc \to abc$ is a congruence by the SAS Postulate. Thus $\overline{bc} \cong \overline{bc'}$. We now know that each of a, b is equidistant from c and c'. Further c, c' are on opposite sides of ab since $c' \subset \overrightarrow{ad} \subset L/c$. Thus $c \neq c'$ and ab is the perpendicular bisector of $\overline{cc'}$ in plane Lc (Th. 21). Certainly $cc' \perp L$ and there exists at least one line with the desired properties.

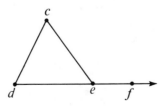

FIGURE 14.19

To prove uniqueness, suppose there are two such lines that are distinct. Then there are distinct points d, $e \subset ab$ such that $cd \perp ab$, $ce \perp ab$ (Fig. 14.19). Let f satisfy (*def*). Show $\angle cde$, $\angle cef$ are right angles and obtain a contradiction of the Exterior Angle Theorem. Then uniqueness is justified.

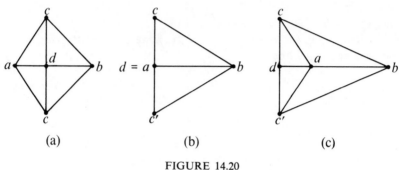

(a)   (b)   (c)

FIGURE 14.20

NOTE. It is not essential to use the equidistance concept in the proof above to show $cc' \perp L$. But it is hard to see how to give a unitary proof without it. For example, if d is the intersection of cc' and ab (Fig. 14.20) we can show $\angle adc \cong \angle adc'$ or $\angle bdc \cong \angle bdc'$ directly and infer $cc' \perp ab$, but several cases are involved.

16. Alternate Interior Angles and Parallel Lines

We introduce a definition of "alternate interior angles" which frees the idea from dependence on diagrams and we relate it to parallelism of lines.

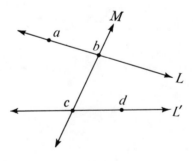

FIGURE 14.21

DEFINITION. Let lines L, L' and M be distinct and coplanar (Fig. 14.21). Let M meet L, L' in b, c; $b \neq c$. Let $a \subset L$, $d \subset L'$ be such that a and d

are on opposite sides of M. Then $\angle abc$, $\angle bcd$ are a pair of *alternate interior angles formed by* M *and* L, L'.

THEOREM 25.   Let lines M and L, L' form a pair of alternate interior angles that are congruent. Then $L \parallel L'$.

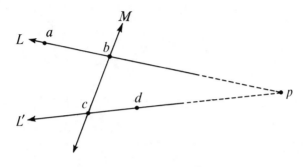

FIGURE 14.22

Proof.   Let $\angle abc$, $\angle bcd$ be a pair of congruent alternate interior angles formed by M and L, L', the notation chosen so that $L = ab$, $L' = cd$, $M = bc$ and a, d are on opposite sides of M. Let plane $P \supset L$, L', M.

Since a, d are opposite sides of M, $\overline{ad}$ meets M and a, $d \notin M$. Hence the Plane Separation Theorem applies and yields

$$P = M/a \cup M \cup M/d.$$

Suppose ab, cd meet, say in p. Then $p \subset P$, $p \notin M$ so that $p \subset M/a$ or $p \subset M/d$.

Suppose $p \subset M/a$. Then $\overline{pa}$ meets M in a point which must be b since b is the intersection of $pa = ab$ and M. Thus (pba) and $\angle abc$ is an exterior angle of $\triangle pbc$. Further, p and d are in M/a and so on the same side of M. Hence $p \subset cd$ implies $p \subset \overrightarrow{cd}$ (Ch. 12, Th. 3, Cor. 3). Thus $\angle bcd = \angle bcp$ and is an interior angle of $\triangle bcp$. Hence $\angle abc \cong \angle bcd$ contradicts the Exterior Angle Theorem. The case $p \subset M/d$ is symmetrical and we conclude $ab \parallel cd$, that is $L \parallel L'$, and the proof is complete.

Now we can easily prove the existence of parallel lines.

THEOREM 26.   If $p \notin L$ there exists at least one line M such that $M \supset p$ and $M \parallel L$.

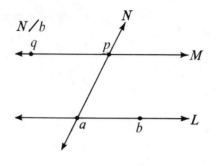

FIGURE 14.23

Proof. Let $L = ab$. Let $N = pa$ and consider the half-plane N/b. By the Ray Location Principle (Th. 11 (v)) there exists $\overrightarrow{pq} \subset N/b$ such that $\angle qpa \cong \angle pab$. Let $M = pq$. Then $\angle qpa, \angle pab$ are a pair of alternate interior angles formed by N and M, L; and $M \parallel L$ by Theorem 25.

We have now completed our basic study of the theories of incidence, of order, and of congruence. All the results of Chapter 7 and Chapters 10–14 hold equally well in Euclidean and Lobachevskian geometries. As we observed above (Sec. 11) the set of postulates I1–I6, B1–B6, C1–C8 characterize the concept of Neutral Geometry (Ch. 3) which studies the common substratum of these two geometries. Euclidean geometry and Lobachevskian geometry can be formally characterized by adjoining a suitable parallel postulate to this set of postulates.

17. Euclidean Geometry

We conclude this chapter with a brief introduction to Euclidean geometry. We introduce Euclid's parallel postulate (in the form of Playfair's Postulate) and deduce two important theorems.

POSTULATE E. *If point a is not in line L there exists one and only one line M such that M $\supset$ a and M $\parallel$ L.*

NOTE: In view of Theorem 26 we need only assume uniqueness in Postulate E and could omit the existence requirement.

THEOREM 27.   Let $L \parallel L'$, and let M intersect L, L'. Then each pair of alternate interior angles that M forms with L and L' are congruent.

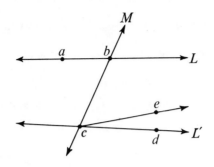

FIGURE 14.24

Proof.   Let $\angle abc$, $\angle bcd$ be a pair of alternate interior angles formed by M and L, L', the notation chosen so that $L = ab$, $L' = cd$, $M = bc$, and a, d are on opposite sides of M. By the Ray Location Principle (Th. 11 (v)) there exists $\overrightarrow{ce} \subset M/a$ such that $\angle abc \cong \angle bce$. Then $\angle abc$, $\angle bce$ are alternate interior angles formed by M with L and ce. Thus $ce \parallel L$ by Theorem 25. Since $cd \parallel L$, Postulate E implies $cd = ce$. Points d, e are on the same side of M since both are in M/a. Hence $d \subset ce$ implies $d \subset \overrightarrow{ce}$ (Ch. 12, Th. 3, Cor. 3) and $\overrightarrow{cd} = \overrightarrow{ce}$ follows. Thus $\angle bcd = \angle bce$ and we conclude $\angle abc \cong \angle bcd$.

THEOREM 28.   The angles of any triangle are supplementary.

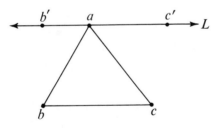

FIGURE 14.25

Proof.   Given $\triangle abc$, let L be the line which contains a and is parallel to bc. Let b', $c' \subset L$ and satisfy $(b'ac')$. Then $\overrightarrow{ab'}$, $\overrightarrow{ac'}$ are opposite. Also b', c' are on opposite sides of ac. Thus one, say b', is on the same side of ac as b. We show first that $\angle b'ab$, $\angle bac$, $\angle c'ac$ form a linear triple (Ch. 12, Sec. 6).

The fact that b, b' are on the same side of ac implies (Ch. 12, Th. 11, Cor. 2)

$$(\overrightarrow{ac}\ \overrightarrow{ab}\ \overrightarrow{ab'}),\ (\overrightarrow{ac}\ \overrightarrow{ab'}\ \overrightarrow{ab})\ \text{or}\ \overrightarrow{ab} = \overrightarrow{ab'}.$$

Suppose $(\overrightarrow{ac}\ \overrightarrow{ab'}\ \overrightarrow{ab})$. Then $\overline{bc}$ meets $\overrightarrow{ab'}$ (Ch. 12, Th. 7) and so bc meets $ab' = L$, contradicting $L \parallel bc$. Suppose $\overrightarrow{ab} = \overrightarrow{ab'}$. Then $b \subset ab' = L$, again contradicting $L \parallel bc$. We conclude $(\overrightarrow{ac}\ \overrightarrow{ab}\ \overrightarrow{ab'})$ and so

(1) $(\overrightarrow{ab'}\ \overrightarrow{ab}\ \overrightarrow{ac})$.

Thus $\angle b'ab$, $\angle bac$, $\angle cac'$ form a linear triple by definition. Note, for use below that (1) implies

(2) $(\overrightarrow{ab}\ \overrightarrow{ac}\ \overrightarrow{ac'})$

since $\overrightarrow{ac'}$ is opposite $\overrightarrow{ab'}$ (Ch. 12, Th. 10).

Now we show $\angle abc$, $\angle bac$, $\angle acb$ supplementary. Relation (1) implies that b', c are on opposite sides of ab (Ch. 12, Th. 6). Hence $\angle b'ab$ and $\angle abc$ are alternate interior angles of L and bc. Thus $\angle abc \cong \angle b'ab$ by Theorem 27. Similarly, using (2) we get $\angle acb \cong \angle c'ac$. By definition $\angle abc$, $\angle bac$, $\angle acb$ are supplementary and the theorem is proved.

COROLLARY.   The sum of the measures of the angles of any triangle is 180.

Proof.   By Theorem 16.

E X E R C I S E S

Prove without assuming any parallel postulate:
1. In $\triangle abc$, if $Dab < Dbc$, then $m\angle acb < m\angle bac$.
2. $\triangle abc$, if $m\angle acb < m\angle bac$, then $Dab < Dbc$.
3. In $\triangle abc$, $Dac < Dab + Dbc$.
†4. If $\triangle abc$ and $\triangle a'b'c'$ are such that $Dab = Da'b'$, $Dbc = Db'c'$ and $m\angle abc < m\angle a'b'c'$, then $Dac < Da'c'$.
5. If $\triangle abc$ and $\triangle a'b'c'$ are such that $Dab = Da'b'$, $Dbc = Db'c'$ and $Dac < Da'c'$, then $m\angle abc < m\angle a'b'c'$.
6. If $p \subset I(\triangle abc)$, then $Dap + Dbp < Dac + Dbc$, and $m\angle acb < m\angle apb$.
7. If two angles and a side of one triangle are congruent to the corresponding angles and side of a second triangle, then the triangles are congruent.
8. If $\triangle abc$ and $\triangle a'b'c'$ are such that $\angle bca$ and $\angle b'c'a'$ are right angles, $\overline{ab} \cong \overline{a'b'}$, and $\overline{ac} \cong \overline{a'c'}$, then the triangles are congruent.

9. A point cannot be equidistant from three distinct points of a line.

10. If $ap \perp pq$, then $Dap < Daq$.

†11. C7 is dependent on the remaining postulates.

†12. C8 is independent of the remaining postulates.

Congruence Without Numbers—
An Introduction

In this chapter we give an introduction, somewhat brief, to an intrinsically geometric theory of congruence. In contrast with our treatment in the last chapter, we take congruence of segments and congruence of angles as primitive or undefined notions and derive their properties from suitable postulates. Numbers, as measures of segments or angles, are not introduced in the postulates; we show how to establish a theory of congruence which is devoid of numerical assumptions. We cover the same ground as in Chapter 14 and conclude with an indication of how numerical measures for segments can be constructed.

We introduce the ideas of order ("less than") and addition for segments and for angles, and state their basic properties. However, to keep the discussion within reasonable limits we do not prove these properties and consequently do not use them in proving theorems. Nevertheless the basic properties of order and addition of segments and angles are indispensible in the more advanced theory of congruence, and we have included several graded sets of exercises to make their proofs more accessible.

After completing Section 1 you can continue with Sections 2–5 on the "algebraic" properties of segments and angles, or independently study Sections 6–10 which treat the material of Chapter 14 on a purely geometrical basis.

1. Postulates for Congruence

We introduce first a new primitive notion, the relation of *congruence* of two *segments*, and denote the relation, $\overline{ab}$ is congruent to $\overline{cd}$, by the notation

$\overline{ab} \cong \overline{cd}$. We assume for an ordered incidence geometry the following postulates.

C(i). (REFLEXIVENESS) $\overline{ab} \cong \overline{ab}$.

C(ii). (SYMMETRY) If $\overline{ab} \cong \overline{cd}$ then $\overline{cd} \cong \overline{ab}$.

C(iii). (TRANSITIVENESS) If $\overline{ab} \cong \overline{cd}$, $\overline{cd} \cong \overline{ef}$ then $\overline{ab} \cong \overline{ef}$.

C(iv). (ADDITIVITY PRINCIPLE) If $\overline{ab} \cong \overline{a'b'}$, $\overline{bc} \cong \overline{b'c'}$, (abc) and $(a'b'c')$ then $\overline{ac} \cong \overline{a'c'}$.

C(v). (POINT LOCATION PRINCIPLE) Let $\overrightarrow{ab}$ be any ray and $\overline{cd}$ any segment. Then there is exactly one point $p \subset \overrightarrow{ab}$ such that $\overline{ap} \cong \overline{cd}$.

It should be noted that $\overline{ab} \cong \overline{cd}$ is a relation between two segments conceived as sets of points and is independent of the notation used to describe them. Thus we shall not distinguish between the statements $\overline{ab} \cong \overline{cd}$, $\overline{ba} \cong \overline{cd}$, $\overline{ab} = \overline{dc}$, et cetera. For example C(i) may be written $\overline{ab} \cong \overline{ba}$.

Observe how different our viewpoint is from that in the last chapter. There $\overline{ab} \cong \overline{cd}$ was *defined* to mean $Dab = Dcd$ (Ch. 14, Sec. 7), and was studied on the basis of postulates for distance (Ch. 14, Sec. 2). Although we use the same notation in the present treatment, $\overline{ab} \cong \overline{cd}$ is literally *undefined* and is to be studied on the basis of the postulates independently adopted here.

Postulates C(i)–C(v) are not unfamiliar—they were derived (Ch. 14, Th. 9) as the basic properties of congruent segments in the former theory, and so form a natural basis for the present theory. They also are naturally motivated by the physical applications that are made of Euclidean geometry. Interpreted physically $\overline{ab} \cong \overline{cd}$ can be taken to mean that $\overline{ab}$ and $\overline{cd}$ give equivalent results when "measured" with an unmarked ruler or compared with a pair of compasses.

Our treatment of the second primitive notion, congruence of angles, parallels that for congruence of segments. The postulates are the basic properties of congruence of angles derived in Theorem 11 of Chapter 14. We denote the relation $\angle abc$ is congruent to $\angle def$, by $\angle abc \cong \angle def$ and assume the following postulates.

C(i'). (REFLEXIVENESS) $\angle abc \cong \angle abc$.

C(ii'). (SYMMETRY) If $\angle abc \cong \angle def$ then $\angle def \cong \angle abc$.

C(iii'). (TRANSITIVENESS) If $\angle abc \cong \angle def$, $\angle def \cong \angle ghk$ then $\angle abc \cong \angle ghk$.

C(iv') (ADDITIVITY PRINCIPLE) If $\angle aob \cong \angle a'o'b'$, $\angle boc \cong \angle b'o'c'$, $(\overrightarrow{oa}\ \overrightarrow{ob}\ \overrightarrow{oc})$ and $(\overrightarrow{o'a'}\ \overrightarrow{o'b'}\ \overrightarrow{o'c'})$ then $\angle aoc \cong \angle a'o'c'$.

C(v′). (RAY LOCATION PRINCIPLE)   *Given a half-plane H, a ray $\overrightarrow{ab}$ contained in its edge, and an angle $\angle pqr$. Then there exists exactly one ray $\overrightarrow{ac}$ such that $\overrightarrow{ac} \subset H$ and $\angle bac \cong \angle pqr$.*

After having introduced congruence of segments and congruence of angles we proceed exactly as in the last chapter (Ch. 14, Sec. 11) to define a *congruence* between two triangles. Then we relate congruence of segments and congruence of angles as in the first theory by adopting the

SAS Postulate. *If a correspondence between the vertices of two triangles is such that two sides and the included angle of one triangle are respectively congruent to the corresponding two sides and angle of the other, then the correspondence is a congruence.*

This completes our list of postulates.

2. Order of Segments

The betweenness concept can be used to define a two-term order relation or "precedence" for segments. So to speak segments are compared by laying one off on the other.

DEFINITION.   Let $\overline{ab}$, $\overline{cd}$ be given. If there exists a point p such that $\overline{ab} \cong \overline{cp}$ and (cpd) we say $\overline{ab}$ *is less than* $\overline{cd}$ and write $\overline{ab} < \overline{cd}$.

On the basis of this definition the following properties can be proved.

(1) (Irreflexiveness) $\overline{ab} < \overline{ab}$ *is always false.*
(2) (Transitiveness) *If* $\overline{ab} < \overline{cd}$ *and* $\overline{cd} < \overline{ef}$ *then* $\overline{ab} < \overline{ef}$.
(3) (Comparability) *For any two segments* $\overline{ab}$, $\overline{cd}$ *one of the following holds:*

$$\overline{ab} \cong \overline{cd}, \overline{ab} < \overline{cd} \text{ or } \overline{cd} < \overline{ab}.$$

(4) (Substitution Principle) *If* $\overline{ab} < \overline{cd}$ *and* $\overline{ab} \cong \overline{a'b'}$ *then* $\overline{a'b'} < \overline{cd}$; *similarly if* $\overline{cd} \cong \overline{c'd'}$ *then* $\overline{ab} < \overline{c'd'}$.

Thus our segments form an ordered system in which congruence plays the role of equality. This is a partial substitute in the present theory for the order relations of the real numbers as used in Chapter 14.

Note that the definition of $\overline{ab} < \overline{cd}$ as stated involves the order in which the points a, b and c, d are given. The Substitution Principle shows that this is only apparent, for it implies that if $\overline{ab} < \overline{cd}$ then $\overline{ba} < \overline{cd}$, $\overline{ab} < \overline{dc}$ and $\overline{ba} < \overline{dc}$. Thus the relation "less than" for segments is intrinsic—it depends on the segments related, not on the notation used to describe them.

Properties (1)–(4) are justified in the following set of exercises.

EXERCISES

Prove the following statements.

1. $\overline{ab} < \overline{ab}$ is always false.
2. If (abc) then $\overline{ab} < \overline{ac}$.
3. If $\overline{a'b'} \cong \overline{ab}$ and $\overline{ab} < \overline{cd}$, then $\overline{a'b'} < cd$.
4. If (abc) and $\overline{ac} \cong \overline{a'c'}$ then there exists a point b' such that $\overline{ab} \cong \overline{a'b'}$ and $(a'b'c')$.
5. If $\overline{ab} < \overline{cd}$ and $\overline{cd} \cong \overline{c'd'}$ then $\overline{ab} < \overline{c'd'}$.
6. If $\overline{ab} < \overline{cd}$ and $\overline{cd} < \overline{ef}$ then $\overline{ab} < \overline{ef}$.
7. Given segments $\overline{ab}$ and $\overline{cd}$, one and only one of the following holds: $\overline{ab} \cong \overline{cd}$, $\overline{ab} < \overline{cd}$, $\overline{cd} < \overline{ab}$.
8. If (abc), $(a'b'c')$, $\overline{ab} \cong \overline{a'b'}$ and $\overline{ac} \cong \overline{a'c'}$ then $\overline{bc} \cong \overline{b'c'}$.

3. Addition of Segments

To add two segments interpreted as physical objects we line them up and take the segment formed as their sum—as when children add numbers geometrically by putting together rods of appropriate sizes. This suggests the following

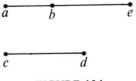

FIGURE 15.1

DEFINITION. Let segments $\overline{ab}$, $\overline{cd}$ be given (Fig. 15.1). Let e be such that $\overline{cd} \cong \overline{be}$ and (abe). Then we say $\overline{ae}$ is the *sum* of $\overline{ab}$ and $\overline{cd}$ in that order and write $\overline{ab} + \overline{cd} = \overline{ae}$.

Note that $\overline{ab} + \overline{cd}$ as defined depends on the notation for segment $\overline{ab}$, for $\overline{ba} + \overline{cd}$ would yield a segment with endpoint b which would be distinct from $\overline{ab} + \overline{cd}$. However as we shall see, $\overline{ba} + \overline{cd}$ is congruent to $\overline{ab} + \overline{cd}$ and the definition is not as strange as it seems. Observe that if (abc) then $\overline{ab} + \overline{bc} = \overline{ac}$ as one would expect.

The basic properties of addition of segments are the following.

(1) (Closure Law) $\overline{ab} + \overline{cd}$ *is a uniquely determined segment.*

(2) (Commutative Law) $\overline{ab} + \overline{cd} \cong \overline{cd} + \overline{ab}$.

(3) (Associative Law) $(\overline{ab} + \overline{cd}) + \overline{ef} = \overline{ab} + (\overline{cd} + \overline{ef})$.

(4) (Substitution Principle) *If* $\overline{ab} \cong \overline{a'b'}, \overline{cd} \cong \overline{c'd'}$ *then* $\overline{ab} + \overline{cd} \cong \overline{a'b'} + \overline{c'd'}$.

(5) (Cancellation Principle) *If* $\overline{ab} + \overline{cd} \cong \overline{a'b'} + \overline{c'd'}$ *and* $\overline{ab} \cong \overline{a'b'}$ *then* $\overline{cd} \cong \overline{c'd'}$.

(6) (Monotonic Principle) *If* $\overline{ab} < \overline{cd}$ *then* $\overline{ab} + \overline{pq} < \overline{cd} + \overline{pq}$.

(7) (Subtraction Principle) $\overline{ab} < \overline{cd}$ *if and only if* $\overline{ab} + \overline{pq} \cong \overline{cd}$, *for some segment* $\overline{pq}$.

We may mention that Properties (1)–(4) characterize the set of segments, with $+$ as defined, as a *commutative semigroup*, in which $\cong$ plays the role of equality. By Property (6), in view of Section 2, the segments form an *ordered semigroup*.

Properties (1)–(7) are justified in the set of exercises at the end of this section.

The Substitution Principle throws light on our definition of addition of segments. For it implies, since $\overline{ab} \cong \overline{ba}, \overline{cd} \cong \overline{dc}$, that

$$\overline{ab} + \overline{cd} \cong \overline{ba} + \overline{cd} \cong \overline{ab} + \overline{dc} \cong \overline{ba} + \overline{dc}.$$

Thus, although the sum of the two segments depends on the order in which the endpoints are written, it does not depend on the order in a pathological way. Interchanging endpoints in a segment may change the sum segment, but merely to a congruent segment. This is not distressing since we are constructing this algebra of segments for use as a sort of "measure" of segments, and congruent segments will have, so to speak, the "same" measure. Phrased differently, two segments do not have a unique sum, but the different sums are congruent or "equivalent." Actually, congruence of segments is an *equivalence relation* (Ch. 9, Sec. 3) by Postulates C(i), C(ii), C(iii). So if we think of congruence as the basic equality in the system of segments, a property such as (2) will neither seem nor look unnatural—it is merely an "equation" in which "$\cong$" is the equality symbol.

The slight awkwardness involved in the definition of $\overline{ab} + \overline{cd}$ can be eliminated in an elegant but complicated way that is frequently used in modern algebra. The Substitution Principle shows that if we replace a segment in a sum by a congruent or equivalent segment, the sum segment is not effectively altered—it is merely replaced by a congruent or "equivalent" segment. This suggests that we consider congruent or equivalent segments as essentially the same in defining addition, and that addition be defined, not for individual segments, but for whole classes of congruent ones. Thus we define the *congruence class* (or *equivalence class*) determined by segment $\overline{ab}$ to be the class of segments $\overline{xy}$ such that $\overline{xy} \cong \overline{ab}$, and we denote this class by $\{\overline{ab}\}$. Then we define addition for such congruence classes:

$$\{\overline{ab}\} + \{\overline{cd}\} = \{\overline{ab} + \overline{cd}\}.$$

That is, to form the sum of two congruence classes, we add their determining segments and take the congruence class determined by the result. This yields an algebraic system of the familiar sort in which congruence plays no explicit role, and the role of equality is taken by equality of congruence classes, that is, equality of sets.

E X E R C I S E S

Prove the following statements. (Make free use of the exercises at the end of Sec. 2.)

1. $\overline{ab} + \overline{cd}$ is a uniquely determined segment.
2. $\overline{ab} + \overline{cd} \cong \overline{cd} + \overline{ab}$.
3. If $\overline{ab} \cong \overline{a'b'}$ and $\overline{cd} \cong \overline{c'd'}$ then $\overline{ab} + \overline{cd} \cong \overline{a'b'} + \overline{c'd'}$.
4. $(\overline{ab} + \overline{cd}) + \overline{ef} = \overline{ab} + (\overline{cd} + \overline{ef})$.
5. If $\overline{ab} < \overline{cd}$ then $\overline{ab} + \overline{pq} < \overline{cd} + \overline{pq}$.
6. If $\overline{ab} \cong \overline{a'b'}$, $\overline{cd} < \overline{c'd'}$ then $\overline{ab} + \overline{cd} < \overline{a'b'} + \overline{c'd'}$.
7. If $\overline{ab} + \overline{cd} \cong \overline{a'b'} + \overline{c'd'}$ and $\overline{ab} \cong \overline{a'b'}$ then $\overline{cd} \cong \overline{c'd'}$.
8. If $\overline{ab} + \overline{cd} < \overline{a'b'} + \overline{c'd'}$ and $\overline{ab} \cong \overline{a'b'}$ then $\overline{cd} < \overline{c'd'}$.
9. $\overline{ab} < \overline{cd}$ if and only if $\overline{ab} + \overline{pq} \cong \overline{cd}$, for some segment $\overline{pq}$.

4. Order of Angles

The treatment parallels that in Section 2 for order of segments.

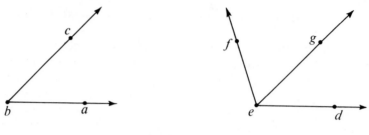

FIGURE 15.2

DEFINITION. Let $\angle abc$, $\angle def$ be given (Fig. 15.2). If there exists a ray $\overrightarrow{eg}$ such that $\angle abc \cong \angle deg$ and $(\overrightarrow{ed}\ \overrightarrow{eg}\ \overrightarrow{ef})$ we say $\angle abc < \angle def$.

The basic properties of order of angles are the same as those for segments:

(1) (Irreflexiveness) $\angle abc < \angle abc$ is always false.

(2) (Transitiveness) If $\angle abc < \angle def$ and $\angle def < \angle ghk$ then $\angle abc < \angle ghk$.

(3) (Comparability) For any two angles $\angle abc$, $\angle def$ one of the following holds:

$$\angle abc \cong \angle def, \angle abc < \angle def \quad or \quad \angle def < \angle abc.$$

(4) (Substitution Principle) If $\angle abc < \angle def$ and $\angle abc \cong \angle a'b'c'$ then $\angle a'b'c' < \angle def$; similarly if $\angle def \cong \angle d'e'f'$ then $\angle abc < \angle d'e'f'$.

Note by the Substitution Principle that the relation "less than" is an intrinsic property of two angles—it does not depend on the notation used to describe the angles.

EXERCISES

Prove the following statements.

1. $\angle abc < \angle abc$ is always false.

2. If $(\overrightarrow{oa}\ \overrightarrow{ob}\ \overrightarrow{oc})$ then $\angle aob < \angle aoc$.

3. If $\angle a'b'c' \cong \angle abc$ and $\angle abc < \angle def$ then $\angle a'b'c' < \angle def$.

4. If $(\overrightarrow{oa}\ \overrightarrow{ob}\ \overrightarrow{oc})$ and $\angle aoc \cong \angle a'o'c'$ then there exists a ray $\overrightarrow{o'b'}$ such that $\angle aob \cong \angle a'o'b'$ and $(\overrightarrow{o'a'}\ \overrightarrow{o'b'}\ \overrightarrow{o'c'})$. (Hint: Use SAS.)

5. If $\angle abc < \angle def$ and $\angle def \cong \angle d'e'f'$ then $\angle abc < \angle d'e'f'$.

6. If $\angle abc < \angle def$ and $\angle def < \angle ghk$ then $\angle abc < \angle ghk$.

7. Given angles $\angle abc$ and $\angle def$, one and only one of the following holds:
$\angle abc \cong \angle def$, $\angle abc < \angle def$, $\angle def < \angle abc$.

8. If $(\overrightarrow{oa}\ \overrightarrow{ob}\ \overrightarrow{oc})$, $(\overrightarrow{o'a'}\ \overrightarrow{o'b'}\ \overrightarrow{o'c'})$, $\angle aob \cong \angle a'o'b'$ and $\angle aoc \cong \angle a'o'c'$ then
$\angle boc \cong \angle b'o'c'$.

5. Addition of Angles

The treatment is analogous to that for addition of segments, but there is an important difference—the closure law does not hold. Sometimes, so to speak, two angles are too "big" to have a sum.

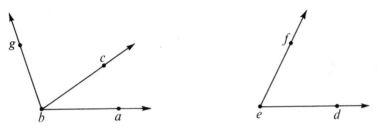

FIGURE 15.3

DEFINITION.   Let angles $\angle abc$, $\angle def$ be given (Fig. 15.3). Let there exist a ray $\overrightarrow{bg}$ such that $\angle def \cong \angle cbg$ and $(\overrightarrow{ba}\ \overrightarrow{bc}\ \overrightarrow{bg})$. Then we say $\angle abg$ is the sum of $\angle abc$ and $\angle def$ in that order and write $\angle abc + \angle def = \angle abg$.

Note that $(\overrightarrow{oa}\ \overrightarrow{ob}\ \overrightarrow{oc})$ implies $\angle aob + \angle boc = \angle aoc$.

The basic properties of addition of angles are the same as Properties (1)–(7) of Section 3 on addition of segments, with suitable assumptions on the existence of the sums involved. For example the commutative law may be stated:

$$\angle abc + \angle def \cong \angle def + \angle abc,$$

provided either side exists. In other words if either sum exists so does the other and they are congruent. The basic properties are considered in the set of exercises below.

Note finally that the discussion at the end of Section 3 concerning the dependence of the sum of two segments on the notation for the segments applies equally well to the sum of angles.

This concludes our discussion of the algebraic properties of segments and angles.

EXERCISES

Prove the following statements.

1. If $\angle abc + \angle def$ exists, it is a uniquely determined angle.
2. If $\angle abc + \angle def$ exists, so does $\angle def + \angle abc$ and $\angle abc + \angle def \cong \angle def + \angle abc$.
3. If $\angle abc \cong \angle a'b'c'$ and $\angle def \cong \angle d'e'f'$ and $\angle abc + \angle def$ exists, then $\angle a'b'c' + \angle d'e'f'$ exists and is congruent to $\angle abc + \angle def$.
4. If $(\angle abc + \angle def) + \angle ghk$ exists, then $\angle abc + (\angle def + \angle ghk)$ exists and is equal to $(\angle abc + \angle def) + \angle ghk$.
5. If $\angle abc < \angle def$ and $\angle def + \angle pqr$ exists, then $\angle abc + \angle pqr$ exists and $\angle abc + \angle pqr < \angle def + \angle pqr$.
6. If $\angle abc \cong \angle a'b'c'$, $\angle def < \angle d'e'f'$ and $\angle a'b'c' + \angle d'e'f'$ exists, $\angle abc + \angle def$ exists, and $\angle abc + \angle def < \angle a'b'c' + \angle d'e'f'$.
7. If $\angle abc + \angle def \cong \angle a'b'c' + \angle d'e'f'$ and $\angle abc \cong \angle a'b'c'$ then $\angle def \cong \angle d'e'f'$.
8. If $\angle abc + \angle def < \angle a'b'c' + \angle d'e'f'$ and $\angle abc \cong \angle a'b'c'$ then $\angle def < \angle d'e'f'$.
9. $\angle abc < \angle def$ if and only if $\angle abc + \angle pqr \cong \angle def$, for some angle $\angle pqr$.

6. Congruence of Triangles

Now we begin to study, on the basis of the postulates assumed in Section 1, the familiar geometrical material of Chapter 14: angles, triangles, midpoints and perpendicular bisectors of segments, et cetera. We base the treatment on Postulates C(i)–C(v), C(i')–C(v') and the SAS Postulate, and make no use of order or addition properties of segments or angles. We have to reorganize some of the material, for example, properties of segments and angles will often have to be proved by congruent triangles, rather than by a quick appeal to the measure postulates of Chapter 14. Consequently, some of the proofs will be longer and more difficult. We have tried nevertheless to preserve the methods of Chapter 14 and in particular have made heavy use of the "equidistance—perpendicular bisector" theme.

In Section 1 we introduced, in order to state the SAS Postulate, the idea of a congruence between two triangles (Ch. 14, Sec. 11). We continue by adopting the notation $\triangle abc \cong \triangle def$ to indicate that the correspondence $abc \rightarrow def$ is a congruence (Ch. 14, Sec. 11). This convenient notation was not employed in Chapter 14 because we wanted to stress there the basic idea of a congruence between triangles as a correspondence.

The results of Section 12 of Chapter 14 are valid in the present theory. Specifically we have the following theorems.

THEOREM 1. (Ch. 14, Th. 19)   If two sides of a triangle are congruent, then the angles opposite these sides are congruent.

Proof.   Use the proof of Theorem 19 in Chapter 14.

THEOREM 2. (ASA Theorem—Ch. 14, Th. 20) If a correspondence between the vertices of two triangles is such that two angles and the included side of the first triangle are congruent respectively to the corresponding two angles and side of the second triangle, then the correspondence is a congruence.

Proof.   Use the method of Theorem 20 in Chapter 14.

COROLLARY.   If two angles of a triangle are congruent, then the sides opposite these angles are congruent.

7. Supplementary Angles, Right Angles, Vertical Angles

We maintain the familiar definitions, adopted earlier, of supplementary angles (Ch. 14, Sec. 9), right angles (Ch. 14, Sec. 10) and vertical angles (Ch. 12, Sec. 6). To obtain familiar results we have to work a bit harder.

THEOREM 3. (Ch. 14, Th. 14)   Supplements of congruent angles are congruent.

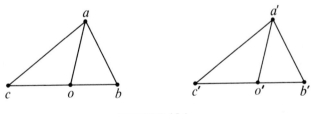

FIGURE 15.4

Proof.   It suffices to consider the case where the pairs of supplementary angles are linear pairs, since congruence of angles is an equivalence relation. Let us suppose then $\angle aob$, $\angle aoc$, and $\angle a'o'b'$, $\angle a'o'c'$ are two linear pairs such that $\overrightarrow{ob}$, $\overrightarrow{oc}$ are opposite and $\overrightarrow{o'b'}$, $\overrightarrow{o'c'}$ are opposite and $\angle aob \cong \angle a'o'b'$.

We show $\angle aoc \cong \angle a'o'c'$. It is not restrictive to assume

$$\overline{oa}, \overline{ob}, \overline{oc} \cong \overline{o'a'}, \overline{o'b'}, \overline{o'c'}.$$

Then the correspondence $oab \rightarrow o'a'b'$ is a congruence by the SAS Postulate, so that $\triangle oab \cong \triangle o'a'b'$. Hence $\overline{ab} \cong \overline{a'b'}$ and $\angle oba \cong \angle o'b'a'$. Further we have (boc), $(b'o'c')$. Hence $\overline{bo} \cong \overline{b'o'}$ and $\overline{oc} \cong \overline{o'c'}$ yield $\overline{bc} \cong \overline{b'c'}$ by the Segment Additivity Principle C(iv). Note $\angle abc \cong a'b'c'$. Then $\triangle abc \cong \triangle a'b'c'$ by the SAS Postulate. Hence $\overline{ac} \cong \overline{a'c'}$ and $\angle acb \cong \angle a'c'b'$. Similarly $\triangle aco \cong \triangle a'c'o'$ by SAS. Thus $\angle aoc \cong \angle a'o'c'$ and the proof is complete.

COROLLARY 1. (Ch. 14, Th. 13)   Vertical angles are congruent.

Proof. Two vertical angles are supplements of the same angle and so congruent by the theorem.

COROLLARY 2. (Ch. 14, Th. 15)   Let $\angle aob$, $\angle aoc$ be supplementary, where b and c are on opposite sides of oa. Then $\overrightarrow{ob}$, $\overrightarrow{oc}$ are opposite and $\angle aob$, $\angle aoc$ form a linear pair.

Proof. Use the proof of Theorem 15 in Chapter 14.

COROLLARY 3.   A right angle is congruent to each of its supplements.

COROLLARY 4.   Any angle congruent to a right angle is a right angle.

Proof. Let A be a right angle; suppose $B \cong A$. By definition $A \cong A'$, where A' is a supplement of A. Let B' be a supplement of B. By the theorem $B \cong A$ implies $B' \cong A'$. We have

$$B \cong A, \quad A \cong A', \quad A' \cong B'$$

so that $B \cong B'$ and B is a right angle by definition.

COROLLARY 5.   Any supplement of a right angle is a right angle.

Proof. Apply Corollaries 3 and 4.

In the next section we derive an important converse of Corollary 4: *Any two right angles are congruent.*

8. Equidistance and Perpendicular Bisectors of Segments

The terms *perpendicular lines, perpendicular bisector* of a segment and *equidistance* are used exactly as defined in Sections 10 and 13 of Chapter 14. Our main result (Th. 8, 9) asserts that, in a plane, a segment has one and only one perpendicular bisector. This principle is rather deep and not easily obtained. In Chapter 14 its derivation was trivial because the measure postulates are such heavily weighted assumptions.

We begin with several familiar properties.

THEOREM 4. (Ch. 14, Th. 17)   If one of the angles formed by two intersecting lines is a right angle, then all of the angles formed by them are right angles.

Proof.   The proof given in Chapter 14 is valid in view of the corollaries to Theorem 3.

COROLLARY.   If two lines are perpendicular, all the angles they form are right angles.

THEOREM 5.   Any point in a perpendicular bisector of a segment is equidistant from the endpoints of the segment.

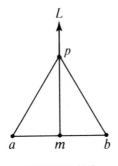

FIGURE 15.5

Proof.   Let m be a midpoint of $\overline{ab}$. Suppose $L \supset m$ and $L \perp ab$. Let $p \subset L$. If $p = m$ then $\overline{pa} \cong \overline{pb}$. Suppose $p \neq m$; then $p \not\subset ab$. By the last corollary $\angle pma$ is a right angle. Since (amb), $\angle pma$ and $\angle pmb$ form a linear pair and are supplementary. Thus $\angle pma \simeq \angle pmb$, since a right angle is congruent to each supplement (Th. 3, Cor. 3). It follows that $\triangle pma \simeq \triangle pmb$ by SAS and we infer $\overline{pa} \simeq \overline{pb}$.

THEOREM 6.   Let m be a midpoint of $\overline{ab}$, p be equidistant from a, b, and $p \nsubseteq ab$. Then pm is a perpendicular bisector of $\overline{ab}$.

Proof.   Use Theorem 1 and SAS.

The following lemma enables us to "construct" a point equidistant from two given points. (We have no compasses!) It leads eventually to the existence of a perpendicular bisector of a segment (Th. 8).

LEMMA.   Suppose c, d are on the same side of ab, $\angle cab \simeq \angle dba$, $\overline{ac} \simeq \overline{bd}$, and $\overrightarrow{ac}$, $\overrightarrow{bd}$ do not meet. Then $\overrightarrow{ad}$ meets $\overrightarrow{bc}$ in a point p which is equidistant from a and b.

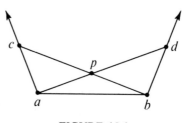

FIGURE 15.6

Proof.   Since c, d are on the same side of ab we have (Ch. 12, Th. 11 Cor. 2)

(1) $(\overrightarrow{ab}\ \overrightarrow{ac}\ \overrightarrow{ad})$, $(\overrightarrow{ab}\ \overrightarrow{ad}\ \overrightarrow{ac})$ or $\overrightarrow{ac} = \overrightarrow{ad}$.

Suppose $(\overrightarrow{ab}\ \overrightarrow{ac}\ \overrightarrow{ad})$. Then $\overrightarrow{ac}$ meets $\overline{bd}$ (Ch. 12, Th. 7) and $\overrightarrow{ac}$ meets $\overrightarrow{bd}$ contrary to hypothesis. Also $\overrightarrow{ac} = \overrightarrow{ad}$ implies $\overrightarrow{ac}$ meets $\overrightarrow{bd}$. Hence (1) implies $(\overrightarrow{ab}\ \overrightarrow{ad}\ \overrightarrow{ac})$, so that $\overrightarrow{ad}$ meets $\overline{bc}$. Thus $\overrightarrow{ad}$ meets $\overrightarrow{bc}$; let p be their intersection. We have

$$\overline{ac} \simeq \overline{bd}, \quad \angle cab \simeq \angle dba, \quad \overline{ab} \simeq \overline{ba}$$

so that $\triangle abc \simeq \triangle bad$ by SAS. Hence $\angle abc \simeq \angle bad$. Thus $\angle abp \simeq \angle bap$ and $\overline{pa} \simeq \overline{pb}$ (Th. 2, Cor.).

THEOREM 7. In a plane, if two distinct points are given, there is a point equidistant from and not collinear with them.

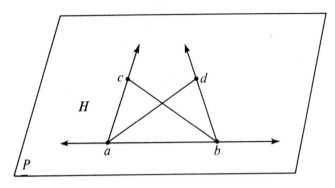

FIGURE 15.7

Proof. Let $a, b \subset P$; $a \neq b$. Let $c \subset P$, $c \nsubseteq ab$. Let H be the half-plane with edge ab that contains c. By C(v′) the Ray Location Principle, there exists ray $\overrightarrow{bd} \subset H$ such that $\angle cab \cong \angle dba$. If $\overrightarrow{ac}, \overrightarrow{bd}$ meet let p be their intersection. Then $\overline{pa} \cong \overline{pb}$ (Th. 2, Cor.).

Suppose $\overrightarrow{ac}, \overrightarrow{bd}$ do not meet. Then the Lemma applies, since we may consider d chosen such that $\overline{ac} \cong \overline{bd}$. Thus $\overrightarrow{ad}, \overrightarrow{bc}$ meet, say in p, and $\overline{pa} \cong \overline{pb}$. Note $\overrightarrow{ac}, \overrightarrow{ad} \subset H$. Hence in either case $p \subset H$ and $p \nsubseteq ab$.

COROLLARY. In a plane, a segment has a perpendicular bisector which contains a given midpoint of the segment.

Proof. Let $P \supset \overline{ab}$ and m be a midpoint of $\overline{ab}$. Note $P \supset a, b$. By the theorem there is a point $p \subset P$ such that $\overline{pa} \cong \overline{pb}$ and $p \nsubseteq ab$. Then pm is a perpendicular bisector of $\overline{ab}$ (Th. 6).

We have now reached one of the principal theorems. The proof, though simple in outline, is much more complicated than that given in Chapter 14 (Sec. 13, Par. 4).

THEOREM 8. In a plane, any segment has a perpendicular bisector.

Proof. Suppose $\overline{ab} \subset P$. Then $ab \subset P$. Our procedure is to find points $p, p' \subset P$ which are equidistant from a, b and on opposite sides of ab. Then

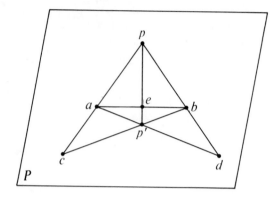

FIGURE 15.8

pp' will be the desired perpendicular bisector. By the preceding theorem there is a point $p \subset P$, $p \not\subset ab$ such that $\overline{pa} \cong \overline{pb}$. Choose c, d to satisfy

$$c \subset a|p, \quad d \subset b|p, \quad \overline{ac} \cong \overline{bd}.$$

Then (cap), so that $\angle pab$, $\angle cab$ form a linear pair and are supplementary. Similarly $\angle pba$, $\angle dba$ are supplementary. Since $\angle pab \cong \angle pba$ (Th. 1) we have $\angle cab \cong \angle dba$ (Th. 3). Further $\overrightarrow{ac}$, $\overrightarrow{bd}$ can not meet since $a|c$, $b|d$ meet in p. Thus the Lemma implies that $\overrightarrow{ad}$ meets $\overrightarrow{bc}$ in a point p' such that $\overline{p'a} \cong \overline{p'b}$. We show

(1) $(\overrightarrow{pa}\,\overrightarrow{pp'}\,\overrightarrow{pb})$.

Since $p' \subset \overrightarrow{ad}$ it follows that p', d are on the same side of pc. Similarly p', c are on the same side of pd. Thus $(\overrightarrow{pc}\,\overrightarrow{pp'}\,\overrightarrow{pd})$ (Ch. 12, Th. 11, Cor. 3) and (1) is immediate since $\overrightarrow{pc} = \overrightarrow{pa}$, $\overrightarrow{pd} = \overrightarrow{pb}$. (1) implies (Ch. 12, Th. 7) that $\overrightarrow{pp'}$ meets $\overline{ab}$, say in e. Note p, p' are on opposite sides of ab, since p, c are on opposite sides of ab and p', c are on the same side of ab. Hence $\overline{pp'}$ meets ab in a point which must be e. Hence $\overline{pp'}$ meets $\overline{ab}$ in e and $e \neq a$, b. Thus p, a, p' are noncollinear; similarly for p, b, p'. We show $\triangle pap' \cong \triangle pbp'$.

We have $\overline{pa} \cong \overline{pb}$, $\overline{p'a} \cong \overline{p'b}$. Hence (Th. 1)

(2) $\angle pab \cong \angle pba, \quad \angle p'ab \cong \angle p'ba.$

Further note (pep') and $e \subset \overrightarrow{ab}$. Hence by definition

(3) $(\overrightarrow{ap}\,\overrightarrow{ab}\,\overrightarrow{ap'})$.

Similarly (pep') and $e \subset \overrightarrow{ba}$ imply

(4) $(\overrightarrow{bp}\ \overrightarrow{ba}\ \overrightarrow{bp'})$.

By C(iv'), the Angle Additivity Principle, (2), (3), (4) yield $\angle pap' \cong \angle pbp'$. Thus $\triangle pap' \cong \triangle pbp'$ by SAS and we have $\angle app' \cong \angle bpp'$ or $\angle ape \cong \angle bpe$. By SAS $\triangle ape \cong \triangle bpe$. It easily follows that e is a midpoint of $\overline{ab}$ and $pe \perp ab$ and the proof is complete.

COROLLARY 1.   In a plane, any segment has a midpoint.

This leaves open the possibility of a system satisfying our postulates in which there is only one line and not every segment has a midpoint.

COROLLARY 2.   If a point is equidistant from the endpoints of a segment and not collinear with them, it is in a perpendicular bisector of the segment.

Proof.   Suppose $\overline{pa} = \overline{pb}$, $p \notin ab$. By the proof of the theorem, there is a line pp' which is a perpendicular bisector of $\overline{ab}$.

THEOREM 9.   In a plane, any segment has a unique perpendicular bisector.

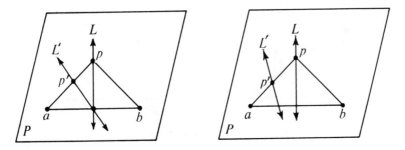

FIGURE 15.9

Proof.   We have $\overline{ab} \subset P$, and $L, L' \subset P$ are perpendicular bisectors of $\overline{ab}$. Suppose $L \neq L'$. Choose point p such that $p \subset L$, $p \notin ab$, $p \notin L'$. L' coplanes with p, a, b; $L' \not\supset p, a, b$; and L' meets $\overline{ab}$. Hence by B6, L' meets $\overline{pa}$ or $\overline{pb}$. Let us say L' meets $\overline{pa}$ in p'. Since L is a perpendicular bisector of $\overline{ab}, \overline{pa} \cong \overline{pb}$ (Th. 5). Hence $\angle pab \cong \angle pba$ (Th. 1). Similarly, $\angle p'ab \cong \angle p'ba$.

Since $\angle pab = \angle p'ab$ we have $\angle pba \cong \angle p'ba$. But p, p' and so $\overrightarrow{bp}$, $\overrightarrow{bp'}$ are on the same side of ab. Hence $\overrightarrow{bp} = \overrightarrow{bp'}$ by the uniqueness condition of $C(v')$, the Ray Location Principle. This implies $p = p' \subset \overline{pa}$, which is impossible. Thus $L = L'$ and the proof is complete.

COROLLARY 1.   In a plane, any segment has a unique midpoint.

Proof. Suppose $P \supset \overline{ab}$ and $\overline{ab}$ has midpoints m, m'. Then $\overline{ab}$ has perpendicular bisectors in P which contain m, m' respectively (Th. 7, Cor.). We infer $m = m'$, since in P, $\overline{ab}$ has a unique perpendicular bisector which is distinct from ab.

COROLLARY 2.   In a plane, there is a unique line perpendicular to a given line at a given point of the line.

Proof. Let $P \supset L \supset p$. Choose a, $b \subset L$ such that p is the midpoint of $\overline{ab}$. Then $M \perp L$ at p if and only if M is a perpendicular bisector of $\overline{ab}$. The result follows by the theorem.

COROLLARY 3.   Any two right angles are congruent.

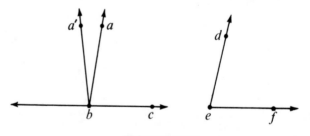

FIGURE 15.10

Proof. Let $\angle abc$, $\angle def$ be right angles. By the Ray Location Principle there exists $\angle a'bc$ such that $\angle a'bc \cong \angle def$ and $\overrightarrow{ba'}$ is on the same side of bc as a. Hence $\angle a'bc$ is a right angle, since it is congruent to one (Th. 3, Cor. 4). Thus $ab \perp bc$, $a'b \perp bc$ so that $ab = a'b$ by Corollary 2. Then $a \subset \overrightarrow{ba'}$, since $\overrightarrow{ba'}$ contains all points of $a'b$ on the same side of bc as a' (Ch. 12, Th. 3, Cor. 3). Thus $\overrightarrow{ba} = \overrightarrow{ba'}$ and $\angle abc = \angle a'bc \cong \angle def$.

THEOREM 10. (Ch. 14, Th. 21)   In a plane, a point is equidistant from the endpoints of a segment if and only if it is in the perpendicular bisector of the segment.

Proof.   Theorem 5, and the second part of the proof of Theorem 21 in Chapter 14.

The SSS Theorem (Ch. 14, Th. 22) follows as in Chapter 14.

9. The Exterior Angle Theorem and its Consequences

We can now prove a weakened form of the Exterior Angle Theorem (Ch. 14, Th. 23) which is sufficient to yield the remaining results of the chapter. We define *exterior angle* and *remote interior angle* of a triangle as in Section 14 of Chapter 14.

THEOREM 11.   An exterior angle of a triangle is never congruent to a remote interior angle.

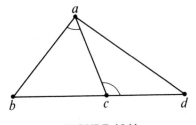

FIGURE 15.11

Proof.   Given $\triangle abc$, suppose (bcd). We show the exterior angle $\angle acd$ is not congruent to $\angle bac$. Suppose $\angle acd \cong \angle bac$. We may consider d chosen so that $\overline{cd} \cong \overline{ab}$. Then $\overline{ac} \cong \overline{ca}$ and $\triangle acd \cong \triangle cab$ by SAS. Hence $\angle cad \cong \angle acb$. But $\angle acb$, $\angle acd$ form a linear pair. Hence $\angle cad$, $\angle cab$ are supplementary by definition. Since b, d are on opposite sides of ac, rays $\overrightarrow{ad}$, $\overrightarrow{ab}$ are opposite (Th. 3, Cor. 2). Thus b, a, d are collinear; this is impossible and our supposition is false. To complete the proof show by a symmetrical argument that the angle vertical to $\angle acd$ is not congruent to $\angle abc$.

We now can continue as in Section 15 of Chapter 14 and prove Theorem 24, that there is a unique perpendicular to a line through an external point.

Similarly, the results in Section 16 on alternate interior angles and parallel lines are justified. Finally the results of Section 17 on Euclidean geometry through Theorem 28, that the angles of a triangle are supplementary, are valid as proved. The Corollary to Theorem 28 on the sum of the measures of the angles of a triangle can not be derived without introducing a theory of measures of angles.

10. Construction of Distance Functions

We conclude by indicating how numerical measures can be introduced on the basis we have adopted. We assume that every segment has a unique midpoint—this certainly is satisfied if a plane exists (Th. 9, Cor. 1).

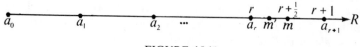

FIGURE 15.12

We begin with the problem of setting up a "coordinate system" on a ray. Let ray R be given (Fig. 15.12). Let a_0 denote its endpoint. Choose point a_1 arbitrarily in R to play the role of a "unit point." We proceed to construct the "integral points" $a_2, a_3, \ldots$ of R relative to this choice. We determine a_2 by the conditions

$$(a_0 a_1 a_2), \quad \overline{a_0 a_1} \cong \overline{a_1 a_2}.$$

Similarly, a_3 satisfies $(a_0 a_2 a_3)$, $\overline{a_0 a_1} \cong \overline{a_2 a_3}$. By continuing in this way we determine a sequence $a_1, a_2, \ldots, a_n, \ldots$ such that

$$(a_0 a_i a_{i+1}), \quad \overline{a_0 a_1} \cong \overline{a_i a_{i+1}} \quad (i \geq 1).$$

We call the sequence $a_1, a_2, \ldots, a_n, \ldots$ an *equicongruent sequence* of points of R, and say the sequence is *determined* by a_1. We assign to the points $a_1, a_2, \ldots, a_n, \ldots$ the numbers $1, 2, \ldots, n, \ldots$ as their *coordinates* (relative to a_1 as unit point).

FIGURE 15.13

Now we consider how to assign coordinates to the other points of R (Fig. 15.13). Suppose for example p falls between two consecutive points of the sequence $a_0, a_1, a_2, \ldots, a_n, \ldots$. Say $(a_r p a_{r+1})$. Then we shall assign to p a real number x between r and $r + 1$. To determine x we employ a bisection process which gives rise to a binary representation:

$$(1) \qquad x = r + b_1/2^1 + b_2/2^2 + \cdots + b_n/2^n + \cdots.$$

Let m be the midpoint of $\overline{a_r a_{r+1}}$. Then

$$p \subset \overline{a_r a_{r+1}} = \overline{a_r m} \cup m \cup \overline{m a_{r+1}}.$$

If $p = m$, assign to p the number $x = r + \frac{1}{2}$ (that is, in (1), we set $b_1 = 1$, and $b_n = 0$ for $n > 1$) and the process terminates. If $p \subset \overline{a_r m}$, we require $r < x < r + \frac{1}{2}$, so in (1) we set $b_1 = 0$. Similarly, if $p \subset \overline{m a_{r+1}}$, we set $b_1 = 1$. We continue by applying the bisection process to whichever of $\overline{a_r m}$ or $\overline{m a_{r+1}}$ contains p. Suppose $p \subset \overline{a_r m}$. Then if p is the midpoint m' of $\overline{a_r m}$, we assign to p the number $r + 0/2^1 + 1/2^2$, and the process terminates. Or, for example, if $p \subset \overline{m' m}$, we set $b_2 = 1$ in (1) so that x has the form

$$x = r + 0/2^1 + 1/2^2 + \cdots.$$

The process may terminate after a finite number of steps. This will happen if p is obtained by applying a finite number of bisections to a segment $\overline{a_{n-1} a_n}$. In this case the number x assigned to p has a terminating binary representation, that is, $b_n = 0$ in (1) for n sufficiently large.

If the process does not terminate, p will not be an integral point or a point obtained by a finite number of bisections but will always be between two consecutive "points of division" at each stage of the process. In this case the number x assigned to p does not have a terminating binary representation. In either case there is assigned to p a uniquely determined positive real number, which is called the *coordinate* of p (relative to unit point a_1).

A question immediately arises: How do we know that the procedure employed above is applicable to *any* point p of R? Specifically, how do we know that p will fall between two consecutive points a_r, a_{r+1} of the sequence $a_0, a_1, \ldots, a_n, \ldots$? This or an equivalent must be postulated. Consider the following statement:

THE POSTULATE OF ARCHIMEDES. *Let R be a ray with endpoint a_0; let $a_1, a_2, \ldots, a_n, \ldots$ be an equicongruent sequence of points of R. Then if $p \subset R$ there is a point a_j such that $(a_0 p a_j)$.*

It can be shown, using the Postulate of Archimedes, that every point of R has a uniquely determined coordinate (relative to the unit point a_1), and that distinct points have distinct coordinates.

The converse question forces itself on our attention: Given a positive real number x, can we "plot" it on ray R, that is, is x the coordinate of some point p or R? Let us represent x in binary notation

(2) $$x = r + b_1/2^1 + b_2/2^2 + \cdots + b_n/2^n + \cdots,$$

where r is a positive integer and each b_n is zero or 1. Further we may assume that $b_n = 1$ for all $n > k$ is false. Otherwise we could express x more simply by a terminating representation, using the relation

$$1/2^{k+1} + 1/2^{k+2} + \cdots = 1/2^k.$$

Under the given assumption x is uniquely representable in the form (2).

First suppose the expression for x in (2) terminates. Then it is not hard to see that x is the coordinate of a point p of R—in fact p will be an integral point $a_n, n \geq 1$, or p will be obtained by applying a finite number of bisections to a segment $\overline{a_{n-1}a_n}$.

Now suppose the expression for x does not terminate. Let us assume that there *is* a point p whose coordinate is x and try to locate p. By the process used above to determine the coordinate of a point, we have from (2) that p is between a_r and a_{r+1}. Similarly, p is between the points with coordinates $r + b_1/2$ and $r + (b_1 + 1)/2$. In general p is between the points with coordinates

$$r + b_1/2^1 + \cdots + b_n/2^n, \quad r + b_1/2^1 + \cdots + (b_n + 1)/2^n.$$

Thus (2) determines an infinite sequence of segments $S_1, \ldots, S_n, \ldots$ each of which contains the next—and p, the point we are seeking, must be common to all segments of the sequence. This suggests a geometric principle that there exists a point common to the segments of any sequence of the given type. But such a principle, unless qualified, necessarily fails. For example, the segments which join the points of R with coordinates 1 and $1 + 1/2^n$ ($n \geq 1$) have no common point. This difficulty however is easily obviated if the principle is stated not for segments but for closed segments, that is segments united with their endpoints. This suggests the

POSTULATE OF CANTOR. *Let* $S_1', \ldots, S_n', \ldots$ *be a sequence of closed segments, nested in the sense that* $S_n' \supset S_{n+1}'$ ($n \geq 1$). *Then there exists a point that is common to all segments of the sequence.*

By using the Postulate of Cantor it can be shown that any positive real number x is the coordinate of some point p of ray R. The argument follows the analysis above. Cantor's Postulate is employed in the case where x has a nonterminating binary representation to show that there is a point common to the segments of the sequence $S_1, S_2, \ldots, S_n, \ldots$ determined by relation (2) above.

Summary.   The Postulates of Archimedes and Cantor imply

(i) each point of ray R has a uniquely determined coordinate x;

(ii) $p \to x$ is a one-to-one correspondence between R and the set of positive real numbers.

We use this result to construct a distance function in our geometry. Choose a segment $\overline{uv}$ which is to be fixed in the discussion and is to play the role of a "unit segment." Let p, q be any two distinct points. Consider $\overrightarrow{pq}$. Let $a_1 \subset \overrightarrow{pq}$ and satisfy $\overline{pa_1} \simeq \overline{uv}$. Let x be the coordinate of q in $\overrightarrow{pq}$ (relative to a_1 as unit point). Then we define the distance from p to q (relative to unit segment $\overline{uv}$) to be x, and we write functionally $d(p, q) = x$.

It can be shown that $d(p, q)$ is a distance function in the sense of Chapter 14, Section 4. That is the following properties hold:

(i) $d(p, q) = d(q, p) > 0$.

(ii) If $\overrightarrow{ab}$ is any ray and x any positive real number, then there is exactly one point $p \subset \overrightarrow{ab}$ such that $d(a, p) = x$.

(iii) If (pqr) then $d(p, q) + d(q, r) = d(p, r)$.

Further the distance function $d(p, q)$ has an intimate relation to the congruence concept which was employed in its construction: *Let $\overline{pq}, \overline{rs}$ be any two segments. Then $\overline{pq} = \overline{rs}$ if and only if $d(p, q) = d(r, s)$.*

Finally, angle measure functions (Ch. 14, Sec. 5) can be constructed by similar procedures without introducing additional postulates.

E X E R C I S E S

1. Let $p, q \subset \overrightarrow{ab}$. Prove $\overline{ap} < \overline{aq}$ if and only if (apq).

2. Let $p, q, r \subset \overrightarrow{ab}$. Prove (pqr) if and only if $\overline{ap} < \overline{aq} < \overline{ar}$ or $\overline{ar} < \overline{aq} < \overline{ap}$.

3. Let a, b, c be distinct and collinear. Prove that $\overline{ab} + \overline{bc} \simeq \overline{ac}$ if and only if (abc).

4. Let b, c be on the same side of oa. Prove $(\overrightarrow{oa}\,\overrightarrow{ob}\,\overrightarrow{oc})$ if and only if $\angle aob < \angle aoc$.

DEFINITION.   Ray $\overrightarrow{ox}$ is a *bisector* of $\angle aob$ if $(\overrightarrow{oa}\,\overrightarrow{ox}\,\overrightarrow{ob})$ and $\angle aox \simeq \angle box$.

5. Prove: Any angle has one and only one bisector.

6. Prove: If $a \not\subset L$, then there is one and only one line which contains a and is perpendicular to L.

7. Let $\angle p$ be an exterior angle of a triangle, and $\angle q$ one of its remote interior angles. Prove $\angle q < \angle p$.

8. In $\triangle abc$, if $\overline{ab} < \overline{bc}$, prove $\angle bca < \angle cab$.

9. In $\triangle abc$, if $\angle bca < \angle cab$, prove $\overline{ab} < \overline{bc}$.

10. In $\triangle abc$, prove $\overline{ab} < \overline{bc} + \overline{ca}$.

11. If $\triangle abc$ and $\triangle a'b'c'$ are such that $\overline{ab} \cong \overline{a'b'}$, $\overline{bc} \cong \overline{b'c'}$, and $\angle abc < \angle a'b'c'$, prove that $\overline{ac} < \overline{a'c'}$.

12. If $\triangle abc$ and $\triangle a'b'c'$ are such that $\overline{ab} \cong \overline{a'b'}$, $\overline{bc} \cong \overline{b'c'}$, and $\overline{ac} < \overline{a'c'}$, prove that $\angle abc < \angle a'b'c'$.

13. If two angles and a side of one triangle are congruent to the corresponding angles and side of a second triangle, prove that the triangles are congruent.

List of Postulates

Postulates for Incidence

I1. A line is a set of points, containing at least two points.

I2. Two distinct points are contained in one and only one line.

I3. A plane is a set of points, containing at least three points which do not belong to the same line.

I4. Three distinct points which do not belong to the same line are contained in one and only one plane.

I5. If a plane contains two distinct points of a line, it contains the line (that is, it contains all points of the line).

I6. If two planes have one point in common, they have a second point in common.

Euclid's Parallel Postulate (Playfair Form)

E. If point a is not in line L there exists one and only one line M such that M contains a and $M \parallel L$.

Postulates for Precedence

P1. $a < a$ is always false.

P2. $a < b$, $b < c$ imply $a < c$.

P3. If a, b are distinct then one of the relations $a < b$, $b < a$ holds.

Postulates for Betweenness

B1. (abc) implies (cba).

B2. (abc) implies the falsity of (bca).

B3. a, b, c are distinct and collinear if and only if (abc), (bca) or (cab).

B4. Let p colline with and be distinct from a, b, c. Then (apb) implies (bpc) or (apc) but not both.

B5. If $a \neq b$ there exist x, y, z such that (xab), (ayb), (abz).

B6. Let L coplane with and not contain a, b, c. Then if L meets $\overline{ab}$, it meets $\overline{bc}$ or $\overline{ac}$ but not both.

Postulates for Distance

C1. To each pair of distinct points (a, b) there corresponds a unique positive real number Dab called the distance between a and b, or the length of $\overline{ab}$, or the measure of $\overline{ab}$.

C2. Let $\overrightarrow{ab}$ be any ray and x any positive real number. Then there is exactly one point $p \subset \overrightarrow{ab}$ such that $Dap = x$.

C3. If (abc) then $Dab + Dbc = Dac$.

Postulates for Angle Measure

C4. To each angle $\angle abc$ there corresponds a unique real number x, $0 < x < 180$, called the measure of the angle, and denoted $m \angle abc$.

C5. Given a half-plane H, a ray $\overrightarrow{ab}$ contained in its edge, and a real number x, $0 < x < 180$. Then there exists exactly one ray $\overrightarrow{ac}$ such that $\overrightarrow{ac} \subset H$ and $m \angle bac = x$.

C6. If $(\overrightarrow{oa}\ \overrightarrow{ob}\ \overrightarrow{oc})$ then $m \angle aob + m \angle boc = m \angle aoc$.

C7. If $\overrightarrow{oa}$ is opposite $\overrightarrow{ob}$, and $d \not\subset oa$ then $m \angle aod + m \angle bod = 180$.

SAS Postulate

If a correspondence between the vertices of two triangles is such that two sides and the included angle of one triangle are respectively congruent to the corresponding two sides and angle of the other, then the correspondence is a congruence.

Postulates for Congruence

C(i). $\overline{ab} \cong \overline{ab}$.

C(ii). If $\overline{ab} \cong \overline{cd}$ then $\overline{cd} \cong \overline{ab}$.

C(iii). If $\overline{ab} \cong \overline{cd}$, $\overline{cd} \cong \overline{ef}$ then $\overline{ab} \cong \overline{ef}$.

C(iv). If $\overline{ab} \cong \overline{a'b'}$, $\overline{bc} \cong \overline{b'c'}$, (abc) and $(a'b'c')$ then $\overline{ac} \cong \overline{a'c'}$.

C(v). Let $\overrightarrow{ab}$ be any ray and $\overline{cd}$ any segment. Then there is exactly one point $p \subset \overrightarrow{ab}$ such that $\overline{ap} \cong \overline{cd}$.

C(i'). $\angle abc \cong \angle abc$.

C(ii'). If $\angle abc \cong \angle def$ then $\angle def \cong \angle abc$.

C(iii'). If $\angle abc \cong \angle def$, $\angle def \cong \angle ghk$ then $\angle abc \cong \angle ghk$.

C(iv'). If $\angle aob \cong \angle a'o'b'$, $\angle boc \cong \angle b'o'c'$, $(\overrightarrow{oa}\ \overrightarrow{ob}\ \overrightarrow{oc})$ and $(\overrightarrow{o'a'}\ \overrightarrow{o'b'}\ \overrightarrow{o'c'})$ then $\angle aoc \cong \angle a'o'c'$.

C(v'). Given a half-plane H, a ray $\overrightarrow{ab}$ contained in its edge, and an angle $\angle pqr$. Then there exists exactly one ray $\overrightarrow{ac}$ such that $\overrightarrow{ac} \subset H$ and $\angle bac \cong \angle pqr$.

The Postulate of Archimedes

Let R be a ray with endpoint a_0; let $a_1, a_2, \ldots, a_n, \ldots$ be an equicongruent sequence of points of R. Then if $p \subset R$ there is a point a_j such that $(a_0 p a_j)$.

The Postulate of Cantor

Let $S_1', \ldots, S_n', \ldots$ be a sequence of closed segments, nested in the sense that $S_n' \supset S_{n+1}' (n \geq 1)$. Then there exists a point that is common to all segments of the sequence.

Index

343

Printed in the United States of America

E F G H I J   5 4 3 2 1 7 0

ABOUT THE AUTHORS

Walter Prenowitz is a Professor of Mathematics at Brooklyn College. He received the degrees of AB, AM, and, in 1936, Ph.D. from Columbia University. His field of specialization is geometry. A frequent lecturer at Summer Institutes for high school teachers and other high school teacher groups, and at meetings of the Mathematical Association of America, the author has published research and expository papers in the field of geometry. Dr. Prenowitz was a member of the Institute for Advanced Study at Princeton from 1936–37 and 1953–54, and held a Ford Foundation Faculty Fellowship for a project on the advanced undergraduate geometry curriculum from 1953–54; he held a National Science Foundation Science Faculty Fellowship from 1960–61. A member of the Mathematics Steering Committee of the African Education Program of Educational Services Incorporated, Dr. Prenowitz was also Chairman of the Entebbe Mathematics Workshop in 1962 and 1963. In 1959 and 1960 he was a member of the tenth grade geometry writing team of the School Mathematics Study Group and is now on their Monograph Panel, which is responsible for the publication of the New Mathematical Library.

Meyer Jordan is an Assistant Professor of Mathematics at Brooklyn College. He obtained the Ph.D. from New York University in 1956, and specializes in the area of geometry. Dr. Jordan was a member of the College Geometry Course Project writing team of the 1964 Summer Conference of the Minnesota School Mathematics and Science Center. He has directed and taught in a number of programs sponsored by the National Science Foundation: Summer and In-Service Institutes for Secondary School Teachers of Mathematics, Mathematics Programs for High-Ability Secondary School Students, and Undergraduate Independent Study Programs.

THIS BOOK WAS SET IN

TIMES ROMAN AND PERPETUA TYPES BY

UNIVERSITIES PRESS.

TYPOGRAPHY AND DESIGN

ARE BY THE STAFF OF

BLAISDELL PUBLISHING COMPANY.